ENERGY SCIENCE, ENGINEERING AND TECHNOLOGY

LIGHTNING IN A BOTTLE: ELECTRICAL ENERGY STORAGE

ENERGY SCIENCE, ENGINEERING AND TECHNOLOGY

Additional books in this series can be found on Nova's website under the Series tab.

Additional E-books in this series can be found on Nova's website under the E-books tab.

ENERGY SCIENCE, ENGINEERING AND TECHNOLOGY

LIGHTNING IN A BOTTLE: ELECTRICAL ENERGY STORAGE

FEDOR NOVIKOV
EDITOR

Nova Science Publishers, Inc.
New York

Copyright © 2011 by Nova Science Publishers, Inc.

All rights reserved. No part of this book may be reproduced, stored in a retrieval system or transmitted in any form or by any means: electronic, electrostatic, magnetic, tape, mechanical photocopying, recording or otherwise without the written permission of the Publisher.

For permission to use material from this book please contact us:
Telephone 631-231-7269; Fax 631-231-8175
Web Site: http://www.novapublishers.com

NOTICE TO THE READER

The Publisher has taken reasonable care in the preparation of this book, but makes no expressed or implied warranty of any kind and assumes no responsibility for any errors or omissions. No liability is assumed for incidental or consequential damages in connection with or arising out of information contained in this book. The Publisher shall not be liable for any special, consequential, or exemplary damages resulting, in whole or in part, from the readers' use of, or reliance upon, this material. Any parts of this book based on government reports are so indicated and copyright is claimed for those parts to the extent applicable to compilations of such works.

Independent verification should be sought for any data, advice or recommendations contained in this book. In addition, no responsibility is assumed by the publisher for any injury and/or damage to persons or property arising from any methods, products, instructions, ideas or otherwise contained in this publication.

This publication is designed to provide accurate and authoritative information with regard to the subject matter covered herein. It is sold with the clear understanding that the Publisher is not engaged in rendering legal or any other professional services. If legal or any other expert assistance is required, the services of a competent person should be sought. FROM A DECLARATION OF PARTICIPANTS JOINTLY ADOPTED BY A COMMITTEE OF THE AMERICAN BAR ASSOCIATION AND A COMMITTEE OF PUBLISHERS.

Additional color graphics may be available in the e-book version of this book.

Library of Congress Cataloging-in-Publication Data

Lightning in a bottle : electrical energy storage / [edited by] Fedor Novikov.
   p. cm.
 Includes index.
 ISBN 978-1-61470-481-2 (hardcover)
 1. Energy storage. I. Novikov, Fedor.
 TK2980.L54 2011
 333.793'2--dc23
                        2011023235

Published by Nova Science Publishers, Inc. † New York

CONTENTS

Preface		vii
Chapter 1	Energy Storage for the Electricity Grid: Benefits and Market Potential Assessment Guide *Jim Eyer and Garth Corey*	1
Chapter 2	Advanced Materials and Devices for Stationary Electrical Energy Storage Applications *United States Department of Energy*	193
Chapter 3	Electric Power Industry Needs for Grid-Scale Storage Applications *United States Department of Energy*	231
Chapter 4	Energy Storage: Program Planning Docoument *United States Department of Energy*	261
Chapter 5	Solar Energy Grid Integration Systems: Energy Storage (SEGIS-ES) *Dan T. Ton, Charles J. Hanley, Georgianne H. Peek, and John D. Boyes*	291
Chapter Sources		315
Index		317

PREFACE

This book describes a high-level, technology-neutral framework for assessing potential benefits from and economic market potential for energy storage used for electric-utility-related applications. The overarching theme addressed is the concept of combining applications/benefits into attractive value propositions that include use of energy storage, possibly including distributed and/or modular systems. Other topics addressed include high-level estimates of application-specific lifecycle benefits and maximum market potential. Combined, these criteria indicate the economic potential for a given energy storage applications and benefits.

Chapter 1- Electric energy storage is poised to become an important element of the electricity infrastructure of the future. The storage opportunity is multifaceted – involving numerous stakeholders and interests – and could involve potentially rich value propositions. Those rich value propositions are possible because, as described in this report, there are numerous potentially complementary and significant benefits associated with storage use that could be aggregated into attractive value propositions. In addition, proven storage technologies are in use today, while emerging storage technologies are expected to have improved performance and/or lower cost. In fact, recent improvements in energy storage and power electronics technologies, coupled with changes in the electricity marketplace, indicate an era of expanding opportunity for electricity storage as a cost-effective electric energy resource.

Chapter 2- Reliable access to cost-effective electricity is the backbone of the U.S. economy, and electrical energy storage is an integral element in this system. Without significant investments in stationary electrical energy storage, the current electric grid infrastructure will increasingly struggle to provide reliable, affordable electricity, jeopardizing the transformational changes envisioned for a modernized grid. Investment in energy storage is essential for keeping pace with the increasing demands for electricity arising from continued growth in U.S. productivity, shifts in and continued expansion of national cultural imperatives (e.g., the distributed grid and electric vehicles), and the projected increase in renewable energy sources.

Chapter 3- Reliable access to cost-effective electricity is the backbone of the U.S. economy, and electrical energy storage is an integral element in this system. Without significant investments in stationary electrical energy storage, the current electric grid infrastructure will increasingly struggle to provide reliable, affordable electricity, and will jeopardize the transformational changes envisioned for a modernized grid. Investment in energy storage is essential for keeping pace with the increasing demands for electricity arising from continued growth in U.S. productivity, shifts and continued expansion of national cultural imperatives (e.g., emergence of the distributed grid and electric vehicles), and the projected increase in renewable energy sources.

Chapter 4- Energy storage systems have the potential to extend and optimize the operating capabilities of the grid, since power can be stored and used at a later time. This allows for flexibility in generation and distribution, improving the economic efficiency and utilization of the entire system while making the grid more reliable and robust. Additionally, alternatives to traditional power generation, including variable wind and solar energy technologies, may require back-up power storage. Thus, modernizing the power grid may require a substantial volume of electrical energy storage (EES).

Chapter 5- In late 2007, the U.S. Department of Energy (DOE) initiated a series of studies to address issues related to potential high penetration of distributed photovoltaic (PV) generation systems on our nation's electric grid. This Renewable Systems Interconnection (RSI) initiative resulted in the publication of 14 reports and an Executive Summary that defined needs in areas related to utility planning tools and business models, new grid architectures and PV systems configurations, and models to assess market penetration and the effects of high-penetration PV systems. As a result of this effort, the Solar Energy Grid Integration Systems Program (SEGIS) was initiated in early 2008. SEGIS is an industry-led effort to develop new PV inverters, controllers, and energy management systems that will greatly enhance the utility of distributed PV systems.

In: Lightning in a Bottle: Electrical Energy Storage
Editor: Fedor Novikov
ISBN: 978-1-61470-481-2
© 2011 Nova Science Publishers, Inc.

Chapter 1

ENERGY STORAGE FOR THE ELECTRICITY GRID: BENEFITS AND MARKET POTENTIAL ASSESSMENT GUIDE[*]

Jim Eyer and Garth Corey

NOTICE

Prepared by
Sandia National Laboratories
Albuquerque, New Mexico 87185 and Livermore, California 94550

Sandia is a multiprogram laboratory operated by Sandia Corporation, a Lockheed Martin Company, for the United States Department of Energy's National Nuclear Security Administration under Contract DE-AC04-94AL85000.

Approved for public release; further dissemination unlimited.

Issued by Sandia National Laboratories, operated for the United States Department of Energy by Sandia Corporation.

This report was prepared as an account of work sponsored by an agency of the United States Government. Neither the United States Government, nor any agency thereof, nor any of their employees, nor any of their contractors, subcontractors, or their employees, make any warranty, express or implied, or assume any legal liability or responsibility for the accuracy, completeness, or usefulness of any information, apparatus, product, or process disclosed, or represent that its use would not infringe privately owned rights. Reference herein to any specific commercial product, process, or service by trade name, trademark, manufacturer, or otherwise, does not necessarily constitute or imply its endorsement, recommendation, or favoring by the United States Government, any agency thereof, or any of their contractors or subcontractors. The views and opinions expressed herein do not necessarily state or reflect those of the United States Government, any agency thereof, or any of their contractors.

---

[*] This is an edited, reformatted and augmented version of a Sandia National Laboratories publication, Sandia Report Sand2010-0815, dated February 2010.

ABSTRACT

This guide describes a high-level, technology-neutral framework for assessing potential benefits from and economic market potential for energy storage used for electric-utility-related applications. The overarching theme addressed is the concept of combining applications/benefits into attractive value propositions that include use of energy storage, possibly including distributed and/or modular systems. Other topics addressed include: high-level estimates of application-specific lifecycle benefit (10 years) in $/kW and maximum market potential (10 years) in MW. Combined, these criteria indicate the economic potential (in $Millions) for a given energy storage application/benefit.

The benefits and value propositions characterized provide an important indication of storage system cost targets for system and subsystem developers, vendors, and prospective users. Maximum market potential estimates provide developers, vendors, and energy policymakers with an indication of the upper bound of the potential demand for storage. The combination of the value of an individual benefit (in $/kW) and the corresponding maximum market potential estimate (in MW) indicates the possible impact that storage could have on the U.S. economy.

The intended audience for this document includes persons or organizations needing a framework for making first-cut or high-level estimates of benefits for a specific storage project and/or those seeking a high-level estimate of viable price points and/or maximum market potential for their products. Thus, the intended audience includes: electric utility planners, electricity end users, non-utility electric energy and electric services providers, electric utility regulators and policymakers, intermittent renewables advocates and developers, Smart Grid advocates and developers, storage technology and project developers, and energy storage advocates.

ACKNOWLEDGMENTS

The authors give special thanks to Imre Gyuk of the U.S. Department of Energy (DOE) for his support of this work and related research. Thanks also to Dan Borneo and John Boyes of Sandia National Laboratories for their support. Joel Klein and Mike Gravely of the California Energy Commission, Tom Key of the Electric Power Research Institute Power Electronics Applications Center and Susan Schoenung of Longitude 122 West also provided valuable support. Finally, authors are grateful to Paul Butler of Sandia National Laboratories who provided a thoughtful, thorough, and very valuable review.

This work was sponsored by the DOE Energy Storage Systems Program under contract to Sandia National Laboratories. Sandia is a multiprogram laboratory operated by Sandia Corporation, a Lockheed Martin Company for the DOE's National Nuclear Security Administration under Contract DE-AC04-94AL85000.

EXECUTIVE SUMMARY

Introduction

Electric energy storage is poised to become an important element of the electricity infrastructure of the future. The storage opportunity is multifaceted – involving numerous stakeholders and interests – and could involve potentially rich value propositions. Those rich value propositions are possible because, as described in this report, there are numerous potentially complementary and significant benefits associated with storage use that could be aggregated into attractive value propositions. In addition, proven storage technologies are in use today, while emerging storage technologies are expected to have improved performance and/or lower cost. In fact, recent improvements in energy storage and power electronics technologies, coupled with changes in the electricity marketplace, indicate an era of expanding opportunity for electricity storage as a cost-effective electric energy resource.

Scope and Purpose

This guide provides readers with a high-level understanding of important bases for electricutility-related business opportunities involving electric energy storage. More specifically, this guide is intended to give readers a basic understanding of the benefits for electric-utility-related uses of energy storage.

The guide includes characterization of 26 benefits associated with the use of electricity storage for electric-utility-related applications. The 26 storage benefits characterized are categorized as follows: 1) Electric Supply, 2) Ancillary Services, 3) Grid System, 4) End User/Utility Customer, 5) Renewables Integration, and 6) Incidental. For most of these benefits, the financial value and maximum market potential are estimated. An estimate of the potential economic impact associated with each benefit is also provided.

As a complement to characterizations of individual benefits, another key topic addressed is the concept of aggregating benefits to comprise financially attractive value propositions. Value propositions examples are provided.

Also addressed are storage opportunity drivers, challenges, and notable developments affecting storage. Finally, observations and recommendations are provided regarding the needs and opportunities for electric-energy-storage-related research and development.

Intended Audience

The intended audience for this guide includes persons or organizations needing a framework for making first-cut or high-level estimates of benefits for a specific storage project and/or those seeking a high-level estimate of viable price points and/or maximum market potential for their products. Thus, the intended audience includes, in no particular order: electric utility planners and researchers, non-utility electricity service providers and load aggregators, electricity end users, electric utility regulators and policymakers, and storage project and technology developers and vendors.

Value Propositions

As a complement to coverage of *individual* benefits, a key topic addressed in this guide is the *aggregation* of benefits into financially attractive *value propositions*. That is important because, in many cases, the value of a *single* benefit may not exceed storage cost whereas the value of *combined* benefits may be greater than the cost.

Characterizing the full spectrum of possible value propositions is beyond the scope of this guide; however, eight potentially attractive value propositions are characterized as examples:

1) Electric Energy Time-shift Plus Transmission and Distribution Upgrade Deferral
2) Time-of-use Energy Cost Management Plus Demand Charge Management
3) Renewables Energy Time-shift Plus Electric Energy Time-shift
4) Renewables Energy Time-shift plus Electric Energy Time-shift plus Electric Supply Reserve Capacity
5) Transportable Storage for Transmission and Distribution Upgrade Deferral and Electric Service Power Quality/Reliability at Multiple Locations
6) Storage to Serve Small Air Conditioning Loads
7) Distributed Storage *in lieu* of New Transmission Capacity
8) Distributed Storage for Bilateral Contracts with Wind Generators

Notable Challenges for Storage

Clearly, there are important challenges to be addressed before the full potential for storage is realized. At the highest level, in most cases storage cost exceeds *internalizable* benefits[1] for a variety of reasons, primarily the following:

- High storage cost (relative to internalizable benefits) for modular storage.
- To a large extent, pricing of electric energy and services does not enable storage owners to internalize most benefits.
- Limited regulatory 'permission' to use storage and/or to share benefits among stakeholders – especially benefits from distributed/modular storage.
- Key stakeholders have limited or no familiarity with storage technology and/or benefits.
- Infrastructure needed to control and coordinate storage, especially smaller distributed systems, is limited or does not exist.

---

[1] The concept of an internalizable benefit is an important theme for this report. An internalizable benefit is one that can be 'captured', 'realized', or received by a given stakeholder. An internalizable financial benefit takes the form of revenue and/or a cost reduction or avoided cost.

Notable Storage Opportunity Drivers

Some notable recent and emerging developments driving the opportunities for storage include the following (in no particular order):

- Modular storage technology development in response to the growing market for hybrid vehicles and for portable electronic devices.
- Increasing interest in managing peak demand and reliance on 'demand response' programs – due to peaking generation and transmission constraints.
- Expected increased penetration of distributed energy resources.
- Adoption of the Renewables Portfolio Standard, which will drive increased use of renewables generation with intermittent output.
- Financial risk that limits investment in new transmission capacity, coupled with increasing congestion on some transmission lines and the need for new transmission capacity in many regions.
- Increasing emphasis on richer electric energy and services pricing, such as time-of-use energy prices, locational marginal pricing, and increasing exposure of market-based prices for ancillary services.
- The increasing use of distributed energy resources and the emergence of Smart Grid and distributed energy resource and load aggregation.
- Accelerating storage cost reduction and performance improvement.
- Increasing recognition by lawmakers, regulators, and policymakers of the important role that storage should play in the electricity marketplace of the future.

Research and Development Needs and Opportunities

The following R&D needs and opportunities have been identified as ways to address some of the important challenges that limit increased use of storage:

1) Establish consensus about priorities and actions.
2) Identify and characterize attractive value propositions.
3) Identify and characterize important challenges and possible solutions.
4) Identify and develop standards, models, and tools.
5) Ensure robust integration of distributed/modular storage and Smart Grid.
6) Develop more refined market potential estimates.
7) Develop model risk and reward sharing mechanisms.
8) Develop model rules for utility ownership of distributed/modular storage.
9) Characterize, understand, and communicate the *societal* value proposition for storage.

Key Assumptions and Primary Results

Key assumptions and primary results from the guide are provided in Table ES-1. That table contains five criteria for the 17 primary benefits characterized in this report.

Table ES-1. Summary of Key Assumptions and Results

#	Benefit Type	Discharge Duration* Low	Discharge Duration* High	Capacity (Power: kW, MW) Low	Capacity (Power: kW, MW) High	Benefit ($/kW)** Low	Benefit ($/kW)** High	Potential (MW, 10 Years) CA	Potential (MW, 10 Years) U.S.	Economy ($Million)† CA	Economy ($Million)† U.S.
1	Electric Energy Time-shift	2	8	1 MW	500 MW	400	700	1,445	18,417	795	10,129
2	Electric Supply Capacity	4	6	1 MW	500 MW	359	710	1,445	18,417	772	9,838
3	Load Following	2	4	1 MW	500 MW	600	1,000	2,889	36,834	2,312	29,467
4	Area Regulation	15 min.	30 min.	1 MW	40 MW	785	2,010	80	1,012	112	1,415
5	Electric Supply Reserve Capacity	1	2	1 MW	500 MW	57	225	636	5,986	90	844
6	Voltage Support	15 min.	1	1 MW	10 MW	400		722	9,209	433	5,525
7	Transmission Support	2 sec.	5 sec.	10 MW	100 MW	192		1,084	13,813	208	2,646
8	Transmission Congestion Relief	3	6	1 MW	100 MW	31	141	2,889	36,834	248	3,168
9.1	T&D Upgrade Deferral 50th percentile††	3	6	250 kW	5 MW	481	687	386	4,986	226	2,912
9.2	T&D Upgrade Deferral 90th percentile††	3	6	250 kW	2 MW	759	1,079	77	997	71	916
10	Substation On-site Power	8	16	1.5 kW	5 kW	1,800	3,000	20	250	47	600
11	Time-of-use Energy Cost Management	4	6	1 kW	1 MW	1,226		5,038	64,228	6,177	78,743
12	Demand Charge Management	5	11	50 kW	10 MW	582		2,519	32,111	1,466	18,695
13	Electric Service Reliability	5 min.	1	0.2 kW	10 MW	359	978	722	9,209	483	6,154
14	Electric Service Power Quality	10 sec.	1 min.	0.2 kW	10 MW	359	978	722	9,209	483	6,154
15	Renewables Energy Time-shift	3	5	1 kW	500 MW	233	389	2,889	36,834	899	11,455
16	Renewables Capacity Firming	2	4	1 kW	500 MW	709	915	2,889	36,834	2,346	29,909

Table ES-1. (Continued).

#	Benefit Type	Discharge Duration* Low	Discharge Duration* High	Capacity (Power: kW, MW) Low	Capacity (Power: kW, MW) High	Benefit ($/kW)** Low	Benefit ($/kW)** High	Potential (MW, 10 Years) CA	Potential (MW, 10 Years) U.S.	Economy ($Million)† CA	Economy ($Million)† U.S.
17.1	Wind Generation Grid Integration, Short Duration	10 sec.	15 min.	0.2 kW	500 MW	500	1,000	181	2,302	135	1,727
17.2	Wind Generation Grid Integration, Long Duration	1	6	0.2 kW	500 MW	100	782	1,445	18,417	637	8,122

*Hours unless indicated otherwise. min. = minutes. sec. = seconds.
**Lifecycle, 10 years, 2.5% escalation, 10.0% discount rate.
† Based on potential (MW, 10 years) times average of low and high benefit ($/kW).
†† Benefit for *one year*. However, storage could be used at more than one location at different times for similar benefits.

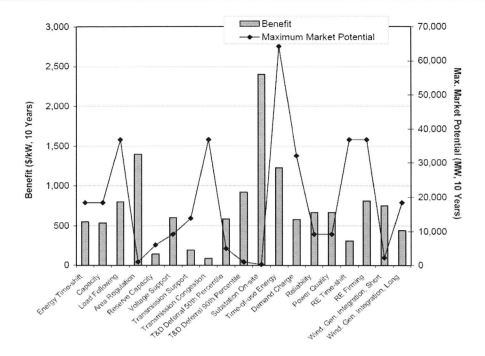

Figure ES-1. Application-specific 10-year benefit and maximum market potential estimates for the U.S.

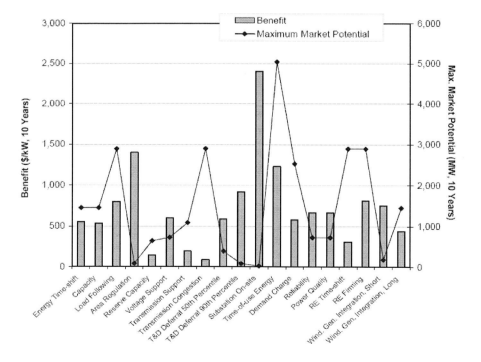

Figure ES-2. Application-specific 10-year benefit and maximum market potential estimates for California.

Discharge duration indicates the amount of time that the storage must discharge at its rated output before charging. Capacity indicates the range of storage system power ratings that apply for a given benefit. The benefit indicates the present worth of the respective benefit type for 10 years (2.5% inflation, 10% discount rate). Potential indicates the maximum market potential for the respective benefit over 10 years. Economy reflects the total value of the benefit given the maximum market potential.

Financial benefits and maximum market potential estimates for the U.S. are provided in Figure ES-1. The same values for California are provided in Figure ES-2.

Care must be used when aggregating specific benefits and market potential values because there may be technical and/or operational conflicts, and/or institutional barriers may hinder or even preclude aggregation, as described in Section 4.4.2.

ACRONYMS AND ABBREVIATIONS

AC	alternating current
A/C	air conditioning
ACE	area control error
AGC	automated generation control
AMI	Advanced Metering Infrastructure
CAES	compressed air energy storage
CAISO	California Independent System Operator
CEC	California Energy Commission
C&I	commercial and industrial (energy users)
DC	direct current
DER	distributed energy resource(s)
DOB	dynamic operating benefit
DOE	U.S. Department of Energy
ELCC	effective load carrying capacity
EPRI	Electric Power Research Institute
EV	electric vehicle
FACTS	flexible AC transmission systems
FERC	Federal Energy Regulatory Commission
kW	kilowatt
kWh	kilowatt-hour
kV	kilovolt
kVA	kilovolt-Ampere (or kilovolt-Amp)
kVAR	kilovolt-Ampere reactive (or kilovolt-Amp reactive)
IEEE	Institute of Electronics and Electrical Engineers
IOU	investor-owned utility
ISO	independent system operator
I^2R	pronounced "I squared R" meaning current squared times electric resistance
LDC	load duration curve
Li-ion	lithium-ion
MES	modular energy storage

MW	megawatt
MWh	megawatt-hour
MVA	megavolt-Ampere (or megavolt-Amp)
Na/S	sodium/sulfur
NERC	North American Electric Reliability Council
NiCad	nickel-cadmium
Ni-MH	nickel-metal hydride
O&M	operation and maintenance
ORNL	Oak Ridge National Laboratory
PCU	power conditioning unit
PEAC	Power Electronics Applications Center
PEV	plug-in electric vehicle
PG&E	Pacific Gas and Electric Company
PHEV	plug-in hybrid electric vehicle
PV	photovoltaic
PW	present worth (factor)
R&D	research and development
RPS	Renewables Portfolio Standard
SCADA	supervisory control and data acquisition
SMES	superconducting magnetic energy storage
SNL	Sandia National Laboratories
StatCom	static synchronous compensator
T&D	transmission and distribution
THD	total harmonic distortion
TOU	time-of-use (energy pricing)
UPS	uninterruptible power supply
VAR	volt-Amperes reactive (or volt-Amps reactive)
VOC	variable operating cost
VOS	value-of-service
Zn/Br	zinc/bromine

GLOSSARY

Area Control Error (ACE) – The momentary difference between electric supply and electric demand within a given part of the electric grid (area).

Automated Generation Control (AGC) – A protocol for dispatching electric supply resources (possibly including demand management) in response to changing demand. AGC resources often respond by changing output at a rate of a few percentage points per minute over a predetermined output range. The AGC signal can vary as frequently as every six seconds though generation is rarely called upon to respond that frequently. Typically, generation responds to an average of that more frequent signal, such that a response (change of output) is required once per minute or perhaps as infrequently as every five minutes.

Application – A specific way or ways that energy storage is used to satisfy a specific need; how/for what energy storage is used.
Arbitrage – Simultaneous purchase and sale of identical or equivalent commodities or other instruments across two or more markets in order to benefit from a discrepancy in their price relationship.
Benefit – See Financial Benefit.
Beneficiaries – Entities to whom financial benefits accrue due to use of a storage system.
Carrying Charges – The annual financial requirements needed to service debt and/or equity capital used to purchase and to install capital equipment (*i.e.*, a storage plant), including tax effects. For utilities, this is the revenue requirement. See also Fixed Charge Rate.
Combined Applications – Energy storage used for two or more *compatible* applications.
Combined Benefits – The sum of all benefits that accrue due to use of an energy storage system, regardless of the purpose for installing the system.
Demand Response – Controlled reduction of power draw by electricity end users accomplished via automated communication and control protocols done to balance demand and supply, possibly *in lieu* of adding generation and/or transmission and distribution (T&D) capacity.
Discharge Duration – Total amount of time that the storage plant can discharge, at its nameplate rating, without recharging. Nameplate rating is the nominal full-load rating, not the emergency, short-duration, or contingency rating.
Discount Rate – The interest rate used to discount future cash flows to account for the time value of money. For this document, the assumed value is 10%.
Dispatchable – Electric power resource whose output can be controlled – increased and/or decreased – as needed. Applies to generation, storage, and load-control resources.
Diurnal – Having a daily cycle or occurring every day.
Diversity – The amount of variability and/or difference there is among members of a group. To the extent that electric resources are diverse – with regard to geography and/or fuel – their reliability is enhanced because diversity limits the chance that failure of one or a few individual resources will cause significant problems.
Economic Benefit – The sum of all financial benefits that accrue to all beneficiaries using storage. For example, if the average financial benefit is $100 for 1 million storage users then the *economic* benefit is $100 × 1 million = $100 Million. See Financial Benefit.
Efficiency (Storage Efficiency) – See Round-trip Efficiency.
Effective Load Carrying Capacity (ELCC) – A characterization of a generator's contribution to planning reserves for a given level of electric supply system reliability. ELCC is a robust and mathematically consistent measure of capacity value. ELCC can be used to establish appropriate payments for resources used to provide capacity needed to meet system reliability goals.
Financial Benefit (Benefit) – Monies received and/or cost avoided by a specific beneficiary, due to use of energy storage.
Financial Life – The plant life assumed when estimating lifecycle costs and benefits. A plant life of 10 years is assumed for lifecycle financial evaluations in this document (*i.e.*, 10 years is the standard assumption value).

Fixed Charge Rate – The rate used to convert capital plant installed cost into an annuity equivalent (payment) representing annual carrying charges for capital equipment. It includes consideration of interest and equity return rates, annual interest payments and return of debt principal, dividends and return of equity principal, income taxes, and property taxes. The standard assumption value is 0.13 for utilities.

Flexible AC Transmission Systems (FACTS) – "A power electronic-based system and other static equipment that provide control of one or more alternating current (AC) transmission system parameters to enhance controllability and increase power transfer capability."[2]

I^2R Energy Losses – Energy losses incurred during transmission and distribution of electric energy, due to heating in an electrical system, caused by electrical currents in the conductors of transformer windings or other electrical equipment. I^2R (pronounced I squared R) indicates that those energy losses are a function of the square of the current (I^2) times the resistance (R) per Joule's Law (which characterizes the amount of heat generated when current flows through a conductor). So, for example, reducing current by 50% reduces I^2R energy losses to one quarter of the original value.

Inflation Rate (Inflation) – The annual average rate at which the price of goods and services increases during a specific time period. For this document, inflation is assumed to be 2.5% per year.

Internalizable Benefit – A benefit (revenue and/or reduced cost) that accrues, in part or in whole, to a *specific* stakeholder or stakeholders. A benefit is most readily internalizable if there is a price associated with it.

Lifecycle – See Financial Life.

Lifecycle Benefit – Present worth (value) of financial benefits that are expected to accrue over the life of a storage plant.

Load Duration Curve (LDC) – Hourly demand values (usually for one year) arranged in order of magnitude, regardless of which hour during the year that the demand occurs. Values to the left represent the highest levels of demand during the year and values to the right represent the lowest demand values during the year.

Loss of Load Expectation – Measure of the electric supply system's reliability that indicates the adequacy of the system to satisfy demand.

Loss of Load Probability – measure of the electric supply system's reliability indicating the likelihood that the system cannot satisfy demand.

Market Estimate – The estimated amount of *energy storage* capacity (MW) that will be installed. For this document, market estimates are made for a 10-year period. Market estimates reflect consideration of prospects for lower cost alternatives to compete for the same applications and benefits. (The Market Estimate is a portion of the Maximum Market Potential.)

Maximum Market Potential – The maximum potential for *actual sale and installation of energy storage,* estimated based on reasonable assumptions about technology and market readiness and trends, and about the persistence of existing institutional challenges. In the context of this document, it is the *plausible market potential* for a given application. (The Maximum Market Potential is a portion of the Market Technical Potential.)

---

[2] Definition provided by the Institute of Electrical and Electronics Engineers (IEEE).

Market Technical Potential – The estimated *maximum possible* amount of energy storage (MW and MWh) that could be installed over 10 years, *given purely technical constraints*.

Plant Rating (Rating) – Storage plant ratings include two primary criteria: 1) *power* – nominal power output and 2) e*nergy* – the maximum amount of energy that the system can deliver to the load without being recharged.

Present Worth Factor (PW Factor) – A value used to estimate the present worth of a stream of annual expenses or revenues. It is a function of a specific combination of investment duration (equipment life), financial escalation rate (*e.g.,* inflation), and an annual discount rate. The PW factor of 7.17 used in this guide is based on the following standard assumption values: a 10-year equipment life, 2.5% annual price/cost inflation rate, 10% annual discount rate, and a mid-year convention.

Price Inflation Rate (Inflation) – See Inflation.

Revenue Requirement – For a utility, the amount of annual revenue required to pay carrying charges for capital equipment and to cover expenses including fuel and maintenance. See also Carrying Charges and Fixed Charge Rate.

Round-trip Efficiency – The amount of electric energy output from a given storage plant/system per unit of electric energy input.

Smart Grid – A concept involving an electricity grid that delivers electric energy using communications, control, and computer technology for lower cost and with superior reliability. As characterized by the U.S. Department of Energy, the following are characteristics or performance features of a Smart Grid: 1) self-healing from power disturbance events; 2) enabling active participation by consumers in demand response; 3) operating resiliently against physical and cyber attack; 4) providing power quality for 21st century needs; 5) accommodating all generation and storage options; 6) enabling new products, services, and markets; and 7) optimizing assets and operating efficiently.

Societal Benefit – A benefit that accrues, in part or in whole, to utility customers as a group and/or to society at large.

Standard Assumption Values (Standard Values) – Standardized/generic values used for example calculations. For example, financial benefits are calculated based on the following standard assumption values: a 10-year lifecycle, 10% discount rate, and 2.5% annual inflation. See also Standard Calculations.

Standard Calculations – Methodologies for calculating benefits and market potential – used in conjunction with Standard Assumption Values.

Storage Discharge Duration – See Discharge Duration.

Storage System Life (System Life) – The period during which the storage system is expected to be operated. For this document, the Storage System Life is equal to the Financial Life.

Supervisory Control and Data Acquisition (SCADA) – A generic term describing various approaches used to automate monitoring and control of T&D equipment and to gather and store data about equipment operation.

1. INTRODUCTION

1.1. About this Document

This document provides high-level characterizations of electric energy storage applications, including key characteristics needed for storage used in electric-grid-related applications. Financial benefits and maximum market potential estimates, in California and the U.S., are provided for those applications.

Financial benefit estimates provide an indication of the financial attractiveness of storage for specific applications. Individual benefits provide bases for value propositions that comprise two or more individual benefits, especially value propositions involving benefits that exceed cost.

Application-specific maximum market potential estimates provide an indication of the potential demand for storage. Values for application-specific benefits are multiplied by the maximum market potential to estimate the potential economic effect ($Millions) for storage used for specific applications.

The goal is to provide 1) bases for first-cut or screening-level evaluation of the benefits and market potential for specific, possibly attractive, storage value propositions and 2) a possible framework for making region-specific or circumstance-specific estimates.

The presentation in this document is storage-technology-neutral, though there is some coverage of storage technology system characteristics as context for coverage of applications, benefits, and value propositions. In fact, value propositions characterized using values and insights in this report may provide a helpful indication of storage system cost and performance targets. Many other existing resources can be used to determine the cost for, and technical viability of, specific storage types.[1][2][3]

1.2. Background and Genesis

The original work underlying this report, supported and funded by the U.S. Department of Energy (DOE), was developed in support of the California Energy Commission (CEC) Public Interest Energy Research (PIER) Program. The purpose of that work – documented in the report *Energy Storage Benefits and Market Analysis Handbook* (Sandia National Laboratories report #SAND2004-6177) – was to provide guidance for organizations seeking CEC co-funding for storage demonstrations. The approach used for selecting co-funding proposals emphasized demonstration of storage to be used for a specific *value proposition*. Furthermore, the CEC gave some preference to value propositions with more potential to have a positive impact.

1.3. Intended Audience

The intended audience for this document includes persons or organizations needing a framework for making first-cut or high-level estimates of benefits for a specific storage project and/or those seeking a high-level estimate of viable price points and/or maximum market potential for their products. Thus, the intended audience includes, in no particular order: electric utility planners and researchers, non-utility electricity service providers and load aggregators, electricity end users, electric utility regulators and policymakers, and storage project and technology developers, and vendors.

1.4. Analysis Philosophy

The methodologies used to estimate application-specific values for benefits and market potential are intended to balance a general preference for precision with the cost to perform rigorous financial assessments and to make rigorous market assessments. Much of the data needed for a more rigorous approach is proprietary or otherwise unavailable; is too expensive, does not exist in a usable form, or does not exist at all. It is also challenging to establish extremely credible generic values for benefits when those values are somewhat-to-very specific to region and circumstances. Similarly, making national estimates of maximum market potential using limited data requires many assumptions that are established using a combination of informal surveys of experts, subjectivity, and authors' familiarity with the subject. Nonetheless, despite those challenges, this report includes just such estimates of generic, application-specific values for benefits and maximum market potential.

Given the diversity of California's generation mix, load types and sizes, regions, weather conditions, *etc.*, it was assumed to be a reasonable basis for estimating national values. The application-specific benefit estimates are especially California-centric. Also, maximum market potential estimates developed for California are extrapolated to estimate values for the entire country. (See Section 4 for details.)

Although the methodology used to estimate benefits and maximum market potential involves some less than rigorous analysis, it was the authors' intention to make reasonable attempts to document assumptions and methodologies used so that the evaluation is as transparent and auditable as is practical. This gives the necessary information to readers and analysts so that they may consider the merits and appropriateness of data and methodologies used in this report. To the extent that superior data or estimates are available, and/or a superior or preferred estimation methodology exists, those should be used *in lieu* of the assumptions and approaches in this report.

Similarly, given the generic nature of the benefit estimates, for specific situations or projects it is prudent to undertake a more circumstance-specific and possibly more detailed evaluation than is possible using the assumptions and estimates in this guide.

1.4.1. Application Versus Benefit

It is important to note the distinction made in this document between applications and benefits. In general terms, an *application* is a *use* whereas a *benefit* connotes a *value*. In many cases, a benefit is quantified in terms of the monetary or financial value. Of course, some qualitative benefits – such as the 'goodness' of reduced noise and improved aesthetics – may not be readily quantifiable and/or expressed in financial terms.

1.4.2. Internalizable Benefits

The concept of an internalizable benefit is an important theme for this report. An internalizable benefit is one that can be 'captured', 'realized', or received by a given stakeholder or stakeholders. An internalizable financial benefit takes the form of revenue or reduced cost. A benefit is most readily internalizable if there is a price associated with it. (Some refer to a benefit for which there is an established financial value – especially in the form of a price – as a benefit that is 'monetized'.)

An example of a readily internalized benefit is electricity bill reduction that accrues to a utility customer who uses storage to reduce on-peak a) energy cost and b) demand charges. In that example, the benefit is a function of a) the amount of energy and the level of demand involved and b) the on-peak and the off-peak prices for energy and the on-peak demand charge.

Continuing with the example; consider that the same customer-owned and -operated storage could also reduce or delay the need (*and cost*) for additional utility-owned transmission and distribution (T&D) capacity. The resulting 'T&D upgrade deferral' benefit (*i.e.,* reduced, deferred or avoided cost) though real, cannot be directly internalized by the utility customer who installs the storage. That is because there is no established 'price' associated with reducing the need for a *specific* T&D capacity upgrade (*i.e.,* the utility's avoided cost cannot be shared with end users who take actions that defer/reduce the need and cost for a T&D upgrade). Rather, the resulting T&D upgrade deferral benefit is internalized by the utility and/or the utility's ratepayers as a group (in the form of reduced, deferred, or avoided price increase).

1.4.3. Societal Benefits

Although not addressed in detail in this report, it is important to consider some important storage-related benefits that accrue, in part or in whole, to electric utility customers as a group and/or to society at large. Three examples of possible storage-related societal benefits are the integration of more renewables, more effectively; reduced air emissions from generation; and improved utilization of grid assets (*i.e.,* generation and T&D equipment).

In most cases, societal benefits are accompanied by an internalizable or partially internalizable benefit. Consider an example: A utility customer uses storage to reduce on-peak energy use. An internalizable benefit accrues to that customer in the form of reduced cost; however, other societal benefits may accrue to utility customers as a group and/or to society as a whole. For example, reduced peak demand could lead to reduced need for generation and transmission capacity, reduced air emissions, and a general improvement of businesses' cost competitiveness.

This topic is especially important for lawmakers, electric utility regulators, energy and electricity policymakers and policy analysts, and storage advocates as laws, regulations, and policies that could affect prospects for increased storage use are developed.

1.5. Grid and Utility-related General Considerations

Applications described in this report affect the electric supply system and the T&D system – known collectively as 'the grid'. This subsection characterizes several important considerations and topics related to the electric grid. Those topics are presented here as context for results presented throughout the rest of this report.

1.5.1. Real Power versus Apparent Power

For the purposes of this document, units of kW and MW (real or true power) are used universally when kVA and MVA (apparent power) may be the more technically correct units. Given the degree of precision possible for market potential and financial benefit estimation, the distinction between these units has relatively little impact on most results.[3]

1.5.2. Ancillary Services

Some possible uses of storage are typically classified as ancillary services. The electric utility industry has a specific definition of ancillary services. (See Appendix A for brief overview of ancillary services.)

Three specific ancillary services are explicitly addressed in this report: 1) area regulation, 2) electric supply reserve capacity, and 3) voltage support. Although not always categorized as an ancillary service, in this guide load following is also included in the ancillary services category.

1.5.3. Electricity Transmission and Distribution

The electric utility transmission and distribution (T&D) system comprises three primary subsystems: 1) transmission, 2) subtransmission, and 3) distribution, as described below. Several storage applications involve benefits associated with one or more of these subsystems.

Electricity Transmission

Electricity transmission is the backbone of the electric grid. Transmission wires, transformers, and control systems transfer electricity from supply sources (generation or electricity storage) to utility distribution systems. Often, the transmission system is used to send large amounts of electricity over relatively long distances. In the U.S., transmission system operating voltages generally range from 200 kV (200,000 V) to 500 kV (500,000 V). Transmission systems typically transfer the equivalent of 200 MW to 500 MW. Most transmission systems use alternating current (AC), though some larger, longer transmission corridors employ high-voltage direct current (DC).

Electricity Subtransmission

Relative to transmission, subtransmission transfers smaller amounts of electricity, at lower operating voltages, over shorter distances. Normally, subtransmission voltages fall within the range of 50 kV (50,000 V) to 100 kV (100,000 V) with 69 kV (69,000 V) being somewhat common.

Electricity Distribution

Electricity distribution is the part of the electric grid that delivers electricity to end users. It is connected to the subtransmission system which, in turn, is connected to the transmission

---

[3] In practice, there are important technical and cost differences between true power (kW or MW) and apparent power (kVA or MVA). Various load types reduce the effectiveness of the grid by, for example, injecting harmonic currents or by increasing reactive power flows. As a general indication of the magnitude of the difference, consider this example: a power system serves 10 MW of peak load (true power). During times when load is at its peak, the 'power factor' may drop to 0.85. Given that power factor, the T&D equipment should have an apparent power rating of at least 10 MW/0.85 = 11.76 MVA.

system and the electric supply system (generation). Relative to electricity transmission, the distribution system is used to send relatively small amounts of electricity over relatively short distances. In the U.S., distribution system operating voltages generally range from a few thousand volts to 50 kV. Typical power transfer capacities range from a few tens of MW for substation transformers to as few as tens of kW for very small circuits.

Two applications addressed in this report apply only to the transmission system: 1) transmission support and 2) transmission congestion relief.

1.5.4. Utility Regulations and Rules

Some of the benefits characterized in this report may not apply in any particular circumstance because provisions of applicable rules or regulations may not provide the means for a given stakeholder to internalize the benefit. For example, one application characterized is demand charge reduction for utility customers; but, if the customer is not eligible for demand charges, then that application does not apply. Consider another example: A utility customer with 100 kW may not be allowed to participate in the market for ancillary services (without some type of 'load aggregation') because the minimum capacity required is 1 MW.

1.5.5. Utility Financials: Fixed Charge Rate

Some important applications involve storage used to reduce the need to own other utility equipment – generation, transmission, and/or distribution. The cost *reduction* is often referred to as an *avoided cost*.

For investor-owned utilities (IOUs), the avoided cost of equipment ownership is primarily consists of six elements: 1) interest payments for bond holders, 2) equity returns (dividends) for stock owners, 3) annual return of principal or depreciation, 4) income taxes, 5) property taxes, and 6) insurance.

Though circumstances can vary, the avoided cost for municipal utilities (munis) and co-operative utilities (coops) includes annual interest payments and 'return of capital' (*i.e.*, amortization). Cooperatives' cost may also be subject to property taxes and insurance.

When estimating benefits related to deferred or avoided cost for utility equipment ownership, it is usually necessary to first estimate the *annual* cost. Utilities often refer to this annual avoided cost as the *annual revenue requirement* because it is equal to the annual revenue needed (from utility customers) to cover the full cost of owning the equipment.

> Although the topic is beyond the scope of this guide, readers should note the important distinction between—
> 1) avoided cost for ownership of a capital *investment* (in this case, utility equipment) and
> 2) avoided cost for an *expense* incurred due to equipment operation, such as the cost for fuel or variable maintenance.
> The distinction is important because investor-owned utilities' profit is based on investments made in equipment, whereas expenses are pass throughs to end users as-is (*i.e.*, without profit).

In this guide, a fixed charge rate is used to estimate *annual* avoided cost of equipment ownership. The fixed charge rate reflects the six elements of utility equipment cost listed above (annual interest and equity payments, *etc.*) as applicable for a given utility.

Annual avoided cost is calculated by multiplying the equipment's total *installed* cost by a utility-specific fixed charge rate. (Installed cost includes all costs incurred until equipment enters service, including equipment purchase price, design, installation, commissioning, *etc.*)

Note that the annual avoided cost calculated using the fixed charge rate is equivalent to an annuity payment involving a series of equal annual payments over the equipment's life, similar to a mortgage. Given that the annual avoided cost is expressed as equal annual payments, it is often referred to as a 'levelized' cost.

Consider an example: A new storage system costing $500,000 is installed. Given the utility financial structure and the expected life of the storage system, the utility financial group calculates the fixed charge rate for the equipment to be 0.11. So, the full 'capital carrying charges' incurred to own the storage plant (without regard to energy charging cost and other variable expenses) is $500,000 × 0.11 = $55,000 per year for each year during the expected life of the storage plant. (A fixed charge rate of 0.11 is the standard value used in this guide.)

1.6. Standard Assumption Values

Standard assumption values established for this guide are used to make high-level, generic estimates of financial benefits and maximum market potential for storage. Key standard assumption values are those provided for financial criteria and for storage discharge duration, power rating, and maximum market potential.

Certainly, to one extent or another, establishing such generic values requires subjectivity, speculation, simplifying assumptions, and/or generalizations. So, for any particular circumstance or situation, analysts are encouraged to use circumstance-specific assumptions and/or additional or superior information to establish superior values instead of the generic assumptions, as appropriate. To the extent possible, the rationale and underlying assumptions used to establish standard assumption values are presented and described in this report.

1.6.1. Standard Assumption Values for Financial Calculations
The following standard assumption values are used in this report to generalize and to simplify the calculations used as examples.

1.6.1.1. Storage Project Life
A storage project life of 10 years is assumed for lifecycle financial evaluations. That is an especially important standard assumption value for a variety of reasons. Clearly, using any one value is suboptimal because, if nothing else, each storage type and system may have a different life and each circumstance is different. Important factors affecting storage life also include the way(s) and amount that storage is used and the frequency and quality of storage system maintenance.

Given such considerations, without selecting one standard assumption for storage project life, it is conceivable that many estimates would have to be made for each benefit. Estimating benefits for various timeframes would add complexity to the evaluations and would yield results that are unwieldy and challenging to report. Furthermore, making numerous estimates for each benefit would require more resources than were allocated for this report.

Although the selection of 10 years is may seem somewhat arbitrary, there was a rationale for doing so. First, though a 10-year life is too short for compressed-air energy storage (CAES) and pumped hydro, it may be generous for the other storage types, given their somewhat-to-very limited record. Additionally, estimates of benefits accruing over periods of 10 to 20 years may not be credible and/or precise, given expected changes to and increasing uncertainty in the electricity marketplace. In fact, given that uncertainty, there is even a chance that some of the benefits may not even exist 10 or 20 years from now. Finally, when accounting for the time value of money, a significant majority of benefits accrue in the first 10 years.

Consider also that, for most benefits, there may be fairly straightforward ways to adjust benefit estimates to accommodate timeframes that are longer than the 10 years assumed. Section 1.6.1.4 provides an indication of a simplified way to accommodate a lifecycle other than 10 years.

1.6.1.2. Price Escalation

A general price escalation of 2.5% per year is assumed for the analysis in this guide. Electric energy and capacity costs and prices are assumed to escalate at that rate during the storage plant's financial life.

1.6.1.3. Discount Rate for Present Worth Calculations

An annual discount rate of 10.0% is used for making present worth (PW) calculations to estimate lifecycle benefits.

1.6.1.4. Present Worth Factor

The simplified approach described below for estimating the present worth (PW) of a stream of annual expenses or revenues is used throughout this guide. It is intended to provide a simple, auditable, and flexible way to estimate PW. Detailed treatment of more sophisticated financial calculations is beyond the scope of this guide.

Present worth calculations are made using these standard assumptions:

- 2.5% per year annual price/cost escalation
- 10.0% per year discount rate
- 10-year storage equipment life
- Mid-year convention

The PW factor is calculated based on these assumptions. That value is used to estimate present worth based on the value in the first year of operation. Given the standard assumption values of 2.5% cost/price escalation rate, 10% discount rate, and 10-year storage system life, the standard assumption value for the PW factor is 7.17.

Consider an example of how the PW factor is used: For an annual/first year benefit of $100,000, the estimated lifecycle benefit is $100,000 · 7.17 = $717,000 (present worth) for 10 years.

The equation for the PW factor for a 10-year service life is as follows:

$$\text{PW Factor} = \sum_{i=1}^{10} \frac{(1+e)^{i-.5}}{(1+d)^{i-.5}}$$

e = annual price escalation rate (%/year)
d = discount rate (%/year)
i = year

Figure 1 shows PW factors for three discount rates, assuming a cost escalation of 2.5% per year. (Note that the value of 'I' is calculated at mid-year.) For a given life/discount rate combination, the PW factor represents the present worth for a sum of a stream of annual values. Table 1 includes PW factors for Years 5 to 20 for a discount rate of 10% (shown with the solid line). The figure allows for quick comparisons of annually recurring costs and benefits for various storage project lifecycles and discount rates.

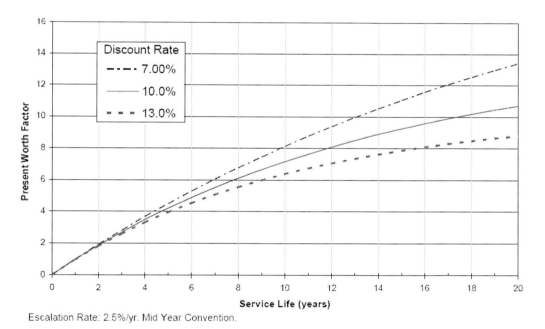

Figure 1. Present worth factors.

Table 1. Present Worth Factors, 2.5% Escalation, 10% Discount Rate

Year	5	6	7	8	9	10	11	12	13	14	15	16	17	18	19	20
PW Factor	4.21	4.89	5.52	6.11	6.66	7.17	7.65	8.09	8.5	8.89	9.25	9.58	9.9	10.2	10.5	10.7

Consider another example: Assume that a storage plant will operate for 20 years and that it has a first-year operating cost of $100,000 which is expected to escalate at a rate of 2.5% per year. If the owner uses a 13% discount rate, then the PW factor is about 8.80 (as shown in Figure 1). So, the 20-year present worth of all operating costs (before taxes) is

$100,000 · 8.80 = $880,000.

Implicit in this approach is the assumption that annual benefits for all years considered (10 in this case) are the same as the first year, except that the cost or price escalates at 2.5%. If that approach is not appropriate, then an actual cash flow evaluation may be required to estimate the lifecycle benefits.

1.6.1.5. Fixed Charge Rate

The standard assumption value for fixed charge rate – which applies to utilities only – is 0.11. The fixed charge rate is used as follows: Consider utility equipment whose installed cost is $500,000. The utility's *annual* revenue requirement (and avoided cost) is

$500,000 · 0.11 = $55,000/year.

1.7. Results Summary

Key study results are summarized in Table 2. The table contains three criteria for the 17 primary benefits characterized in this guide, for California and for the U.S.: 1) benefit, 2) potential, and 3) economy. The 'benefit' value indicates the present worth of the respective benefit type for 10 years (assuming 2.5% inflation and 10% discount rate). 'Potential' indicates the maximum market potential for the respective benefit over 10 years. 'Economy' reflects the total value of the benefit given the maximum market potential.

Table 2. Primary Results Summary — Benefits, Maximum Market Potential, and Maximum Economic Value

#	Benefit Type	Benefit ($/kW)** Low	High	Potential (MW, 10 Years) CA	U.S.	Economy ($Million)† CA	U.S.
1	Electric Energy Time-shift	400	700	1,445	18,417	795	10,129
2	Electric Supply Capacity	359	710	1,445	18,417	772	9,838
3	Load Following	600	1,000	2,889	36,834	2,312	29,467
4	Area Regulation	785	2,010	80	1,012	112	1,415
5	Electric Supply Reserve Capacity	57	225	636	5,986	90	844
6	Voltage Support	400		722	9,209	433	5,525

Table 2. (Continued)

	Benefit Type	Benefit ($/kW)** Low	Benefit ($/kW)** High	Potential (MW, 10 Years) CA	Potential (MW, 10 Years) U.S.	Economy ($Million)† CA	Economy ($Million)† U.S.
7	Transmission Support	192		1,084	13,813	208	2,646
8	Transmission Congestion Relief	31	141	2,889	36,834	248	3,168
9.1	T&D Upgrade Deferral 50th percentile††	481	687	386	4,986	226	2,912
9.2	T&D Upgrade Deferral 90th percentile††	759	1,079	77	997	71	916
10	Substation On-site Power	1,800	3,000	20	250	47	600
11	Time-of-use Energy Cost Management	1,226		5,038	64,228	6,177	78,743
12	Demand Charge Management	582		2,519	32,111	1,466	18,695
13	Electric Service Reliability	359	978	722	9,209	483	6,154
14	Electric Service Power Quality	359	978	722	9,209	483	6,154
15	Renewables Energy Time-shift	233	389	2,889	36,834	899	11,455
16	Renewables Capacity Firming	709	915	2,889	36,834	2,346	29,909
17.1	Wind Generation Grid Integration, Short Duration	500	1,000	181	2,302	135	1,727
17.2	Wind Generation Grid Integration, Long Duration	100	782	1,445	18,417	637	8,122

*Hours unless indicated otherwise. min. = minutes. sec. = seconds.
**Lifecycle, 10 years, 2.5% escalation, 10.0% discount rate.
†Based on potential (MW, 10 years) times average of low and high benefit ($/kW).
†† Benefit for *one year*. However, storage could be used at more than one location at different times for similar b.

2. ELECTRIC ENERGY STORAGE TECHNOLOGY OVERVIEW

A general introduction to energy storage technology is provided as context for the applications and benefits addressed in this guide. Storage technology and subsystems are subjects covered in detail by other studies and reports. Section 2.1 provides a brief description of storage types. Sections 2.2 through 2.20 briefly describe important storage characteristics. Note that the order in which these characteristics are presented is not meant to imply order of importance.

2.1. Overview of Storage Types

2.1.1. Electrochemical Batteries

Electrochemical batteries consist of two or more electrochemical cells. The cells use chemical reaction(s) to create a flow of electrons – electric current. Primary elements of a cell include the container, two electrodes (anode and cathode), and electrolyte material. The electrolyte is in contact with the electrodes. Current is created by the oxidation-reduction process involving chemical reactions between the cell's electrolyte and electrodes.

When a battery discharges through a connected load, electrically charged ions in the electrolyte that are near one of the cell's electrodes supply electrons (oxidation) while ions near the cell's other electrode accept electrons (reduction), to complete the process. The process is reversed to charge the battery, which involves ionizing of the electrolyte.

An increasing number of chemistries are used for this process. More familiar ones include lead-acid, nickel-cadmium (NiCad), lithium-ion (Li-ion), sodium/sulfur (Na/S), zinc/bromine (Zn/Br), vanadium-redox, nickel-metal hydride (Ni-MH), and others.

2.1.1.1. Flow Batteries

Some electrochemical batteries (*e.g.,* automobile batteries) contain electrolyte in the same container as the cells (where the electrochemical reactions occur). Other battery types – called flow batteries – use electrolyte that is stored in a separate container (*e.g.,* a tank) outside of the battery cell container. Flow battery cells are said to be configured as a 'stack'.

When flow batteries are charging or discharging, the electrolyte is transported (*i.e.,* pumped) between the electrolyte container and the cell stack. Vanadium redox and Zn/Br are two of the more familiar types of flow batteries. A key advantage to flow batteries is that the storage system's discharge duration can be increased by adding more electrolyte (and, if needed to hold the added electrolyte, additional electrolyte containers). It is also relatively easy to replace a flow battery's electrolyte when it degrades.

2.1.2. Capacitors

Capacitors store electric energy as an electrostatic charge. An increasing array of larger capacity capacitors have characteristics that make them well-suited for use as energy storage.[1] They store significantly more electric energy than conventional capacitors. They are especially well-suited to being discharged quite rapidly, to deliver a significant amount of energy over a short period of time (*i.e.,* they are attractive for high-power applications that require short or very short discharge durations).

2.1.3. Compressed Air Energy Storage

Compressed air energy storage (CAES) involves compressing air using inexpensive energy so that the compressed air may be used to generate electricity when the energy is worth more. To convert the stored energy into electric energy, the compressed air is released into a combustion turbine generator system. Typically, as the air is released, it is heated and then sent through the system's turbine. As the turbine spins, it turns the generator to generate electricity.

---

[1] Trade names for such devices include Supercapacitor and Ultracapacitor.

For larger CAES plants, compressed air is stored in underground geologic formations, such as salt formations, aquifers, and depleted natural gas fields. For smaller CAES plants, compressed air is stored in tanks or large on-site pipes such as those designed for high-pressure natural gas transmission (in most cases, tanks or pipes are above ground).

2.1.4. Flywheel Energy Storage

Flywheel electric energy storage systems (flywheel storage or flywheels) include a cylinder with a shaft that can spin rapidly within a robust enclosure. A magnet levitates the cylinder, thus limiting friction-related losses and wear. The shaft is connected to a motor/generator. Electric energy is converted by the motor/generator to kinetic energy. That kinetic energy is stored by increasing the flywheel's rotational speed. The stored (kinetic) energy is converted back to electric energy via the motor/generator, slowing the flywheel's rotational speed.

2.1.5. Pumped Hydroelectric

Key elements of a pumped hydroelectric (pumped hydro) system include turbine/generator equipment, a waterway, an upper reservoir, and a lower reservoir. The turbine/generator is similar to equipment used for normal hydroelectric power plants that do not incorporate storage.

Pumped hydro systems store energy by operating the turbine/generator in reserve to pump water uphill or into an elevated vessel when inexpensive energy is available. The water is later released when energy is more valuable. When the water is released, it goes through the turbine which turns the generator to produce electric power.

2.1.6. Superconducting Magnetic Energy Storage

The storage medium in a superconducting magnetic energy storage (SMES) system consists of a coil made of superconducting material. Additional SMES system components include power conditioning equipment and a cryogenically cooled refrigeration system.

The coil is cooled to a temperature below the temperature needed for superconductivity (the material's 'critical' temperature). Energy is stored in the magnetic field created by the flow of direct current in the coil. Once energy is stored, the current will not degrade, so energy can be stored indefinitely (as long as the refrigeration is operational).

2.1.7. Thermal Energy Storage

There are various ways to store thermal energy. One somewhat common way that thermal energy storage is used involves making ice when energy prices are low so the cold that is stored can be used to reduce cooling needs – especially compressor-based cooling – when energy is expensive.

2.2. Storage System Power and Discharge Duration

When characterizing the rating of a storage system, the two key criteria to address are power and energy. *Power* indicates the *rate* at which the system can supply energy. *Energy* relates to the *amount* of energy that can be delivered to loads. In practical terms, the amount of energy stored determines the amount of time that the system can discharge at its rated power (output), hence the term *discharge duration*.

Storage power and energy are described in more detail below. For detailed coverage of the topic, readers should refer to a report developed by the Electric Power Research Institute (EPRI) and the DOE entitled *Estimating Electricity Storage Power Rating and Discharge Duration for Utility Transmission and Distribution Deferral, a Study for the DOE Energy Storage Program*.[4]

2.2.1. Storage Power

A storage system's power rating is assumed to be the system's nameplate power rating under normal operating conditions. Furthermore, that rating is assumed to represent the storage system's *maximum* power output under *normal* operating conditions. In this guide, the normal discharge rate used is commonly referred to as the system's 'design' or 'nominal' (power) rating. Generic application-specific power requirements are summarized in Table 4 (in Section 3).

2.2.1.1. Storage 'Emergency' Power Capability

Some types of storage systems can discharge at a relatively high rate (*e.g.,* 1.5 to 2 times their nominal rating) for relatively short periods of time (*e.g.,* several minutes to as much as 30 minutes). One example is storage systems involving an Na/S battery, which is capable of producing two times its rated (normal) output for relatively short durations.[5]

That feature – often referred to as the equipment's 'emergency' rating – is valuable if there are circumstances that occur infrequently that involve an urgent need for relatively high power output, for relatively short durations.

Importantly, while discharging at the higher rate, storage efficiency is reduced (relative to efficiency during discharge at the nominal discharge rate), and storage equipment damage increases (compared to damage incurred at the normal discharge rate).

So, in simple terms, storage with emergency power capability could be used to provide the nominal amount of power required to serve a regularly occurring need (*e.g.,* peak demand reduction) while the same storage could provide additional power for urgent needs that occur infrequently and that last for a few to several minutes at a time.

2.2.2. Storage Discharge Duration

Discharge duration is the amount of time that storage can discharge at its rated output (power) without recharging. Discharge duration is an important criterion affecting the technical viability of a given storage system for a given application and storage plant cost.

To the extent possible, this document includes generalized guidance about the necessary discharge duration for specific applications. Application-specific guidance and standard assumption values are provided in their respective subsections, below. Application-specific discharge durations and the assumptions used to establish them are summarized in Table 5 (in Section 3).

2.3. Energy and Power Density

Power density is the amount of power that can be delivered from a storage system with a given volume or mass. Similarly, energy density is the amount of energy that can be stored in a storage device that has a given volume or mass. These criteria are important in situations for which space is valuable or limited and/or if weight is important.

2.4. Storage System Footprint and Space Requirements

Closely related to energy and power density are footprint and space requirements for energy storage. Depending on the storage technology, floor area and/or space constraints may indeed be a challenge, especially in heavily urbanized areas.

2.5. Storage System Round-Trip Efficiency

All energy transfer and conversion processes have losses. Energy storage is no different. Storage system round-trip efficiency (efficiency) reflects the amount of energy that comes out of storage relative to the amount put into the storage.

Typical values for efficiency include the following: 60% to 75% for conventional electrochemical batteries; 75% to 85% for advanced electrochemical batteries; 73% to 80% for CAES; 75% to 78% for pumped hydro; 80% to 90% for flywheel storage; and 95% for capacitors and SMES.[6][7]

2.6. Storage Operating Cost

Storage total operating cost (as distinct from plant capital cost or plant financial carrying charges) consists of two key components: 1) energy-related costs and 2) operating costs not related to energy. Non-energy operating costs include at least four elements: 1) labor associated with plant operation, 2) plant maintenance, 3) equipment wear leading to loss-of-life, and 4) decommissioning and disposal cost (addressed in Section 2.20).

2.6.1. Charging Energy-Related Costs

The energy cost for storage consists of all costs incurred to purchase energy used to charge the storage, including the cost to purchase energy needed to make up for (round trip) energy losses. An example: For a storage system with 75% efficiency, if the unit price for energy used for charging is 4¢/kWh, then the plant energy cost is

4¢/kWh ÷ 0.75 = 5.33¢/kWh.

2.6.2. Labor for Plant Operation

In some cases, labor may be required for storage plant operation. *Fixed* labor costs are the same magnitude irrespective of how much the storage is used. *Variable* labor costs are proportional to the frequency and duration of storage use. In many cases, labor is required to operate larger storage facilities and/or 'blocks' of aggregated storage capacity whereas little or no labor may be needed for smaller/distributed systems that tend to be designed for autonomous operation. No explicit value is ascribed to this criterion, due in part to the wide range of labor costs that are possible given the spectrum of storage types and storage system sizes.

2.6.3. Plant Maintenance

Plant maintenance costs are incurred to undertake normal, scheduled, and unplanned repairs and replacements for equipment, buildings, grounds, and infrastructure. *Fixed* maintenance costs are the same magnitude irrespective of how much the storage is used. *Variable* maintenance costs are proportional to the frequency and duration of storage use. Plant maintenance costs are highly circumstance-specific and are not addressed explicitly in this report.

2.6.4. Replacement Cost

If specific equipment or subsystems within a storage system are expected to wear out during the expected life of the system, then a 'replacement cost' will be incurred. In such circumstances, a 'sinking fund' is needed to accumulate funds to pay for replacements when needed. That replacement cost is treated as a variable cost (*i.e.,* the total cost is spread out over each unit of energy output from the storage plant). Replacement cost is highly technology- and circumstance-specific and is not addressed explicitly in this report. (See Appendix B for an example calculation of equipment replacement cost.)

2.6.5. Variable Operating Cost

A storage system's total *variable* operating cost consists of applicable non-energy-related variable operating costs *plus* plant energy cost, possibly including charging energy, labor for plant operation, variable maintenance, and replacement costs. Variable operating cost is a key factor affecting the cost-effectiveness of storage. It is especially important for 'high-use' value propositions involving many charge-discharge cycles.

Ideally, storage for high-use applications should have relatively high or very high efficiency and relatively low variable operating cost. Otherwise, the total cost to charge then discharge the storage is somewhat-to-very likely to be higher than the benefit. That can be a significant challenge for some storage types and value propositions.

Consider the example illustrated in Figure 2, which involves a 75% efficient storage system with a non-energy-related variable operating cost of 4Ø/kWh$_{out}$. If that storage system is charged with energy costing 4Ø/kWh$_{in}$, then the total variable operating cost – for energy output – is about 9.33Ø/kWh$_{out}$.

2.7. Lifetime Discharges

To one extent or another, most energy storage media degrade with use (*i.e.,* during each charge-discharge cycle). The rate of degradation depends on the type of storage technology, operating conditions, and other variables. This is especially important for electrochemical batteries.

For some storage technologies – especially batteries – the extent to which the system is emptied (discharged) also affects the storage media's useful life. Discharging a small portion of stored energy is a 'shallow' discharge and discharging most or all of the stored energy is a 'deep' discharge. For these technologies, a shallow discharge is less damaging to the storage medium than a deep discharge.

Note that many battery vendors can produce storage media with extra service life (relative to the baseline product) to accommodate additional charge-discharge cycles and/or deeper discharges. Of course, there is usually a corresponding incremental cost for the superior performance. To the extent that the storage medium degrades and must be replaced during the expected useful life of the storage system, the cost for that replacement must be added to the variable operating cost of the storage system.

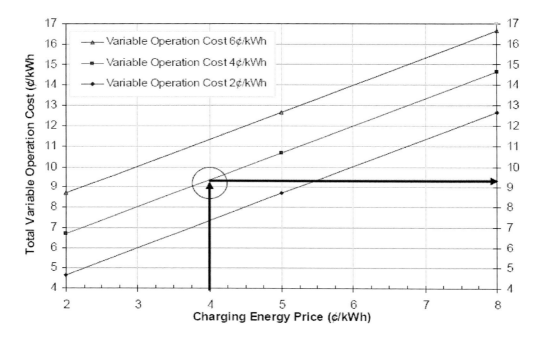

Figure 2. Storage total variable operation cost for 75% storage efficiency.

2.8. Reliability

Like power rating and discharge duration, storage system reliability requirements are circumstance-specific. Little guidance is possible. The project design engineer is responsible for designing a plant that provides enough power and that is as reliable as necessary to serve the specific application.

2.9. Response Time

Storage response time is the amount of time required to go from no discharge to full discharge. At one extreme, under almost all conditions, storage has to respond quite rapidly if used to provide capacity on the margin *in lieu* of T&D capacity. That is because the output from T&D equipment (*i.e.,* wires and transformers) changes nearly instantaneously in response to demand.

In contrast, consider storage used *in lieu* of generation capacity. That storage does not need to respond as quickly because generation tends to respond relatively slowly to demand changes. Specifically, some types of generation – such as engines and combustion turbines – take several seconds to many minutes before generating at full output. For other generation types, such as those fueled by coal and nuclear energy, the response time may be hours.

Most types of storage have a response time of several seconds or less. CAES and pumped hydroelectric storage tend to have a slower response, though they still respond quickly enough to serve several important applications.

2.10. Ramp Rate

An important storage system characteristic for some applications is the ramp rate – the rate at which power output can change. Generally, storage ramp rates are rapid (*i.e.,* output can change quite rapidly); pumped hydro is the exception. Power devices with a slow response time tend also to have a slow ramp rate.

2.11. Charge Rate

Charge rate – the rate at which storage can be charged – is an important criterion because, often, modular energy storage (MES) must be recharged so it can serve load during the next day. If storage cannot recharge quickly enough, then it will not have enough energy to provide the necessary service. In most cases, storage charges at a rate that is similar to the rate at which it discharges. In some cases, storage may charge more rapidly or more slowly, depending on the capacity of the power conditioning equipment and the condition and/or chemistry and/or physics of the energy storage medium.

2.12. Energy Retention and Standby Losses

Energy retention time is the amount of time that storage retains its charge. The concept of energy retention is important because of the tendency for some types of storage to self-discharge or to otherwise dissipate energy while the storage is not in use. In general terms, energy losses could be referred to as *standby* losses.

Storage that depends on chemical media is prone to self-discharge. This self-discharge is due to chemical reactions that occur while the energy is stored. Each type of chemistry is different, both in terms of the chemical reactions involved and the rate of self-discharge. Storage that uses mechanical means to store energy tends to be prone to energy dissipation. For example, energy stored using pumped hydroelectric storage may be lost to evaporation. CAES may lose energy due to air escaping from the reservoir.

To the extent that storage is prone to self-discharge or energy dissipation, retention time is reduced. This characteristic tends to be less important for storage that is used frequently.

For storage that is used infrequently (*i.e.*, is in standby mode for a significant amount of time between uses), this criterion may be very important.

2.13. Transportability

Transportability can be an especially valuable feature of storage systems for at least two reasons. First, transportable storage can be (re)located where it is needed most and/or where benefits are most significant. Second, some locational benefits only last for one or two years. Perhaps the most compelling example of the latter is T&D deferral, as discussed in detail in Section 3. Given those considerations, transportability may significantly enhance the prospects that lifecycle benefits will exceed lifecycle cost.

2.14. Modularity

One attractive feature of modular energy storage is the flexibility that system 'building blocks' provide. Modularity allows for more optimal levels and types of capacity and/or discharge duration because modular resources allow utilities to increase or decrease storage capacity, when and where needed, in response to changing conditions. Among other attractive effects, modular capacity provides attractive means for utilities to address uncertainty and to manage risk associated with large, 'lumpy' utility T&D investments.

2.15. Power Conditioning

To one extent or another, most storage types require some type of power conditioning (*i.e.*, conversion) subsystem. Equipment used for power conditioning – the power conditioning unit (PCU) – modifies electricity so that the electricity has the necessary voltage and the necessary form; either alternating current (AC) or direct current (DC). The PCU, in concert with an included control system, must also synchronize storage output with the oscillations of AC power from the grid.

Output from storage with relatively low-voltage DC output must be converted to AC with higher voltage before being discharged into the grid and/or before being used by most load types. In most cases, conversion from DC to AC is accomplished using a device known as an *inverter*.

For storage requiring DC input, the electricity used for charging must be converted from the form available from the grid (*i.e.*, AC at relatively high voltage) to the form needed by the storage system (*e.g.*, DC at lower voltage). That is often accomplished via a PCU that can function as a DC 'power supply'.

2.16. Power Quality

Although requirements for applications vary, the following storage characteristics may or may not be important. To one extent or another, they are affected by the PCU used and/or they drive the specifications for the PCU. In general, higher quality power (output) costs more.

2.16.1. Power Factor
Although detailed coverage of the concept of power factor is beyond the scope of this report, it is important to be aware of the importance of this criterion. At a minimum, the power output from storage should have an acceptable power factor, where acceptable is somewhat circumstance-specific. For some applications, the storage system may be called upon to provide power with a variable power factor. (See Appendix C for more details about this consideration.)

2.16.2. Voltage Stability
In most cases, it is important for storage output voltage to remain somewhat-to-very constant. Depending on the circumstances, voltage can vary; though, it should probably remain within about 5% to 8% of the rated value.

2.16.3. Waveform
Assuming that storage output is AC, in most cases, the waveform should be as close as possible to that of a sine wave. In general, higher quality PCUs tend to have waveforms that are quite close to that of a sine wave whereas output from lower quality PCUs tends to have a waveform that is somewhat square.

2.16.4. Harmonics
Harmonic currents in distribution equipment can pose a significant challenge. Harmonic currents are components of a periodic wave whose frequency is an integral multiple of the fundamental frequency. In this case, the fundamental frequency is the utility power line frequency of 60 Hz. So, for example, harmonic currents might exist with frequencies of 3 ̣60 Hz (180 Hz) or 7 X 60 Hz (420 Hz). Total harmonic distortion (THD) is the contribution of all the individual harmonic currents to the fundamental.

2.17. Storage System Reactive Power Capability

One application (Voltage Support) and one incidental benefit (Power Factor Correction) described in this guide involve storage whose capabilities include absorbing and injecting reactive power (expressed in units of volt-Amperes reactive or VARs). This feature is commonly referred as VAR support. In most cases, storage systems by themselves do not have reactive power capability. For a relatively modest incremental cost, however, reactive power capability can be added to most storage system types. (See Appendix C for more details.)

2.18. Communications and Control

Storage used for most applications addressed in this report must receive and respond to appropriate control signals. In some cases, storage may have to respond to a dispatch control signal. In other cases, the signal may be driven by a price or prices. Storage response to a control signal may be a simple ramp up or ramp down of power output in proportion to the control signal. A more sophisticated response, requiring one or more control algorithms, may be needed. An example of that is storage used to respond to price signals or to accommodate more than one application.

2.19. Interconnection

If storage will be charged with energy from the grid or will inject energy into the grid, it must meet applicable interconnection requirements. At the distribution level, an important point of reference is the Institute of Electronics and Electrical Engineers (IEEE) Standard 1547.[8] Some states and utilities have more specific interconnection rules and requirements.

2.20. Decommissioning and Disposal Needs and Cost

Although not addressed explicitly in this report, in most cases there will be non-trivial decommissioning costs associated with almost any storage system. For example, eventually batteries must be dismantled and the chemicals must be removed. Ideally, dismantled batteries and their chemicals can be recycled, as is the case for the materials in lead-acid batteries. Ultimately, decommissioning-related costs should be included in the total cost to own and to operate storage.

3. ELECTRIC ENERGY STORAGE APPLICATIONS

3.1. Introduction

This section characterizes 17 electric grid-related energy storage applications. Included in each characterization are a description of the application, an overview of application-specific technical considerations, and a summary of possible synergies with other applications. (Section 2 includes a brief characterization of several important storage system characteristics.) The 17 applications are grouped into five categories as shown in Table 3.

Table 3. Five Categories of Energy Storage Applications

Category 1 — Electric Supply
1. Electric Energy Time-shift
2. Electric Supply Capacity
Category 2 — Ancillary Services

Category 1 — Electric Supply
3. Load Following
4. Area Regulation
5. Electric Supply Reserve Capacity
6. Voltage Support
Category 3 — Grid System
7. Transmission Support
8. Transmission Congestion Relief
9. Transmission & Distribution (T&D) Upgrade Deferral
10. Substation On-site Power
Category 4 — End User/Utility Customer
11. Time-of-use (TOU) Energy Cost Management
12. Demand Charge Management
13. Electric Service Reliability
14. Electric Service Power Quality
Category 5 — Renewables Integration
15. Renewables Energy Time-shift
16. Renewables Capacity Firming
17. Wind Generation Grid Integration

3.1.1. Power Applications Versus Energy Applications

Although this report does not focus on specific storage technologies, it is helpful to be aware of the distinction between storage technologies classified as those that are best suited for *power* applications and those best suited to *energy* applications.

Power applications require high power output, usually for relatively short periods of time (a few seconds to a few minutes). Storage used for power applications usually has capacity to store fairly modest amounts of energy per kW of rated power output. Notable storage technologies that are especially well-suited to power applications include capacitors, SMES, and flywheels.

Energy applications are uses of storage requiring relatively large amounts of energy, often for discharge durations of many minutes to hours. So, storage used for energy applications must have a much larger energy storage reservoir than storage used for power applications. Storage technologies that are best suited to energy applications include CAES, pumped hydro, thermal energy storage, and most battery types.

3.1.2. Capacity Applications Versus Energy Applications

Similar to the distinction between power applications and energy applications is the distinction between *capacity* applications and *energy* applications. In simple terms, capacity applications are those involving storage used to defer or to reduce the need for other equipment. For example, storage can be used to reduce the need for generation or T&D equipment. Depending on circumstances, capacity applications tend to require relatively limited amounts of energy discharge throughout the year.

As described above, energy applications involve storing a significant amount of electric energy to offset the need to purchase or to generate the energy when needed. Typically, energy-related applications require a relatively significant amount of energy to be stored and discharged throughout the year. An important consideration is that storage used for energy applications should be relatively efficient, or the cost incurred due to energy losses will offset a significant amount of the benefit. The same applies to non-energy-related variable operation cost.

Importantly, for investor-owned utilities (IOUs) capacity is generally treated like an *investment* whereas purchases of or generation of energy are typically thought of as an *expenses* involving variable operating cost and fuel-related costs. This distinction is especially important for investor-owned utilities given what is sometimes referred to as the *revenue requirement* method for establishing cost-of-service. Under that regulatory scheme utilities earn a rate of return (*i.e.,* profit) on *investments* in capital *equipment* whereas *expenses* are treated as a 'pass-through' to end users without any mark-up (*i.e.,* IOUs do not earn profit for *energy* provided).

3.1.3. Application-specific Power and Discharge Duration

Table 4 and Table 5 list application-specific standard assumption values for two key storage design criteria: 1) power rating and 2) discharge duration. Also shown are key underlying assumptions used when establishing those values. Table 4 lists application-specific, standard assumption values for storage power ratings and notes explaining the rationale used to make the estimates. Table 5 lists application-specific standard assumption values for discharge durations along with notes explaining the rationale used to make the estimates.

The standard assumption values used herein are intended to be generic. They were developed based on varying levels of engineering judgment and simplifying assumptions. Readers are encouraged to use case-specific assumptions and additional information, as needed and available, for more precise estimates of power ratings and discharge durations.

Table 4. Standard Assumption Values for Storage Power

#	Type	Storage Power Low	Storage Power High	Note
1	Electric Energy Time-shift	1 MW	500 MW	Low per ISO transaction min. (Can aggregate smaller capacity.) High = combined cycle gen.
2	Electric Supply Capacity	1 MW	500 MW	Same as above.
3	Load Following	1 MW	500 MW	Same as above.
4	Area Regulation	1 MW	40 MW	Low per ISO transaction min. Max is 50% of estimated CA technical potential of 80 MW.
5	Electric Supply Reserve Capacity	1 MW	500 MW	Low per ISO transaction min. (Can aggregate smaller capacity.) High = combined cycle gen.
6	Voltage Support	1 MW	10 MW	Assume distributed deployment, to serve Voltage support needs locally.

		Storage Power		
#	Type	Low	High	Note
7	Transmission Support	10 MW	100 MW	Low value is for substransmission.
8	Transmission Congestion Relief	1 MW	100 MW	Low per ISO transaction min. (Can aggregate smaller capacity.) High = 20% of high capacity transmission.
9.1	T&D Upgrade Deferral 50th percentile	250 kW	5 MW	Low = smallest likely, High = high end for distribution & subtransmission.
9.2	T&D Upgrade Deferral 90th percentile	250 kW	2 MW	Same as above.
10	Substation On-site Power	1.5 kW	5 kW	Per EPRI/DOE Substation Battery Survey.
11	Time-of-use Energy Cost Management	1 kW	1 MW	Residential to medium sized commercial/industrial users.
12	Demand Charge Management	50 kW	10 MW	Small commercial to large commercial/industrial users.
13	Electric Service Reliability	0.2 kW	10 MW	Low = Under desk UPS. High = facility-wide for commercial/industrial users.
14	Electric Service Power Quality	0.2 kW	10 MW	Same as above.
15	Renewables Energy Time-shift	1 kW	500 MW	Low = small residential PV. High = "bulk" renewable energy fueled generation.
16	Renewables Capacity Firming	1 kW	500 MW	Same as above.
17.1	Wind Generation Grid Integration, Short Duration	0.2 kW	500 MW	Low = small residential turbine. High = larged wind farm boundary.
17.2	Wind Generation Grid Integration, Long Duration	0.2 kW	500 MW	Same as above.

Table 5. Standard Assumption Values for Discharge Duration

		Discharge Duration*		
#	Type	Low	High	Note
1	Electric Energy Time-shift	2	8	Depends on energy price differential, storage efficiency, and storage variable operating cost.
2	Electric Supply Capacity	4	6	Peak demand hours
3	Load Following	2	4	Assume: 1 hour of discharge duration provides approximately 2 hours of load following.
4	Area Regulation	15 min.	30 min.	Based on demonstration of Beacon Flywheel.

Table 5. (Continued)

#	Type	Discharge Duration* Low	Discharge Duration* High	Note
5	Electric Supply Reserve Capacity	1	2	Allow time for generation-based reserves to come on-line.
6	Voltage Support	15 min.	1	Time needed for a) system stabilization or b) orderly load shedding.
7	Transmission Support	2 sec.	5 sec.	Per EPRI-DOE Handbook of Energy Storage for Transmission and Distribution Applications.[17]
8	Transmission Congestion Relief	3	6	Peak demand hours. Low value is for "peaky" loads, high value is for "flatter" load profiles.
9.1	T&D Upgrade Deferral 50th percentile	3	6	Same as Above
9.2	T&D Upgrade Deferral 90th percentile	3	6	Same as Above
10	Substation On-site Power	8	16	Per EPRI/DOE Substation Battery Survey.
11	Time-of-use Energy Cost Management	4	6	Peak demand hours.
12	Demand Charge Management	5	11	Maximum daily demand charge hours, per utility tariff.
13	Electric Service Reliability	5 min.	1	Time needed for a) shorter duration outages or b) orderly load shutdown.
14	Electric Service Power Quality	10 sec.	1 min.	Time needed for events ridethrough depends on the type of PQ challenges addressed.
15	Renewables Energy Time-shift	3	5	Depends on energy cost/price differential and storage efficiency and variable operating cost.
16	Renewables Capacity Firming	2	4	Low & high values for Renewable Gen./Peak Load correlation (>6 hours) of 85% & 50%.
17.1	Wind Generation Grid Integration, Short Duration	10 sec.	15 min.	For a) Power Quality (depends on type of challenge addressed) and b) Wind Intermittency.
17.2	Wind Generation Grid Integration, Long Duration	1	6	Backup, Time Shift, Congestion Relief.

*Hours unless indicated otherwise. Min. = minutes. Sec. = Seconds.

3.2. Electric Supply Applications

3.2.1. Application #1 — *Electric Energy Time-shift*

3.2.1.1. Application Overview

Electric energy time-shift (time-shift) involves purchasing inexpensive electric energy, available during periods when price is low, to charge the storage plant so that the stored energy can be used or sold at a later time when the price is high.

Entities that time-shift may be regulated utilities or non-utility merchants. Importantly, this application tends to involve purchase of inexpensive energy from the *wholesale* electric energy market for storage charging. When the energy is discharged, it could be resold via the wholesale market, or it may offset the need to purchase wholesale energy and/or to generate energy to serve end users' needs.

3.2.1.2. Technical Considerations

For the time-shift application, the plant storage discharge duration is determined based on the incremental benefit associated with being able to make additional buy-low/sellhigh transactions during the year *versus* the incremental cost for additional energy storage (discharge duration).

The standard assumption value for storage *minimum* discharge duration for this application is two hours. The upper boundary for discharge duration is defined by potential CAES or pumped hydroelectric facilities. For storage types that have a high incremental cost to increase the amount of energy that can be stored (*i.e.*, to increase discharge duration), the upper boundary is probably five or six hours — the typical duration of a utility's daily peak demand period.

> It is common for those involved with storage to refer to energy time-shift transactions (using storage) as *arbitrage*. It is important to note, however, what arbitrage means to people involved in *finance*.
>
> A finance-centric definition of arbitrage is *the* simultaneous *purchase and sale of identical or equivalent commodities or other instruments across two or more markets in order to benefit from a discrepancy in their price relationship.*
>
> So, strictly speaking, *from a finance perspective* the term 'arbitrage' may be regarded as a misnomer when it is applied to most energy storage 'buy-low/sell-high' (time-shift) transactions. That is because the purchase and storage of electric energy occurs at a different time than sale or use of the energy. In fact, most often charging and discharging are separated by several hours.

Both storage (non-energy-related) variable operating cost and storage efficiency are especially important for this application because electric energy time-shift involves many possible transactions whose economic merit is based on the difference between the cost to purchase, store, and discharge energy (discharge cost) and the benefit derived when the energy is discharged. Any increase in variable operating cost or reduction of efficiency reduces the number of transactions for which the benefit exceeds the cost. That number of transactions is quite sensitive to the discharge cost, so a modest increase may reduce the number of viable transactions considerably.

Two performance characteristics that have a significant impact on storage variable operating cost are efficiency and the rate at which storage performance declines as it is used.

3.2.1.3. Application Synergies

Although each case is unique, if a plant used for electric energy time-shift is in the right location and if it is discharged at the right times, it could also serve the following applications: electric supply capacity, T&D upgrade deferral, transmission congestion relief, electric service reliability, electric service power quality, and ancillary services.

3.2.2. Application #2 — Electric Supply Capacity

3.2.2.1. Application Overview

Depending on the circumstances in a given electric supply system, energy storage could be used to defer and/or to reduce the need to buy new central station generation capacity and/or to 'rent' generation capacity in the wholesale electricity marketplace.

In many areas of the U.S., the most likely type of new generation plant 'on the margin' is a natural gas-fired combined cycle power plant. For utilities needing additional *peaking* capacity, the conventional proxy or default alternative is usually a relatively clean, simple cycle combustion turbine. Depending on circumstances, however, other peaking resources may be preferred (*e.g.,* other types of central/bulk generation, distributed generation, demand response, and energy efficiency).

The marketplace for electric supply capacity is evolving. In some cases, to one extent or another, generation capacity cost is included in wholesale *energy* prices (as an allocated cost per unit of energy). In other cases, market mechanisms may allow for capacity-related payments. In fact, the price paid for capacity *not* used – under terms of utility demand response programs – may reflect some or all of the marginal cost for generation capacity.

3.2.2.2. Technical Considerations

The operating profile for storage used as supply capacity (characterized by annual hours of operation, frequency of operation, and duration of operation for each use) is circumstance-specific. Consequently, it is challenging to make generalizations about storage discharge duration for this application. Another key criterion affecting discharge duration for this application is the way that generation capacity is priced. For example, if capacity is priced per hour, then storage plant duration is flexible. If prices require that the capacity resource be available for a specified duration for each occurrence (*e.g.,* five hours), or require operation during an entire time period (*e.g.,* 12:00 p.m. to 5:00 p.m.), then the storage plant discharge duration must accommodate those requirements.

3.2.2.3. Application Synergies

Depending on location and other circumstances, storage used for this application may be compatible with the following applications: electric energy time-shift, electric supply reserve capacity, area regulation, voltage support, T&D upgrade deferral, transmission support and congestion relief, electric service power quality, and electric service reliability.

3.3. Ancillary Services Applications

3.3.1. Application #3 — *Load Following*

3.3.1.1. Application Overview

Load following is one of the ancillary services required to operate the electricity grid. (See Appendix A for more detail about ancillary services.) Load following capacity is characterized by power output that changes as frequently as every several minutes. The output changes in response to the changing balance between electric supply (primarily generation) and end user demand (load) within a specific region or area. Output variation is a "...response to changes in system frequency, timeline loading, or the relation of these to each other..." that occurs as needed to "...maintain the scheduled system frequency and/or established interchange with other areas within predetermined limits."[9]

Conventional generation-based load following resources' output *increases* to follow demand *up* as system load increases. Conversely, load following resources' output *decreases* to follow demand *down* as system load decreases. Typically, the amount of load following needed in the up direction (load following up) increases each day as load increases during the morning. In the evening, the amount of load following needed in the down direction (load following down) increases as aggregate load on the grid drops. A simple depiction of load following is shown in Figure 3.

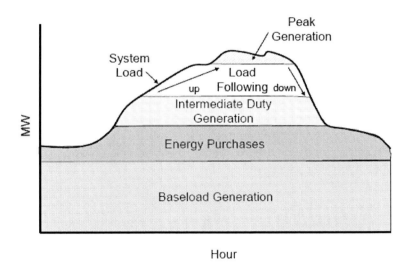

Figure 3. Electric supply resource stack.

Normally, generation is used for load following. For *load following up*, generation is operated such that its output is less than its design or rated output (also referred to as 'part load operation'). That allows operators to increase the generator's output, as needed, to provide load following up to accommodate increasing load. For *load following down*, generation starts at a high output level, perhaps even at design output, and the output is decreased as load decreases.

These operating scenarios are notable because operating generation at part load requires more fuel and results in increased air emissions relative to generation operated at its design output level. Also, varying the output of generators (rather than operating at constant output) may increase fuel use and air emissions, and it increases the need for generator variable maintenance.

Storage is well-suited to load following for several reasons. First, most types of storage can operate at partial output levels with relatively modest performance penalties. Second, most types of storage can respond very quickly (compared to most types of generation) when more or less output is needed for load following. Consider also that storage can be used effectively for both load following up (as load increases) and for load following down (as load decreases), either by discharging or by charging. (See Appendix D for details.)

When *charging* storage for load following, the energy stored must be purchased at the prevailing wholesale price. This is an important consideration – especially for storage with lower efficiency and/or if the energy used for charging is relatively expensive – because the cost of energy used to charge storage (to provide load following) may exceed the value of the load following service.

Conversely, the value of energy *discharged* from storage to provide load following is determined by the prevailing price for wholesale energy. Depending on circumstances (*i.e.,* if the price for the load following service does not include the value of the wholesale energy involved), when discharging for load following, two benefits accrue – one for the load following service and another for the energy.

Storage competes with central and aggregated distributed generation and with aggregated demand response/load management resources including curtailable/interruptible loads and direct load control.

3.3.1.2. Technical Considerations

Storage used for load following should be somewhat-to-very reliable or it cannot be used to meet contractual obligations associated with bidding in the load following market. Storage used for load following will probably need access to automated generation control (AGC) from the respective independent system operator (ISO). Typically, an ISO requires output from an AGC resource to change every minute.

For this application, storage could provide up to two service hours per hour of discharge duration. (See Appendix D for details.)

3.3.1.3. Application Synergies

Large/central storage used for load following may be especially complementary to other applications if charging and discharging for the other applications can be coordinated with charging and discharging to provide load following. For example, storage used to provide generation capacity mid-day could be charged in the evening thus following diminished system demand down during evening hours.

Load following could have good synergies with renewables capacity firming, electric energy time-shift, and possibly electric supply reserve capacity applications. If storage is distributed, then that same storage could also be used for most of the distributed applications and for voltage support.

3.3.2. Application #4 — Area Regulation

3.3.2.1. Application Overview

Area regulation (regulation) is one of the ancillary services for which storage may be especially well-suited. Regulation involves managing "interchange flows with other control areas to match closely the scheduled interchange flows" and moment to moment variations in demand within the control area.[10]

The primary reasons for including regulation in the power system are to maintain the grid frequency and to comply with the North American Electric Reliability Council's (NERC's) Control Performance Standards 1 and 2 (NERC 1999a). Regulation also assists in recovery from disturbances, as measured by compliance with NERC's Disturbance Control Standard.[11]

In more basic terms, regulation is used to reconcile momentary differences between supply and demand. That is, at any given moment, the amount of electric supply capacity that is operating may exceed or may be less than load. Regulation is used for damping of that difference. Consider the example shown in Figure 4. In that figure, the thin (red) plot with numerous fluctuations depicts total system demand without regulation. The thicker (black) plot shows system load after damping of the short-duration fluctuations with regulation.

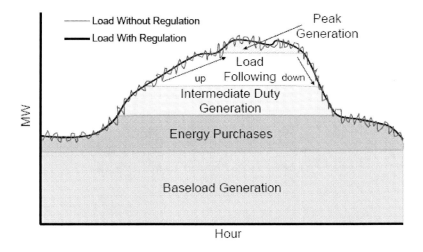

Figure 4. System load without and with area regulation.

Regulation is typically provided by generating units that are online and ready to increase or decrease power as needed. When there is a momentary shortfall of electric supply capacity, output from regulation resources is increased to provide *up regulation*. Conversely, regulation resources' output is reduced to provide *down regulation* when there is a momentary excess of electric supply capacity.

An important consideration for this application is that most thermal/baseload generation used for regulation service is not especially well-suited or designed to provide regulation. This is because most types of thermal/baseload generation are not designed for operation at part load or to provide variable output. Notably, thermal power plant fuel conversion is usually most efficient when power plants operate at a specific and constant (power) output level. Similarly, air emissions and plant wear and tear are usually lowest (per kWh of output) when thermal generation operates at full load and with constant output.

So, storage may be an attractive alternative to most generation-based load following for at least three reasons: 1) in general, storage has superior part-load efficiency, 2) *efficient* storage can be used to provide up to two times its rated capacity (for regulation), and 3) storage output can be varied rapidly (*e.g.*, output can change from none to full or from full to none within seconds rather than minutes).

Two possible operational modes for 1 MW of *storage* used for regulation and three possible operational modes for *generation* used for regulation are shown in Figure 5. The leftmost plot shows how less-efficient storage could be used for regulation. In that case, increased storage discharge is used to provide up regulation and reduced discharge is used to provide down regulation. In essence, one half of the storage's capacity is used for up regulation and the other half of the storage capacity is used for down regulation (similar to the rightmost plot which shows how 1 MW of *generation* is often used for regulation service). Next, consider the second plot which shows how 1 MW of *efficient* storage can be used to provide 2 MW of regulation – 1 MW up and 1 MW down – using discharging and charging, respectively.

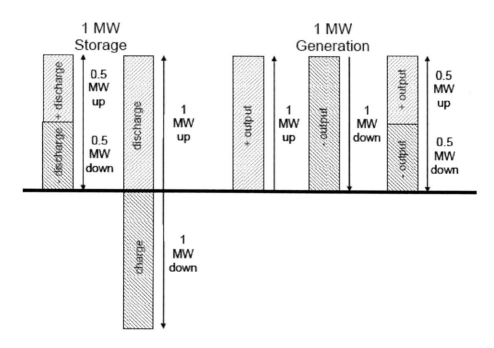

Figure 5. Storage and generation operation for area regulation.

When storage provides down regulation by charging, it *absorbs* energy from the grid, and the storage operator must pay for that energy. That is notable – especially for storage with lower efficiency – because the cost for that energy may exceed the value of the load following service. (Energy stored during load following, however, could be used later for other benefits which, if combined with the load following benefit, may still be attractive.)

3.3.2.2. Technical Considerations

The rapid-response characteristic (*i.e.,* fast ramp rate) of some types of storage makes that storage especially valuable as a regulation resource. In fact, the benefit of regulation from storage with a fast ramp rate (*e.g.,* flywheels, capacitors, and some battery types) is on the order of two times that of regulation provided by generation. (See Appendix E for details.)

Storage used for regulation should have access to and be able to respond to the area control error (ACE) signal which may require a response time of less than five seconds. Resources used to provide regulation should be quite reliable, and they must have high quality, stable (power) output characteristics.

3.3.2.3. Application Synergies

In most cases, storage used to provide area regulation cannot be used *simultaneously* for another application. However, at any given time, storage *could* be used for another more beneficial application *instead* of using it for regulation (*e.g.,* electric energy time-shift, electric supply capacity, electric supply reserve capacity, or T&D upgrade deferral).

3.3.3. Application #5 — Electric Supply Reserve Capacity

3.3.3.1. Application Overview

Prudent operation of an electric grid includes use of electric supply reserve capacity (reserve capacity) that can be called upon when some portion of the normal electric supply resources become unavailable unexpectedly. In the electric utility realm, this reserve capacity is classified as an ancillary service. (See Appendix A and [12] for details about ancillary services.)

At minimum, reserves should be at least as large as the single largest resource (*e.g.,* the single largest generation unit) serving the system. Generally, reserve capacity is equivalent to 15% to 20% of the normal electric supply capacity, although specific reserve margins are designated in rules and/or regulations. In the U.S., the National Electric Reliability Council (NERC) is a key agency involved in establishing reserve capacity requirements.[13]

The three generic types of reserve capacity are:

- **Spinning Reserve** – Generation capacity that is online but unloaded and that can respond within 10 minutes to compensate for generation or transmission outages. 'Frequency-responsive' spinning reserve responds within 10 seconds to maintain system frequency. Spinning reserves are the first type used when a shortfall occurs.
- **Supplemental Reserve** – Generation capacity that may be offline, or that comprises a block of curtailable and/or interruptible loads, and that can be available within 10 minutes. Unlike spinning reserve capacity, supplemental reserve capacity is not synchronized with the grid (frequency). Supplemental reserves are used after all spinning reserves are online.
- **Backup Supply** – Generation that can pick up load within one hour. Its role is, essentially, a backup for spinning and supplemental reserves. Backup supply may also be used as backup for commercial energy sales.

Importantly for storage, *generation* resources used as reserve capacity must be online and operational (*i.e.,* at part load). Unlike generation, in almost all circumstances, *storage* used for reserve capacity does not discharge at all – it just has to be ready and available to discharge *if needed.*

Note that storage can provide two times its capacity as reserve capacity when the storage is charging, because the storage can simultaneously stop charging *and* start discharging.

3.3.3.2. Technical Considerations

Of course, storage used for reserve capacity must have enough stored energy to discharge for the required amount of time (usually at least one hour).

Storage used for this application must be somewhat reliable, though penalties for not providing the service after a bid are not onerous for individual events. Reserve capacity resources must receive and respond to appropriate control signals. Typical discharge durations for this application are between one and two hours. Reserve capacity may have to respond to the ISO's AGC signal.

3.3.3.3. Application Synergies

Electric supply reserve capacity is especially compatible with other applications and application combinations primarily for the following reasons:

- Most times when storage is used for reserves, it does not discharge.
- While charging, storage can provide two times its capacity as reserve capacity.
- If there is an hour-ahead market for reserve capacity, then decisions can be made almost real-time regarding the merits of discharging – if needed – compared to saving the energy to use later, for more benefit.[14]

In most cases, storage cannot serve any other applications while it is providing electric supply reserve capacity. Nevertheless, when storage is not used as electric supply reserve capacity, it could be used for electric energy time-shift, electric supply capacity, other ancillary services, renewables energy time-shift, renewables capacity firming, and wind generation grid integration. Depending on location, it could also be used for transmission congestion relief and T&D upgrade deferral.

3.3.4. Application #6 — Voltage Support

3.3.4.1. Application Overview

An important technical challenge for electric grid system operators is to maintain necessary voltage levels with the required stability. In most cases, meeting that challenge requires management of a phenomenon called 'reactance'. Reactance occurs because equipment that generates, transmits, or uses electricity often has or exhibits characteristics like those of inductors and capacitors in an electric circuit. (See Appendix C for more details.)

To manage reactance at the grid system level, grid system operators rely on an ancillary service called 'voltage support'. The purpose of voltage support is to offset reactive effects so that grid system voltage can be restored or maintained.

Historically, voltage support has been provided by generation resources. Those resources are used to generate reactive power (VAR) that offsets reactance in the grid. New technologies (*e.g.,* modular energy storage, modular generation, power electronics, and communications and control systems) make new alternatives for voltage support increasingly viable.[15][16]

(Conventional 'power factor correction' capacitors are good for managing *localized* reactance that occurs *during normal operating conditions*. Capacitors do not perform well as a voltage support resource, however, because they draw an increasing amount of current as voltage drops – to maintain power – which adds to voltage-related problems affecting the greater grid system. See Section 5.3.6 and Appendix C for more details about power factor correction.)

This is an application for which *distributed* storage may be especially attractive because reactive power cannot be transmitted efficaciously over long distances. Notably, many major power outages are at least partially attributable to problems related to transmitting reactive power to load centers. So, distributed storage – located within load centers where most reactance occurs – provides especially helpful voltage support.[17][18]

One especially notable load type for this application is smaller air conditioning (A/C) equipment like that used for residences and for small businesses. The reactance from motors used for A/C compressors poses a significant voltage-related challenge because, as grid voltage drops – during localized or region-wide grid emergencies – the motors draw an increasing amount of current to maintain power. That exacerbates the voltage problem, in part because air conditioners are most likely to be turned on when the grid is most heavily loaded and possibly when the grid is especially prone to voltage-related problems.

3.3.4.2. Technical Considerations

Storage systems used for voltage support must have VAR support capability if they will be used to inject reactive power. Also, storage used for voltage support must receive and respond quickly to appropriate control signals.

The standard value for discharge duration is assumed to be 30 minutes — time for the grid system to stabilize and, if necessary, to begin orderly load shedding.

3.3.4.3. Application Synergies

In general, storage used for voltage support must be available within a few seconds to serve load for a few minutes to perhaps as much as an hour. Thus, storage serving another application could also provide voltage support if the storage can be available within a few seconds to provide voltage support and if the storage has enough stored energy to discharge for durations ranging from a few minutes to an hour.

Central/bulk storage used for voltage support could also be used for electric energy time-shift, electric supply capacity, other ancillary services, renewables energy time-shift, renewables capacity firming, and wind generation integration.

Distributed storage used for voltage support probably cannot be used for area regulation or transmission support though it probably could be used for most or all of the other applications covered in this report with little or no *technical* conflict, though circumstance-specific dispatch needs may cause *operational* conflicts.

If the same storage is used for voltage support and for another 'must-run' application (*e.g.,* T&D upgrade deferral), then the worst case is that the storage is completely dedicated to serving local demand during the few dozen to few hundred hours per year when the T&D equipment is most heavily loaded, leaving storage available during 95%+ of the year to serve other applications.

3.4. Grid System Applications

3.4.1. Application #7 — Transmission Support

3.4.1.1. Application Overview

Energy storage used for transmission support improves T&D system performance by compensating for electrical anomalies and disturbances such as voltage sag, unstable voltage, and sub-synchronous resonance. The result is a more stable system with improved performance (throughput). It is similar to the ancillary service (not addressed in this guide) referred to as Network Stability. Benefits from transmission support are highly situation-specific and site-specific. Table 6 briefly describes ways that energy storage can provide transmission support.

Table 6. Types of Transmission Support

Type	Description
Transmission Stability Damping	Increase load carrying capacity by improving dynamic stability.
Sub-synchronous Resonance Damping	Increase line capacity by allowing higher levels of series compensation by providing active real and/or reactive power modulation at sub-synchronous resonance modal frequencies.
Voltage Control and Stability	1. Transient Voltage Dip Improvement Increase load carrying capacity by reducing the voltage dip that follows a system disturbance. 2. Dynamic Voltage Stability Improve transfer capability by improving voltage stability.
Under-frequency Load Shedding Reduction	Reduce load shedding needed to manage under-frequency conditions which occur during large system disturbances.

Source: adapted from information provided by EPRI.[19][20][21].

3.4.1.2. Technical Considerations

To be used for transmission support, energy storage must be capable of sub-second response, partial state-of-charge operation, and many charge-discharge cycles. Communication and control systems are important for this application. Also, storage used for transmission support must be very reliable. For storage to be most beneficial as a transmission support resource, it should provide both real and reactive power.[22]

Typical discharge durations for transmission support are between one and twenty seconds. The standard discharge duration assumed for this application is five seconds.

3.4.1.3. Application Synergies

Storage that is used for transmission support probably cannot be used *concurrently* for other applications. Nevertheless, storage used for transmission support during peak demand or peak congestion times could be used at other times for several other applications, if the storage has the necessary discharge duration (*e.g.*, one hour or more for ancillary services).

3.4.2. Application #8 — Transmission Congestion Relief

3.4.2.1. Application Overview

In many areas, transmission capacity additions are not keeping pace with the growth in peak electric demand. Consequently, transmission systems are becoming congested during periods of peak demand, driving the need and cost for more transmission capacity and increased transmission access charges. Additionally, transmission congestion may lead to increased use of congestion charges or locational marginal pricing (LMP) for electric energy.

Storage could be used to avoid congestion-related costs and charges, especially if the charges become onerous due to significant transmission system congestion. In this application, storage systems would be installed at locations that are electrically downstream from the congested portion of the transmission system. Energy would be stored when there is no transmission congestion, and it would be discharged (during peak demand periods) to reduce transmission capacity requirements.

3.4.2.2. Technical Considerations

The discharge duration needed for transmission congestion relief cannot be generalized easily, given all the possible manifestations. As with the T&D upgrade deferral application, it may be that there are just a few individual hours throughout the year when congestion charges apply. Or, there may be a few occurrences during a year when there are several consecutive hours of transmission congestion. Also, congestion charges may be applied like demand charges with payments made for maximum demand during specific times during specific months of the year. Congestion charges may vary from year to year because supply and demand are always changing.

The standard discharge duration assumed for this application is four hours.

3.4.2.3. Application Synergies

Depending on location, the owner, the discharge duration, and other circumstances, storage used for transmission congestion relief may be compatible with most if not all applications described in this report, especially electric energy time-shift, electric supply capacity (peaking), ancillary services, and possibly renewable energy time-shift.

3.4.3. Application #9 — Transmission and Distribution Upgrade Deferral

3.4.3.1. Application Overview

Transmission and distribution (T&D) upgrade deferral involves delaying – and in some cases avoiding entirely – utility investments in transmission and/or distribution system upgrades, using relatively small amounts of storage. Consider a T&D system whose peak electric loading is approaching the system's load carrying capacity (design rating). In some cases, installing a small amount of energy storage downstream from the nearly overloaded T&D node will defer the need for a T&D upgrade.

Consider a more specific example: A 15-MW substation is operating at 3% below its rating and load growth is about 2% per year. In response, engineers plan to upgrade the substation next year by adding 5 MVA of additional capacity. As an alternative, engineers could consider installing enough storage to meet the expected load growth for next year, plus any appropriate engineering contingencies (*i.e.,* it may not be prudent to install 'just enough' storage, especially if there is uncertainty about load growth). For the 15-MW substation in this example: At a 2% load growth rate, the load growth during the next year will be 300 kW (2% × 15 MW). Adding a 25% engineering contingency, the storage plant needed to defer T&D upgrade would be about 375 kW.

The key theme is that a *small* amount of storage can be used provide enough *incremental* capacity to defer the need for a *large* 'lump' investment in T&D equipment. Doing so reduces overall cost to ratepayers; improves utility asset utilization; allows use of the capital for other projects; and reduces the financial risk associated with lump investments.

Notably, for most nodes within a T&D system, the highest loads occur on just a few days per year, for just a few hours per year. Often, the highest annual load occurs on one specific day whose peak is somewhat higher than any other day. One important implication is that storage used for this application can provide a lot of benefit with limited or no need to discharge. Given that most modular storage types have a high variable operating cost, this application may be especially attractive for some storage types.

Although the emphasis for this application is on T&D *upgrade deferral*, a similar rationale applies to T&D equipment *life extension*. That is, if storage use reduces loading on existing equipment that is nearing its expected life, the result could be to extend the life of the existing equipment. This may be especially compelling for T&D equipment that includes aging transformers and underground power cables.

Readers are encouraged to see the Sandia National Laboratories report entitled *Electric Utility Transmission and Distribution Upgrade Deferral Benefits from Modular Electricity Storage* for more details.[23]

3.4.3.2. Technical Considerations

Energy storage must serve sufficient load, for as long as needed, to keep loading on the T&D equipment below a specified maximum. Discharge duration is a critical design criterion that cannot be generalized well. It may require interaction with utility engineers or engineers that design and/or operate distribution systems. The standard discharge duration is assumed to range from three to six hours.

3.4.3.3. Application Synergies

Utility-owned storage used for T&D deferral is also likely to be well-suited for several other applications, especially electric energy time-shift, electric supply capacity (peaking), and electric supply reserve capacity. Depending on location and circumstances, the same utility-owned storage could also be used for voltage support, transmission congestion relief, electric service reliability, electric service power quality, and renewables energy time-shift.

If the storage is customer-owned, it may be especially compatible with TOU energy cost and demand charge management as well as electric service reliability and electric service power quality and for renewables (co-located distributed PV) capacity firming.

3.4.4. Application #10 — Substation On-site Power

3.4.4.1. Application Overview

There are at least 100,000 battery storage systems at utility substations in the U.S. They provide power to switching components and to substation communication and control equipment when the grid is not energized. The vast majority of these systems use lead-acid batteries, mostly vented and to a lesser extent valve-regulated, with 5% of systems being powered by NiCad batteries.[24]

Apparently, users are generally satisfied, though reduced need for routine maintenance, improved reliability, and longer battery life would make alternatives attractive, especially if the cost is comparable to that of the incumbent technologies.

3.4.4.2. Technical Considerations

One important feature that competitive substation on-site power options must have is equal or better reliability than the standard option. Ideally, new options have lower maintenance requirements than the existing systems. Also, competitive options should have a straightforward way to determine the storage system's remaining useful life and ideally its 'state-of-health'.

One feature needed to address an emerging opportunity is the ability to serve the growing number of on-site DC loads (*e.g.*, from DC motors and actuators replacing electro-mechanical systems). Especially important are the capacity to provide inrush currents (*e.g.*, for motor startup) and a faster ramp rate to serve momentary loads including switchgear operation, motor-driven valves, isolating switches, and the field flashing of generators.[25]

IEEE Standard 485, which addresses sizing of battery systems for substation DC loads, groups substation DC loads into three categories: 1) continuous loads, 2) non-continuous loads, and 3) momentary loads. Based on results from a survey of systems, locations serving voltages of about 69 kV are rated at 1.6 kVA; locations serving the grid at 69 kV to 169 kV have storage rated at about 2.9 kVA; and substations serving the grid at voltages exceeding 169 kV have storage systems rated at 8.5 kVA. The standard value assumed is 2.5 kW. The standard discharge duration is assumed to range from 8 to 16 hours.

3.4.4.3. Application Synergies

Conceptually, the same storage used for substation on-site power could be used for other applications. Key considerations include a) use of the storage for other applications cannot degrade reliability and b) the storage must have sufficient discharge duration to serve the substation on-site power application *plus* other applications (*i.e.*, enough energy must be stored to serve the substation on-site power application *and* the other applications). For example, if 8 hours of discharge duration is required for substation on-site power and 5 hours are required for another application then the total discharge duration must be 8 + 5 = 13 hours. Given the high incremental cost for most types of storage that would be used for substation on-site power, use of the same storage system for other applications may be impractical in most circumstances.

3.5. End User/Utility Customer Applications

3.5.1. Application #11 — Time-of-use Energy Cost Management

3.5.1.1. Application Overview

Time-of-use (TOU) energy cost management involves storage used by energy end users (utility customers) to reduce their overall costs for electricity. Customers charge the storage during off-peak time periods when the electric energy price is low, then discharge the energy during times when on-peak TOU energy prices apply. This application is similar to electric energy time-shift, although electric energy prices are based on the customer's *retail* tariff, whereas at any given time the price for electric energy time-shift is the prevailing *wholesale* price.

Pacific Gas and Electric Company's (PG&E's) Small Commercial TOU A-6 tariff was used for the working example. It applies from May to October, Monday through Friday. Commercial and industrial electricity end users whose peak power requirements are less than or equal to 500 kW are eligible for the A-6 tariff.

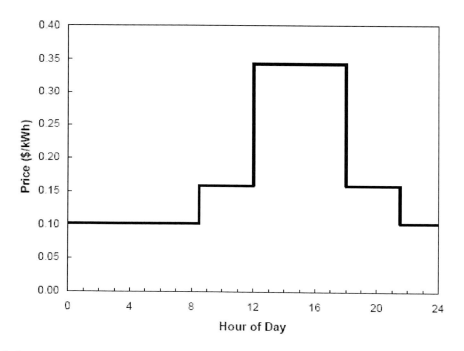

Figure 6. Summer energy prices for PG&E's Small Commercial A-6 TOU rate.

As shown in Figure 6, energy prices are about 32 ¢/kWh on-peak (12:00 p.m. to 6:00 p.m.). Prices during partial-peak (8:30 a.m. to 12:00 p.m. and 6:00 p.m. to 9:30 p.m.) are about 15 ¢/kWh, and during off-peak (9:30 p.m. to 8:30 a.m.) prices are about 10 ¢/kWh.

Although electricity end users receive the benefit for reducing energy cost, it is likely that that storage design, procurement, transaction costs, *etc.* would be too challenging for many potential users, especially those with relatively small energy use. If so, one option is to establish a partnership with an aggregator, as discussed in Section 6.5.4.

3.5.1.2. Technical Considerations

The maximum discharge duration for this application is determined based on the relevant tariff. For example, for the A-6 tariff there are six on-peak hours (12:00 p.m. to 6:00 p.m.). The standard value assumed for this application is five hours of discharge duration.

3.5.1.3. Application Synergies

Depending on overlaps between on-peak energy prices and times when peak demand charges apply, the same storage system use for time-of-use energy cost management might also be compatible with the demand charge management application. It could also provide benefits associated with improved electric service power quality and improved electric service reliability. Similarly, depending on a plant's discharge duration and when discharge occurs, it may be compatible with the T&D upgrade deferral application.

3.5.2. Application #12 — Demand Charge Management

3.5.2.1. Application Overview

Energy storage could be used by electricity end users (*i.e.,* utility customers) to reduce the overall costs for electric service by reducing demand charges, by reducing power draw during specified periods, normally the utility's peak demand periods.

To avoid a demand charge, load[2] must be reduced during all hours of the demand charge period, usually a specified period of time (*e.g.,* 11:00 a.m. to 5:00 p.m.) and on specified days (most often weekdays). In many cases, the demand charge is assessed if load is present during just one 15-minute period, during times of the day and during months when demand charges apply.

The most significant demand charges assessed are those based on the maximum load during the peak demand period (*e.g.,* 12:00 p.m. to 5:00 p.m.) in the respective month. It is somewhat common to also assess additional demand charges for 1) part peak or (partial peak) demand that occurs during times such as 'shoulder hours' in the mornings and evenings and during winter weekdays and 2) 'baseload' or 'facility' demand charges that are based on the peak demand no matter what time (day and month) it occurs. The latter is important for storage because facility demand charges apply at any time, including at night when most storage charging occurs.

Because there is a facility demand charge assessed during charging, the amount paid for *facility* demand charges offsets some of the benefit for reducing demand during times when the higher *peak* demand charges apply. Consider a simple example: The peak demand charge (which applies during summer afternoons, from 12:00 p.m. to 5:00 p.m.) is $10/kW-month, and the annual facility demand charge is $2/kW-month. During the night, when charging

---

[2] In the utility realm, 'demand' often refers to the maximum power draw during a specified period of time (e.g., a month or year). To avoid confusion relative to the more general economics definition, especially regarding demand for energy, in this report 'load' is often used instead of the term demand when referring to power draw.

occurs, the $2/kW facility demand charge is incurred; when storage discharges mid-day (when peak demand charges apply), the $10/kW-month demand charge is avoided. The *net* demand charge reduction in the example is

$10/kW-month − $2/kW-month = $8/kW-month.

Note that the price for electric *energy* is expressed in $/kWh used, whereas demand charges are denominated in $/kW of maximum *power* draw. Tariffs with demand charges have separate prices for energy and for power (demand charges). Furthermore, demand charges are typically assessed for a given month, thus demand charges are often expressed using $/kW per month ($/kW-month).

To reduce load when demand charges are high, storage is charged when there are no or low demand charges. (Presumably, the price for charging energy is low too.) The stored energy is discharged to serve load during times when demand charges apply. Typically, energy storage must discharge for five to six hours for this application, depending on the provisions of the applicable tariff.

Consider the example illustrated in Figure 7. The figure shows a manufacturer's load that is nearly constant at 1 MW for three shifts. During mornings and evenings, the end user's direct load and the facilities' net demand are 1 MW. At night, when the price for energy is low, the facility's net demand doubles as low-priced energy is stored at a rate of 1 MW while the normal load from the end user's operations requires another MW of power. During peak demand times (12:00 p.m. to 5:00 pm in the example), storage discharges (at the rate of 1 MW) to serve the end user's direct load of 1 MW, thus eliminating the real-time demand on the grid.

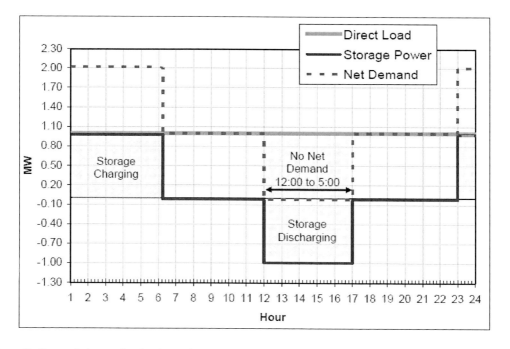

Figure 7. On-peak demand reduction using energy storage.

In the above example, storage is 80% efficient. To discharge for 5 hours, it must be charged for 5 hours ÷ 0.8 = 6.25 hours.

The 'additional' 1.25 hours of charging is needed to offset energy losses. If a facility demand charge applies, it would be assessed on the entire 2 MW (of net demand) used to serve both load and storage charging.

Although it is the electricity customer that internalizes the benefit, for this application, the author presumes that the design, procurement, transaction cost, *etc.* could be challenging for many prospective users, especially those with relatively small peak loads. One possible way for storage to be viable for those prospective users is to partner with an aggregator.

3.5.2.2. Technical Considerations

Given that demand charges apply for an entire month (and perhaps even for an entire year), for maximum load that occurs for even a few minutes, storage must be reliable. It must have acceptable or better power quality for loads served.

For this application, the storage plant discharge duration is based on the applicable tariff. For example, PG&E's E-19 Medium General Demand-Metered TOU tariff defines six on-peak hours (12:00 p.m. to 6:00 p.m.). The standard assumption for this application is five hours of discharge duration.

3.5.2.3. Application Synergies

Although each circumstance is different, storage used for demand charge management may be compatible with the electric energy time-shift application, and it could provide some ancillary services if end users are allowed to participate in the wholesale energy marketplace.

This application may be compatible with the transmission congestion relief and T&D upgrade deferral applications if storage use reduces load on T&D equipment when and where needed. (Note that T&D owners must be motivated and allowed to share related benefits, either by contract or prices.) Storage used for demand charge management is also likely to be compatible with the TOU energy cost management application if storage is discharging during times when energy price is high. Storage used for this application may also be compatible with the electric service power quality, electric service reliability, renewables capacity firming, and electric energy time-shift applications.

3.5.3. Application #13 — Electric Service Reliability

3.5.3.1. Application Overview

The electric service reliability application entails using energy storage to provide highly reliable electric service. In the event of a complete power outage lasting more than a few seconds, the storage system provides enough energy to ride through outages of extended duration; to complete an orderly shutdown of processes; and/or to transfer to on-site generation resources.

3.5.3.2. Technical Considerations

The discharge duration required is based on situation-specific criteria. If an orderly shutdown is the objective, then discharge duration may be an hour or more. If an orderly transfer to a generation device is the objective, then no more than a few minutes of discharge duration are needed. The standard value for discharge duration is 15 minutes.

Storage used for this application must reliably yield power with sufficient quality.

3.5.3.3. Application Synergies

The electric service reliability application may be compatible with most applications described in this report except area regulation and transmission support. It is especially compatible with the electric service power quality application.

If a storage system has sufficient discharge duration to serve the electric service reliability application plus other applications, it could be especially well-suited to serving the TOU energy cost and demand charge management applications as well as renewables (co-located distributed PV) capacity firming.

Depending on circumstances, the same storage system could also be used for electric energy time-shift, electric supply capacity (peaking), ancillary services, voltage support, transmission congestion relief, T&D upgrade deferral, electric service reliability, electric service power quality, and renewables energy time-shift applications.

3.5.4. Application #14 — Electric Service Power Quality

3.5.4.1. Application Overview

The electric service power quality application involves using energy storage to protect on-site loads downstream (from storage) against short-duration events that affect the quality of power delivered to the load. Some manifestations of poor power quality include the following:

- Variations in voltage magnitude (*e.g.*, short-term spikes or dips, longer term surges, or sags).
- Variations in the primary 60-Hz frequency at which power is delivered.
- Low power factor (voltage and current excessively out of phase with each other).
- Harmonics (*i.e.*, the presence of currents or voltages at frequencies other than the primary frequency).
- Interruptions in service, of any duration, ranging from a fraction of a second to several or even many minutes.

3.5.4.2. Technical Considerations

Needless to say, storage used for power quality should produce high-quality power output and should not adversely affect the grid. Typically, the discharge duration required for the power quality application ranges from a few seconds to about one minute.

3.5.4.3. Application Synergies

Given the short discharge duration and distributed deployment of storage for electric service power quality, few if any applications are compatible with storage designed specifically for that application. Nevertheless, the electric service power quality application may be compatible with several other applications if storage is designed for those other applications (*i.e.*, with longer discharge duration), especially time-of-use energy cost management, demand charge management, and electric service reliability.

3.6. Renewables Integration Applications

3.6.1. Application #15 — *Renewables Energy Time-shift*

3.6.1.1. Application Overview

Many renewable energy generation resources produce a significant portion of electric energy when that energy has a low financial value (*e.g.,* at night, on weekends and during holidays) – generally referred to as off-peak times. Energy storage used in conjunction with renewable energy generation could be charged using low-value energy from the renewable energy generation so that energy may be used to offset other purchases or sold when it is more valuable.

The low-value energy is generated off-peak at night and during early mornings when demand is low and supply is adequate. The energy is more valuable on-peak when demand is high and supply is tight. The energy value is especially high during hot summer afternoons when A/C use is most prevalent. The energy that is discharged from the storage could be used by the owner, sold via the wholesale or 'spot' market, or sold under terms of an energy purchase contract (commonly referred to as a 'power purchase agreement' or PPA).

Storage used for renewables energy time-shift could be located at or near the renewable energy generation site or in other parts of the grid, including at or near loads. Energy discharged from storage located at or near the renewable energy generation would have to be transported via the transmission system during on-peak times whereas storage located at or near loads is charged using low-value energy that is transmitted during off-peak times.

Typically, the storage discharge duration needed for energy time-shift ranges from four to six hours, depending mostly on the duration of the region's off-peak and on-peak periods and the on-peak *versus* off-peak energy value or price differential.

Two variations of the renewables energy time-shift application are evaluated in this guide. They are 1) time-shift of energy from *intermittent* renewable energy generation resources and 2) time-shift of energy from *baseload* renewable energy generation resources. Intermittent renewables include solar, wind, ocean wave, tidal and, in some cases, hydroelectric. Baseload renewables – those whose output is somewhat-to-very constant, for several thousand hours per year – include geothermal, biomass, and, in some cases, hydroelectric. The *intermittent* renewable energy generation type evaluated here is wind-fueled generation. The *baseload* renewable energy generation evaluated is generic: It operates 24 hours per day and at a minimum it operates during every weekday during the year.

Storing electric energy from *solar* generation is not addressed in this report for two reasons. First, for situations involving *grid-connected* solar generation, a lot or even most electricity is produced when energy is already valuable, making energy time-shift relatively unattractive. Second, most of the value for storage used with solar generation is for capacity firming. (See Section 3.6.2.) Also not addressed is *seasonal* renewables energy time-shift. That is because storing enough energy for seasonal renewables energy time-shift is either impractical or prohibitively expensive with the possible exception of CAES.

3.6.1.2. Energy Time-shift from Wind Generation

For the case involving wind generation, low-value electric energy from wind generation is stored at night and during early mornings. The stored energy is discharged when it is most valuable — during weekday afternoons when demand for electricity is highest.

Not only does energy from wind generation produced off-peak have a low value, depending on regional circumstances wind generation occurring during off-peak hours can cause operational challenges. Two such operational challenges are minimum load violations and accommodating rapid changes to output from intermittent renewable energy generation. (See Section 3.6.3.)

When minimum load violations occur, the combined output from wind generation capacity plus other 'must-run' generation exceeds demand (must-run generation tends to include that which is fueled by coal, nuclear, baseload renewable energy, and some types of natural-gas-fueled generation). Rapid output changes from intermittent renewable energy generation can lead to 'ramping' of other *dispatchable* generation, which increases wear, fuel use, and emissions (all per kWh).

An example of the daily operation profile for wind generation plus storage on a summer day is shown in Figure 8. For the scenario depicted, wind generation output occurring at night, when the energy's value is low, is used to charge storage. In the example, about one-half of the energy used on-peak is from wind generation that occurs off-peak. The result is constant power for five hours.

For the wind generation case, storage discharge duration required ranges from two and one-half hours to as much as four hours, depending on the amount of energy from wind generation that occurs during on-peak times.

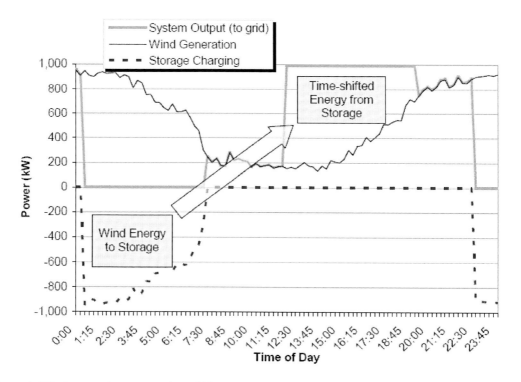

Figure 8. Wind generation energy time-shift.

3.6.1.3. Energy Time-shift from Baseload Renewable Energy Generation

Baseload renewables energy time-shift is accomplished by storing energy at night, during off-peak periods, so the energy can be used when it is most valuable, especially when hot temperatures drive significant air conditioning use.

An example of the concept is illustrated in Figure 9. The example involves storage whose power is equal to that of the generator's (1 MW) and whose discharge duration is five hours. The storage is charged during off-peak times using most or all of the generator's output and the storage discharges during five on-peak hours. Note that time-shift energy from baseload renewable energy generation has the effect of doubling the renewable energy generation's capacity during times when both demand and the value of electric supply *capacity* are highest.

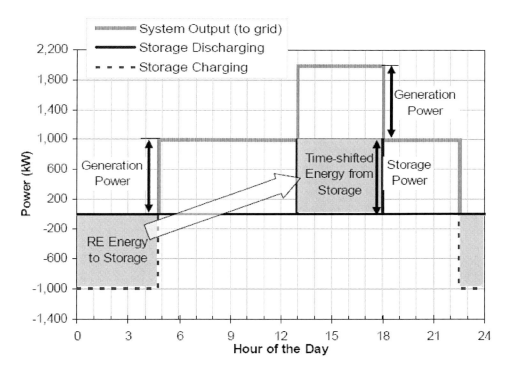

Figure 9. Baseload renewables energy time-shift.

3.6.1.4. Technical Considerations

The discharge duration for this application is circumstance-specific. It depends mostly on expectations about electric energy prices and/or the terms of the energy purchase agreement, especially the price

and timing of purchases. The standard value assumed in this guide for discharge duration is five hours.

For *intermittent* renewable energy generation, another important criterion is the degree to which the renewable energy generation output coincides with times when the price for electric energy is high.

PCUs used in conjunction with many, or even most, renewable energy systems do not have what is needed to facilitate use of storage. Consequently, PCUs used for renewables energy time-shift must have additional hardware and software to accomplish and to manage charging and discharging of the storage.

3.6.1.5. Application Synergies

Depending on the location, the timing of the discharge, storage discharge duration, storage ramp rate, and the owner's flexibility to optimize storage dispatch, storage used to time-shift electric energy from renewables generation could also serve several other applications described in this report.

Renewables energy time-shift is especially compatible with the renewables capacity firming and electric supply capacity applications. Centrally located storage used for this application could also be used for electric supply reserve capacity and area regulation. If the storage is deployed in distributed mode, then the storage could serve most applications (other than area regulation), especially voltage support, transmission congestion relief, T&D upgrade deferral, electric service power quality, electric service reliability, TOU energy cost management, and demand charge management.

3.6.2. Application #16 — Renewables Capacity Firming

3.6.2.1. Application Overview

Renewables capacity firming applies to circumstances involving renewable energy-fueled generation whose output is intermittent. The objective is to use storage to 'fill in' so that the combined output from renewable energy generation plus storage is somewhat-to-very constant.

The resulting firmed capacity offsets the need to purchase or 'rent' additional dispatchable (capacity) electric supply resources. Depending on location, firmed renewable energy output may also offset the need for transmission and/or distribution equipment. Renewables capacity firming is especially valuable when peak demand occurs.

For the purpose of renewables capacity firming, renewable energy generation's output intermittency can be classified as 'short-duration' (*i.e.,* occurring somewhat-to-very randomly over timescales ranging from seconds to minutes) and/or 'diurnal' (*i.e.,* occurring in a regular and/or predictable way during a 24-hour period).

Note the important distinction between renewables *capacity firming*, and renewables *energy time-shift*.

Capacity firming allows use of an intermittent electric supply resource as a nearly constant *power* source. Such use may reduce power-related charges (*e.g., capacity* payments or *demand* charges), or it may offset the need for *equipment* (*e.g.,* wires, transformers, and generation) which is an *investment* with a *fixed* cost.

By contrast, **energy time-shift** involves enhancing the value of *energy* to increase profits and/or reduce fuel, operation, variable operation, and maintenance costs which are e*xpenses*.

In most circumstances, renewables capacity firming is likely to result in a combined benefit comprised of a benefit for renewables energy time-shift and one for the firm capacity.

One important challenge associated with intermittent renewable energy generation is that the generation's power output can change rapidly over short periods of time. Photovoltaic (PV) output can drop quite quickly as clouds pass. Wind generation output can change rapidly during gusty conditions.

These rapid changes (also known as ramping) can lead to the need for dispatchable power sources whose output can also change rapidly. Most new, non-renewable energy generation facilities are best operated at constant output. In some regions, however, there may not be enough dispatchable generation capacity to offset renewable energy generation's ramping. Storage can have an important effect on the amount of dispatchable generation needed to meet the renewable energy generation ramping challenge.

In broad terms, good opportunities for renewables capacity firming tend to involve renewable energy resources whose output is somewhat-to-very coincident with the peak demand and somewhat-to-very constant. Storage used to firm resources with these characteristics needs relatively modest discharge duration. Solar generation's output tends to occur when demand for electricity is highest and varies somewhat modestly, albeit predictably. In some locations, wind-fueled generation output sometimes coincides with peak load and is somewhat stable during peak load periods.

Although, in most cases, wind generation output is not as coincident with peak demand as that from solar generation, non-trivial amounts of wind generation do occur during peak demand periods. Also, wind generation tends to be ramping down as load is increasing, making firming valuable as a way to reduce load following resources. Additionally, wind generation is somewhat to quite predictable.

Given those premises, leading candidates for renewables capacity firming include those fueled with solar energy (especially PV) or with wind energy. Depending on local circumstances, ocean wave generation output could also be firmed with storage, though it is not considered in this report.

3.6.2.2. PV Capacity Firming

Although capacity firming applies somewhat equally to large 'bulk' solar generation facilities and to small systems, *distributed* PV systems are featured here as the solar-fueled generation because, in many circumstances, it is possible for storage to serve other valuable applications if the storage is distributed. And, distributed PV systems are more likely to have suboptimal orientation leading to output that is only somewhat coincident with peak demand periods.

The PV systems are assumed to consist of flat-panel PV modules with a fixed orientation. Fixed-orientation PV remains stationary as the sun's position in the sky changes throughout the day. Output from fixed-orientation PV systems increases as the sun rises during the morning hours; stays somewhat constant (at the daily maximum) for one to two hours during mid-day; and declines as the sun moves across the sky in the afternoon. Consequently, output from PV with a fixed orientation is at a maximum during a portion of the peak load period in many locations. If fixed PV orientation is not optimal, it will produce a modest to significant portion of output before or after the utility's peak demand period.

3.6.2.3. Wind Generation Firming

Large-scale 'bulk' wind generation is featured in this report because a significant portion of wind generation development will involve large wind farms, whereas it seems unlikely that a significant amount of *distributed* wind generation will be added, at least for the foreseeable future. Nonetheless, the capacity firming benefit could apply to distributed wind generation as well as to central/bulk wind farms.

3.6.2.4. Short-duration Intermittency

Solar Generation Short-duration Intermittency

Shading caused by terrestrial obstructions such as trees and buildings can cause relatively short-duration, location-specific intermittency. The most compelling cause of short-duration intermittency from solar generation, however, is clouds. As a cloud passes over solar collectors, power output from the affected solar generation system drops. When the cloud moves away from the collector, the output returns to previous levels. Importantly, when that happens, the *rate* of change (of output from the solar generation plant) can be quite rapid. The resulting ramping increases the need for highly dispatchable and fast-responding generation such as a simple cycle combustion turbine to fill in when clouds pass over the solar collector.

Wind Generation Intermittency

Short-duration intermittency from wind generation is caused by variations of wind speed that occur throughout the day. Although such variations may not be significant during much of the year, it can be a ramping-related challenge if peak demand for electricity coincides with gusty wind conditions. Figure 10 shows a basic example of short-duration intermittency and the implications for storage needed for firming. In the figure, the one-minute average renewable energy output (for a 1-kW renewable energy plant) is plotted. Note the variation from one minute to the next.

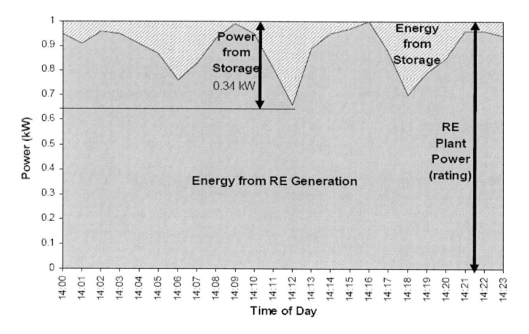

Figure 10. Renewable-fueled generation, short-duration intermittency (example).

As shown in the figure, the power needed from storage to offset the short-duration intermittency is determined based on the maximum difference between the renewable energy plant *rating* and the reduced plant *output* due to short-duration intermittency. In the example, the largest (magnitude) short-duration drop-off of power from the renewable energy generation is about 34% of the renewable energy's plant rating. Consequently, the storage plant would need to have a power rating of at least 0.34 kW per kW of renewable energy generation.

3.6.2.5. Diurnal Intermittency

Solar Generation Diurnal Intermittency

Diurnal intermittency of solar generation is mostly related to the change of insolation throughout the day as the sun rises in the morning and then descends in the evening. Shading (not related to clouds) can also add to solar-energy-fueled generation's diurnal intermittency. Also, the solar energy-to-electricity conversion efficiency for some types of solar generation (especially flat-panel PV) drops as the equipment's temperature increases. Thus, if ambient temperatures are high, then efficiency may drop, reducing output commensurately.

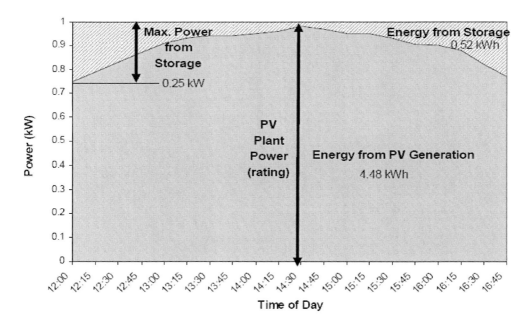

Figure 11. PV generation output variability during peak demand hours (example).

The key source of diurnal intermittency from solar-energy-fueled generation is shown in Figure 11. In that example, storage is discharged when solar generation production is less than the solar plant's rated output. The figure also shows that the lowest output from the solar generation during peak demand hours (about 75% of rated capacity) occurs in the early afternoon as the sun continues to rise. The effects of short-duration intermittency, if any, are not shown. Based on the example (without regard to short-duration intermittency), firming of the PV's output requires storage whose capacity (power) is equivalent to at least 0.25 kW per kW of the solar generation's power rating. The storage must have enough *energy* to deliver 0.52 kWh per day, for each kW of the solar generation's power rating.

Wind Generation Diurnal Intermittency

In most regions, wind tends to be stronger during certain parts of the day than during others. For example, in some regions wind speed is relatively high in the late afternoon and evening and relatively low in the morning and early afternoon. Such a scenario is shown in Figure 12. As shown in Figure 12, storage fills in when wind generation output is less than the wind turbine's rated output. In the figure, the lowest level of output from the wind generation (about 35% of rated capacity) occurs at about 1:45 p.m. (13:45). The effects of short-duration intermittency are not shown. So, for the example described in Figure 12, the storage must provide capacity (power) equal to about 65% of the wind turbine's rating. The storage must be able to deliver 2.36 kWh per kW of wind capacity for firming.

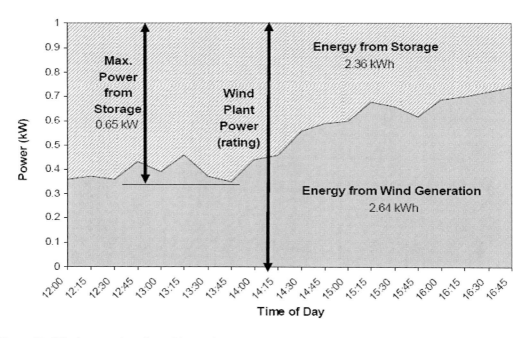

Figure 12. Wind generation diurnal intermittency during peak demand hours.

3.6.2.6. Technical Considerations

Storage power and discharge duration (for renewables capacity firming) are quite circumstance-specific and resource-specific. At the lower end, it is assumed that one-half to as much as two hours of discharge duration is needed to firm solar generation, assuming that much of PV output coincides with peak demand. For the example: To firm wind generation, a somewhat longer discharge duration (two to three hours) is needed.

Storage used for capacity firming should be quite reliable because the primary reason for capacity firming is to provide *constant power*. Also, the price paid for constant power (*i.e.*, demand charges for retail electricity end users or market price for capacity for the wholesale part of the market) is usually accompanied by a significant financial penalty if power is *not* firm.

Power conditioning equipment used for many renewable energy systems does not include the functionality needed for charging and discharging storage, which requires additional hardware and software. Nevertheless, the ability to accommodate storage can be added to the power conditioning equipment used for the renewable energy generation at a relatively low incremental cost.[26]

3.6.2.7. Application Synergies

Although possibilities are circumstance-specific, storage used for renewables capacity firming could also provide benefits related to several other applications. Renewables capacity firming is especially compatible with the renewables energy time-shift and electric supply reserve capacity applications.

For *distributed* renewable energy generation, depending on the location, capacity firming may also be compatible with several other applications including voltage support, transmission congestion relief, T&D upgrade deferral, TOU energy cost management, demand charge management, electric service reliability, and electric service power quality. Incidental benefits that could accrue are those for reduced T&D energy losses and reduced transmission access charges.

One especially attractive synergy for distributed PV plus storage is improved electric service reliability and/or improved electric service power quality. The discharge duration required for reliability-related and quality-related needs varies considerably; it depends on the robustness of the electric grid, T&D quality, and the loads and end uses served. The discharge duration needed for reliability and power quality can range from seconds to hours. For this report, it is assumed that one-quarter to one-half hour of storage (discharge duration) would be added to the PV plus storage system to provide reliability and/or power quality-related benefits.

3.6.3. Application #17 — Wind Generation Grid Integration

3.6.3.1. Application Overview

For all but modest wind generation penetration levels, wind generation is likely to have at least some undesirable impact on the grid. And wind generation *does* seem poised to be a key element of the global move toward increased use of renewable energy. In the U.S., growth of wind generation capacity will be driven, in part, by targets established under the auspices of the Renewables Portfolio Standard (RPS). (See Section 4.3.1.1 for details about RPS.)

To the extent that emphasis on renewable energy *does* increase, wind generation is well-positioned to provide a significant portion of electricity. Wind generation is especially attractive given the relatively low and dropping electricity production cost from wind generation and good or better wind resources in many geographic regions.

As wind generation penetration increases, the electricity grid effects that are unique to wind generation will also increase. Storage could assist with orderly integration of wind generation (wind integration) by managing or mitigating the more challenging and less desirable effects from high wind generation penetration.

The wind generation grid integration application includes six subtypes which are grouped into two categories: 1) short-duration (*i.e.,* lasting for a few seconds to a few minutes) and 2) long-duration (*i.e.,* lasting for many minutes to a few hours). The six subtypes are shown in Table 7.

Table 7. Wind Generation Grid Integration Categories and Subtypes

Short-duration Applications
Reduce Output Volatility
Improve Power Quality
Long-duration Applications
Reduce Output Variability
Transmission Congestion Relief
Backup for Unexpected Wind Generation Shortfall
Reduce Minimum Load Violations

3.6.3.2. Reduce Output Volatility

The reduce output volatility application subtype is related to the need to offset wind power output fluctuations caused by short-duration variation of wind generation output, lasting seconds to a few minutes.

It is important to note that, in most cases, wind turbines' geographical diversity smoothes the aggregate effect of output volatility considerably. If the wind generation is interconnected with a large, well-diversified, electric supply and grid system, then that system can accommodate significant wind generation output fluctuations.[27] Nevertheless, for large wind generation resources, even somewhat modest volatility in the aggregate output may drive a need for a nontrivial supplemental resources to supply capacity and energy. Smaller and/or less diverse wind generation resources may require even more storage capacity (per MW of wind generation capacity).

Although requirements will be different for each location and area, for this report it is assumed that a well-diversified wind generation resource using storage rated at 2% to 3% of the wind generation capacity would reduce aggregate volatility and reduce the need for area regulation significantly.[28][29] That range (2% to 3% of wind generation capacity) applies to wind penetration levels of about 10% (of total generation capacity). Presumably, the capacity needed (per kW of wind generation capacity) will change as wind generation penetration increases.

The benefit for this application is estimated based on avoided need for additional area regulation resources and service. Depending on the amount of output volatility, an alternate approach could involve that described for renewables capacity firming for short-duration intermittency as described in Section 3.6.2.4.

3.6.3.3. Improve Power Quality

The power quality application reflects a category of wind-generation-related challenges that are related to performance standards, interconnection requirements, effects from phenomena such as wind gusts, and changing electrical conditions in parts of the grid affected by and/or with an effect on wind generation operations.[30]

Seven specifically power quality-related challenges are as follows:

- Reactive power
- Harmonics
- Voltage flicker
- Transmission line protection
- Transient stability
- Dynamic stability
- System voltage stability

In most cases, conventional non-storage options are available to address these power quality challenges. For example, capacitors may be used for some reactive-power-related needs. Also, newer wind turbines will, by design, have reduced power quality impacts.[31]

3.6.3.4. Reduce Output Variability

This application is related to the need to offset generation output variability caused by natural wind speed variability over durations of several minutes to a few hours. Increasing wind generation penetration seems likely to increase the need for load following resources beyond what would otherwise be needed for a more dispatchable electric supply mix. It is important to note, however, that large, well-diversified electric supply and transmission systems can accommodate a lot of wind generation variability, especially if the wind generation is geographically diverse and/or comprises a relatively small portion of the electric supply capacity.[32]

This application is somewhat analogous to the 'load following' ancillary service application because of the time scales and operational profiles involved. In fact, at the grid level, system load following resources are used to compensate for such variations. Presumably, reducing aggregate wind generation variability will also reduce the need for central load following.

In more than a few regions, normal wind speed patterns mean that wind generation output drops off just as load picks up (*i.e.*, it decreases as people begin activities in the morning). Similarly, wind generation often increases as load drops off (*i.e.*, generation output rises as people's activity, and the associated electric load, decreases at night). In such a scenario, adding wind generation capacity may also increase the need for load following capacity. In the evening, the grid may need extra load following in the down direction to accommodate increasing wind generation output that occurs during times when load is decreasing. Because wind generation output drops in the morning just as load picks up, more load following in the up direction may be needed as new wind generation capacity is added.

Wind generation variability (and the corresponding need for load following resources) may be an especially compelling challenge during times when load is light. This is because, in many regions, a relatively small amount of *dispatchable* generation is available at those times to accommodate wind generation fluctuations (*i.e.*, the output of most generation online at those times tends to be coal-fired, nuclear, natural gas/steam, 'must-take' energy purchase contracts and some hydroelectric generation that cannot be reduced).[33]

Although requirements will be different for each location and area, for this report, it is assumed that storage capacity whose power rating is 4% to 6% of wind generation capacity could offset the need for a similar amount of system load following resources (*i.e.*, those load following resources *would* be needed to accommodate wind generation's natural variability, without storage).[34]

That range (4% to 6% of wind generation capacity for reducing output variability) applies to a geographically diverse wind resource with wind generation penetration levels of about 10% of total generation capacity. Presumably, the optimal amount of storage would change with wind generation penetrations above 10%.

3.6.3.5. Transmission Congestion Relief

This application reflects an important challenge posed by the installation of significant amounts of wind power capacity. At any given point in time, the transmission system may not have enough capacity to transfer the energy generated by all the wind turbines, causing 'congestion' on the grid (*i.e.,* too much energy to be transferred through the available transmission capacity). Storage could be used *in lieu* of upgrading transmission to accommodate wind generation during times when congestion occurs:

- Storage located *upstream* from the point of congestion could be charged when congestion occurs, so energy can be transmitted when there is no congestion.
- Storage located *downstream* from the point of congestion would allow for transmission of energy for charging when there is no congestion. That energy can be used later when congestion occurs.

3.6.3.6. Backup for Unexpected Wind Generation Shortfall

The need for storage backup for unexpected wind generation shortfall materializes when regional wind velocity is considerably lower than predicted and wind generation is supplying a relatively large portion of total grid power. Although such events are rare, the effect on the grid may be significant. As wind generation penetration increases, the impact from such events may also increase.

Consider one real-world example. On February 27, 2008, the state of Texas experienced an unexpected "drop in wind generation...coupled with colder than expected weather." During the event, wind generation output reportedly dropped from about 1,700 MW to about 300 MW. Grid operators responded by asking grid customers with interruptible electric tariffs to reduce power use by about 1 GW for about 90 minutes.[35] Two key options when this occurs are 1) to call on end users with interruptible or curtailable electric service or 2) to dispatch reserve capacity.

3.6.3.7. Reduce Minimum Load Violations

In some cases, wind generation output occurs when must-run and/or non-dispatchable generation capacity online exceeds demand. In this report, that situation is referred to as a minimum load violation. Possible alternatives for addressing minimum load violations may include 'dumping' or 'spilling' unusable energy or curtailing wind generation output. Storage may be especially helpful to manage those situations, especially if the minimum load violation results in 'negative prices', meaning that energy users get paid to take the energy.

3.6.3.8. Technical Considerations

Storage for wind-generation-related transmission congestion relief and for backup does not have any unique technical requirements. Ramp rate is not especially important, and reliability is not especially important if there are a large number of storage units in service.

Storage used to address wind output intermittency and power quality *must* have a rapid ramp rate. Storage used to address wind output intermittency will likely need to have a very high efficiency and low operation cost because that application involves many charge/discharge cycles per hour.

If reactive power capability is needed for power quality, then the storage system's PCU must have VAR support capability or must be able to produce reactive power.

3.6.3.9. Application Synergies

Generalizing application synergies for wind generation grid integration may not be especially helpful, as technical and operational needs for the six application subtypes vary so much. Nevertheless, there are many possible combinations, some of which may be attractive now or in the future. Especially notable are synergies with the renewables energy time-shift and renewables capacity firming applications; storage used with wind generation for those applications may also reduce grid effects from wind output variability incidentally.

Reducing output volatility is probably not compatible with any other application subtype or with any of the other primary applications described in this report because storage used to manage output volatility is almost always in service. Storage designed for the improved power quality application subtype probably has a short duration and thus may not be compatible with use for other applications.

Depending on the timing of storage output and the storage's location, storage used for the transmission congestion relief, reduce output variability, reduce minimum load violations, and backup for unexpected wind generation shortfall application subtypes may be compatible with each other or with several other primary applications.

If the storage is located at distributed locations (*i.e.,* for small commercial or even residential wind turbines), then storage could also be used for T&D upgrade deferral, electric service reliability, electric service power quality, TOU energy cost management, and demand charge management.

3.7. Distributed Energy Storage Applications

Locating storage near loads opens up opportunities to use the same storage for many more applications than a larger 'central' or 'bulk' resource could address. Depending on the location, storage deployed as a distributed energy resource (DER) may be compatible with all applications listed in this report except for area regulation, transmission support, and some wind integration-related uses.

3.7.1. Locational Distributed Storage Applications

The applications in this subsection are those that are *best* served by *distributed* storage or *cannot* be served unless the storage is deployed in distributed mode (*i.e.,* the storage is located where needed, near to loads). These applications include voltage support, transmission congestion relief, T&D upgrade deferral, TOU energy cost management, demand charge management, electric service reliability, electric service power quality, renewables capacity firming, and wind generation grid integration For example, storage used to defer a T&D capacity upgrade must be located near loads served by the T&D equipment in question. More specifically, the storage must be located downstream (electrically) from the T&D node in question. Another example is storage used to improve localized power quality. That storage must be located where it actually provides the necessary effect(s) on power quality.

3.7.1.1. Voltage Support

For this report, distributed storage (*i.e.,* storage located near loads that most heavily affect voltage) is a viable option for the voltage support application, whereas voltage support provided centrally is assumed to be from large generation facilities. Unless the grid is weak or poor, storage will be used very little, if at all, for this application. Given that consideration, almost any storage located at or near loads that contribute to cascading outages could provide voltage support if it has VAR support capabilities and a discharge duration of 30 minutes or more.

3.7.1.2. Transmission Congestion Relief

If distributed storage is located downstream from congested transmission, then it could be used to store energy when there is no congestion and/or to reduce demand downstream from congestion when the congestion occurs. For distributed storage, this application/benefit may be especially compatible with the following applications/benefits: demand charge management, TOU energy cost management, electric supply reserve capacity, voltage support, electric service reliability, and electric service power quality.

3.7.1.3. T&D Upgrade Deferral

T&D upgrade deferral is one of the richest possibilities for distributed storage because the benefit can be so high. Also, this application/benefit may be compatible with several other applications/benefits, especially the following: electric supply reserve capacity, voltage support, electric service reliability, electric service power quality, TOU energy cost management, demand charge management, and possibly even electric supply reserve capacity and load following.

3.7.1.4. Time-of-use Energy Cost Management and Demand Charge Management

Bill management includes two closely related applications: TOU energy cost management and demand charge management. These applications are notable because storage used for them could also be used for electric service reliability, electric service power quality, electric supply reserve capacity (when charging and when charged but not discharging) and load following (when charging). Storage installed in advantageous locations could also provide voltage support, T&D upgrade deferral, and transmission congestion relief.

3.7.1.5. Electric Service Reliability and Electric Service Power Quality

Electric service reliability and electric service power quality are especially notable applications because significant demand for storage already exists in the form of uninterruptible power supplies (UPSs). They are also notable because, in most cases, storage can provide significant benefit with limited charging/discharging and relatively short discharge durations. In many cases, storage used for several distributed storage applications could also provide backup energy for electric service reliability and could be used to condition power as needed to address power quality problems.

3.7.1.6. Renewables Capacity Firming – Photovoltaics

There are strong synergies when modest storage capacity is coupled with on-site PV. Although PV production may not coincide with *capacity* needs, most PV production occurs during times when most energy is used, and PV alone cannot provide emergency or backup power without sunlight. Distributed storage used to firm PV capacity may also be compatible with other applications, including demand charge management, TOU energy cost management, electric supply reserve capacity, voltage support, electric service reliability, and electric service power quality.

3.7.1.7. Wind Generation Grid Integration

New wind turbine concepts may lead to increasing use of distributed wind generation capacity. As noted in the discussion of the wind generation integration application (Section 3.6), storage may be important if there will be even modest penetration of wind generation capacity at the distribution level. Depending on the circumstances, wind generation's energy could be sold to the grid at a profit or used to reduce TOU energy charges. Also depending on the circumstances, firming wind generation capacity with storage may provide capacity value if the utility has a need for the firm capacity and/or if the end user can use it to reduce demand charges.

3.7.2. Non-locational Distributed Storage Applications

For the following applications, distributed storage may be located anywhere that its operation does not cause operational or technical problems for the grid: electric energy time-shift, electric supply capacity, load following, area regulation, electric supply reserve capacity, and renewables energy time-shift.

3.7.2.1. Electric Energy Time-shift

Assuming that distributed storage is not subject to transmission congestion during charging, distributed storage could be used to store inexpensive off-peak electric energy from the grid so that the energy may be used or sold when value/price is high.

3.7.2.2. Electric Supply Capacity

As with electric energy time-shift, if distributed storage is not subject to transmission congestion when charging occurs, it can be used to store inexpensive off-peak electric energy from the grid so that the energy may be used for electric supply capacity firming when doing so is valuable.

3.7.2.3. Load Following

To the extent that distributed storage can respond to control signals from the ISO, it can be used for load following. Perhaps most interesting is the possibility of providing load following, incidentally, while charging. (See Section 3.3.1 for details.)

3.7.2.4. Area Regulation

Conceptually, area regulation could be provided anywhere within an area if the location does not have any transmission constraints. If the area regulation capacity is located downstream (electrically) from subtransmission or distribution equipment, there may be some back-feed constraints if the equipment cannot accommodate a significant amount of energy flow into the transmission system. If so, then perhaps the area regulation capacity could be matched to local area regulation needs.

3.7.2.5. Electric Supply Reserve Capacity

Distributed storage that is charging or that is in standby mode can provide reserve capacity. Notably, unless the electric supply system served is weak or poorly managed, storage will be used very little for reserve capacity.

3.7.2.6. Renewables Energy Time-shift

As the electricity marketplace evolves, there may be opportunities for using distributed energy storage to store energy generated by large renewable-fueled generation located upstream from transmission and/or distribution system bottlenecks. Key objectives include increasing renewables' energy and capacity value and relieving grid system congestion. This seems especially valuable if distributed storage can be charged when minimum load conditions exist (or even when less severe mismatches between supply and load exist); and/or when charging can be used for load following; and when transmission congestion is not a challenge.

3.7.3. Incidental Applications from Distributed Storage

Distributed storage can serve some applications, incidentally, while charging – most notably load following and electric supply reserve capacity. If the distributed storage (which is charging) has enough *stored* energy then it can also *discharge* to provide *additional* electric supply reserve capacity for other applications including voltage support, electric service reliability, and electric service power quality. Note that reduced storage charging has the same effect as adding reserve capacity. If, after charging is stopped, that same storage then *discharges* into the grid or picks up load, then the storage essentially provides two times its capacity as reserve capacity.

Similarly, distributed storage that is charged can serve several applications, incidentally, while in standby mode (*i.e.,* while not being used for a primary application) including electric supply capacity, voltage support, electric service reliability, and electric service power quality.

3.8. Applications Not Addressed in this Guide

It is important to note that the approach used for this report – involving applications that are defined based on the corresponding electric utility-related benefit – may seem to exclude many possible *uses* of storage. Certainly, that was not the authors' intention. Indeed, the framework developed for this report can be used to estimate the financial benefits associated with many uses of storage, including many not addressed explicitly, because the benefits described are intended to address the various *revenues and avoided costs* that accrue when storage is used.

Consider three examples of storage use: 1) as a backup power source for telecommunications facilities, 2) as part of a rail system to address voltage sags and to recuperate energy using regenerative braking, and 3) for localized reactive power compensation (VAR support) by utilities.

For the first example (backup for telecom facilities), the benefit is related to avoided outages. The magnitude of the benefit can be estimated using an approach similar to that described in this report for the electric service reliability benefit. Specifically, the benefit is either the cost avoided because a more expensive alternative (*e.g.,* diesel engine generators) is not needed if storage is used, or the application-specific value of avoided unserved energy.

The benefit for use of storage in the second example (rail system trackside storage) is some combination of reduced cost for other equipment needed to address the voltage sag challenge; reduced cost to purchase energy; and reduced peak demand charges. In many cases, the equipment purchases that are deferred or avoided are for additional circuits and/or transformers and/or power electronics.

In the third example (utility use of storage for VAR support), the benefit is the avoided cost for equipment that would have to be installed without storage, normally capacitors.

4. MAXIMUM MARKET POTENTIAL ESTIMATION

This section describes a framework for making a high-level, 'first-cut' estimate of the market potential for storage for each of the applications characterized herein (see Figure 13). It entails a generic, three-step process. Estimates for steps one and two are provided in this guide. Taking the estimate to the final step is beyond the scope of this report, as making it requires detailed analysis involving, among other criteria and considerations, 1) a broad array of national and regional market conditions, drivers, and trends; 2) utility regulations and rules; 3) technology cost and performance, existing and trends; 4) the spectrum of benefits (values) for individual applications and for viable application combinations (value propositions); and 5) stakeholder biases and preferences.

4.1. Market Potential Estimation Framework

As indicated by the outer square in Figure 13, the first step required when estimating economic market potential is to ascertain the *technical* market potential. It is the maximum amount (MW) possible given technical constraints. As an upper bound, the technical potential is the peak electric demand.

Next, the *maximum* market potential is established. As shown in Figure 13, maximum market potential is a portion of the technical potential. It is an estimate of the maximum possible demand given constraints that are practical or institutional in nature (*e.g.,* utility regulations and practices). Maximum market potential is also established without regard to storage cost.

Finally, an estimate would be made of the *expected* market potential (market estimate). As shown in Figure 13, the market estimate is some portion of the maximum market potential. The market estimate reflects the amount of storage that an analyst expects to be deployed, over a given period of time (10 years in this document), for the specified application or combination of applications.

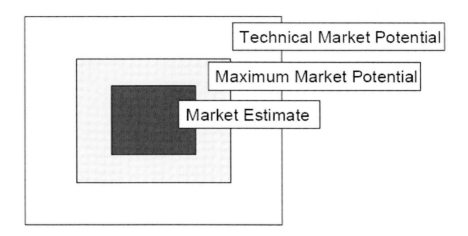

Figure 13. Market potential and estimate.

Market estimates may be as detailed and precise as appropriate. At the very least, various levels of market potential can be tested for reasonableness using a combination of judgment, knowledge, and preliminary product cost estimates. Alternative bases for estimates could include, for example, sales trends and projections, surveys, analysis of utility capital budget plans, detailed product cost estimates, and/or market research or intelligence.

4.1.1. Role of Aggregators

For some applications, and for electricity end users that do not use a lot of energy, the hassle, learning curve, and transaction costs may make using storage and other modular or distributed options too expensive, despite attractive benefits. In a growing number of areas, there may be load and distributed resources aggregators that combine several or many smaller end users in a given area into what could be called power blocks. (See Section 6.5.4 for details.)

4.2. Technical Potential: Peak Electric Load

A key parameter that underlies the maximum possible market size is the total electric load (kW or MW) served by the grid. Market potential is some portion of that peak load. The values in Table 8 include projected peak load in the U.S. and California. The values for the U.S. are based on information from NERC.[36] Visit the NERC website (nerc.com) for details. Values for California are published by the CEC. Visit the CEC website (energy.ca.gov) for details. (Note that the CEC website refers to peak *demand* rather than peak *load.*) The 2008 peak load in California was approximately 62,946 MW, comprising 8% of the total U.S. peak load.[37][38]

Table 8. U.S. and California Peak Load and Peak Load Growth

	California[1]	U.S.[2]
Peak Load, 2008 (MW)	62,946	796,479
Generation Capacity, 2008 (MW)	76,794	925,916
Reserve Margin (%)	22.0%	16.3%
Expected Peak Load Growth Rate (%/year)	1.37%	1.80%
Load Forecast, 2017 (MW)	72,235	920,850
Load Growth Estimate, 2008 to 2017 (MW)	9,289	124,371

[1] Source: California Energy Commission (CEC)
[2] Source: North American Electric Reliability Council (NERC).

4.3. Maximum Market Potential

The maximum market potential for all applications in this guide is the upper bound to the market estimate. It is established by considering constraints (on market potential) that are practical and institutional. Maximum market potential is established without regard to storage cost. For example, given the premise that it is unlikely that storage will displace existing utility equipment, a simplifying assumption (for utility applications) is that the market for new storage to serve electric load is limited to some portion of the annual load growth. For specific applications, other practical or institutional limits on the maximum market potential apply. For example, if the application is for a commercial or industrial customer, then residential customers are not part of the maximum market potential.

4.3.1. Maximum Market Potential Estimates

Maximum market potential estimates for 17 electric-grid-related energy storage applications are shown in Table 9. Estimates for California and U.S. markets are provided, as are the key assumptions and the rationale used to establish those estimates.

4.3.1.1. Caveats about Maximum Market Potential Estimates

The rationale used to establish the above maximum market potential estimates was designed to be transparent (all assumptions used are presented). The values were developed based on a combination of the authors' and supporting analysts' experience and familiarity with the following: energy storage technology; utility loads and supply including costs and prices; utility biases, rules and regulations; electricity market-related business opportunities for energy storage and for modular and distributed resources; and market acceptance of new technologies in the electricity marketplace. Some estimates are based on a relatively high degree of speculation, due to both the dearth of information about the topic and the nascent nature of demand for storage for the applications covered herein. To the extent that analysts have superior and/or newer information, they are encouraged to update or modify these estimates as appropriate.

Table 9. Maximum Market Potential Estimates

#	Type	Maximum Market Potential (MW, 10 Years) CA	U.S.	Note
1	Electric Energy Time-shift	1,445	18,417	10% of peak load is assumed to be in-play, 20% of that, maximum, served by storage.
2	Electric Supply Capacity	1,445	18,417	Same as above.
3	Load Following	2,889	36,834	Total load following = 20% of peak load, 20% of that, maximum, served by storage.
4	Area Regulation	80	1,012	Per CEC/PIER study involving Beacon Power flywheel storage for regulation.
5	Electric Supply Reserve Capacity	636	5,986	20% of peak load is assumed to be in-play, 20% of that, maximum, served by storage.
6	Voltage Support	722	9,209	5% of peak load is assumed to be in-play, 20% of that, maximum, served by storage.
7	Transmission Support	1,084	13,813	1.5% of peak demand, per EPRI/DOE report.
8	Transmission Congestion Relief	2,889	36,834	20% of peak load is assumed to be in-play, 20% of that, maximum, served by storage.
9.1	T&D Upgrade Deferral 50th percentile	386	4,986	T&D upgrade needed for 7.7% of peak load. Of that, a maximum of 50% of qualifying peak load is served by storage. Storage = 3.0% of peak load, on average.
9.2	T&D Upgrade Deferral 90th percentile	77	997	
10	Substation On-site Power	20	250	2.5 kW per system
11	Time-of-use Energy Cost Management	5,038	64,228	67% of peak load is assumed to be in-play. 1%/yr storage adoption rate.
12	Demand Charge Management	2,519	32,111	33% of peak load is assumed to be in-play. 1%/yr storage adoption rate.
13	Electric Service Reliability	722	9,209	10% of peak load is assumed to be in-play, 10% of that, maximum, served by storage.
14	Electric Service Power Quality	722	9,209	Same as above.
15	Renewables Energy Time-shift	2,889	36,834	20% of peak load is assumed to be in-play, 20% of that, maximum, served by storage.
16	Renewables Capacity Firming	2,889	36,834	Same as above.
17.1	Wind Generation Grid Integration, Short Duration	181	2,302	10.0% of peak load is in play. Add storage equal to as much as 2.5% of that amount for intermittency.
17.2	Wind Generation Grid Integration, Long Duration	1,445	18,417	10% of peak load from wind gen., Add storage to a maximum of 20% of that.

The term "in-play" indicates the maximum portion of peak demand that is assumed to be addressable with storage w/o regard to market or technical constraints. Maximum market potential is some portion of that amount.

4.3.2. Renewables Portfolio Standard

Renewable energy seems poised to become a significant fuel source for electric generation. In the U.S., the Renewables Portfolio Standard (RPS) is expected to be a key driver of the trend toward renewables for electricity. Figure 14 indicates RPS-related targets, by state, as of 2008.[39] In this guide, it is assumed that by 2017 15% of electric energy (MWh) in the U.S. will be generated using renewables, and two-thirds of that will be from wind generation.

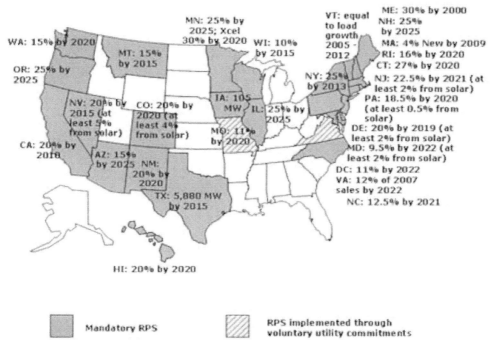

Source: Pew Center Website about Climate Change (as of 2008). http://www.pewclimate.org/

Figure 14. U.S. Renewables Portfolio Standard targets by state.

4.4. Market Estimate

The final step in the market estimation process is to consider the portion of the maximum market potential that will be realized during the target period. The market estimate should be as detailed and precise as appropriate. At the very least, various levels of market potential can be tested for reasonableness using plausible combinations of judgment, knowledge, or preliminary product cost estimates. Alternative bases for estimates could include, for example, sales trends and projections, surveys, analysis of utility capital budget plans, detailed product cost estimates, or market research or intelligence. Note that a market estimate is product-specific and organization-specific, making generic market estimates unhelpful, so none are provided in this report.

4.4.1. Important Considerations
Important criteria affecting market estimates for storage systems include system cost (capital, installation, operation and maintenance, *etc.*), efficiency, marketing costs, market adoption rates, and other considerations discussed in more detail below.

4.4.1.1. Price Signals or Risk and Reward Sharing Mechanisms Must Exist
To include potential demand in the estimate, the region where the demand exists must have price signals or risk and reward sharing mechanisms in order for a given stakeholder to internalize the benefit(s) associated with the targeted value proposition. For example, if utility rules and regulations do not provide adequate incentive for a utility to defer a T&D upgrade, then the T&D deferral application does not apply in that region. Or, if a wind farm developer cannot get a credit for reducing electric service power quality impacts, then that application does not apply in the region.

4.4.1.2. Utility Rules and Regulations Should Give Explicit Permission
It is important to account for utility rules and regulations that forbid use of storage for a given application when making estimates.

4.4.1.3. Storage Must Be Cost Effective
One obvious driver of the market potential for storage systems (used for a given application or applications) is the value proposition to be demonstrated. Specifically, if the cost for storage is higher than the lifecycle benefits, then no storage systems will be sold. If benefits exceed cost by a large margin, then the amount of storage used could be significant.

4.4.1.4. Storage Must Be Cost Competitive
As described in Section 5, benefits associated with the use of energy storage are estimated irrespective of the specific solution being considered. It is important to note that the competitiveness of a given solution (storage or other acceptable substitutes) depends on whether there is a lower cost and/or another viable option.

When establishing the maximum market potential estimate, it is important to account for the fact that solutions whose costs are not competitive are not attractive candidates. Specifically, storage systems whose cost exceeds the cost of another technically viable option are not financially competitive solutions.

4.4.1.5. Changing Electricity Supply and Demand: Effect on On-peak versus Off-peak Electric Energy Price Differential
Two important premises affect the prospects for utility-related use of storage:

1) There are times when electric energy prices are low — because energy *use* is low and because efficient power plants are on the margin, usually at night.
2) There are times when energy prices are high — because energy *use* is high and because inefficient generation is on the margin, usually during the day, especially midday, on weekdays.

Consequently, there is a significant price difference (price delta) between the off-peak price and the on-peak price for electric energy. Nevertheless, there are electric energy supply and demand considerations that could lead to a modest to significant reduction in that price delta. Perhaps most important is the expected increase in the use of plug-in electric vehicles (PEVs) and plug-in hybrid electric vehicles (PHEVs). If a significant number of these vehicles *are* used, then presumably there would be downward pressure on the price delta because more electric energy will be needed during off-peak periods. Similarly, if a lot of energy storage is installed for the applications described in this guide, then additional upward pressure will be exerted on the off-peak price for electric energy. Other possibilities include the increased use of electric energy during off-peak periods to serve loads if, for example, increased economic activity leads to more business and manufacturing activities at night and upward pressure on price for generation fuel used off-peak.

4.4.2. Market Estimates for Combined Applications and Benefits

In many cases, storage may be used for more than one application. When making market estimates for these circumstances, it is important that estimates account for the fact that combining applications may increase storage system benefit ($/kW) while reducing the overall market potential.

Four possible reasons that it may be inappropriate to add the entire market potential for one benefit to the entire market potential for another benefit are as follows:

1) Some benefits accrue to separate stakeholders.
2) Some applications/benefits are region- or location-specific.
3) For most applications the value (magnitude of the benefit) varies among possible beneficiaries.
4) Not all beneficiaries for one benefit ascribe value to the other benefit.

Consider an example: A storage plant is used for the T&D upgrade deferral application. If storage benefits also accrue for electric service reliability, then the estimated market potential is based on the intersection between the market estimate for T&D upgrade deferral alone and the market estimate for electric service reliability alone. The resulting estimate indicates the market potential for customer load that is served by T&D equipment that is due to be upgraded *and* that requires high electric service reliability. This concept of application/benefit intersection is illustrated in Figure 15.

Consider another example: Utility customers will use energy storage for demand charge management, electric service reliability, and electric service power quality. Market estimates would account for the following:

- Technical market potential encompasses all commercial and industrial electricity end users.
- Only a portion of those end users pay demand charges.
- For many commercial and industrial electricity end users that pay demand charges, the benefit associated with increased electric service reliability may be relatively low (depending on the value of the products and/or services involved).
- Only a portion of customers that pay demand charges and that are concerned with electric service reliability will derive a financial benefit from improved power quality.

Similarly, if storage is used for TOU energy cost management *and* for electric service reliability, then some electricity end users who need improved reliability may not pay based on TOU energy prices, and conversely, all end users who pay TOU energy prices may not need improved reliability.

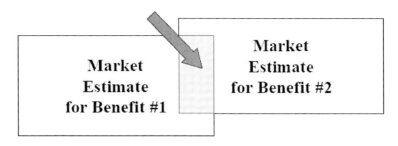

Figure 15. Market intersection.

5. STORAGE BENEFITS

5.1. Introduction

This section discusses the calculation of *application-specific* financial benefits (benefits) associated with using storage for the 17 applications described in Section 3. Also characterized are nine *incidental* benefits that may accrue if storage is used for one or more of the 17 applications. The 26 application-specific and incidental benefits are listed in Table 10.

Table 10. Application-specific and Incidental Benefits of Using Energy Storage

Application-specific Benefits	
1.	Electric Energy Time-shift
2.	Electric Supply Capacity
3.	Load Following
4.	Area Regulation
5.	Electric Supply Reserve Capacity
6.	Voltage Support
7.	Transmission Support
8.	Transmission Congestion Relief
9.	Transmission and Distribution (T&D) Upgrade Deferral
10.	Substation On-site Power
11.	Time-of-use (TOU) Energy Cost Management
12.	Demand Charge Management

	Application-specific Benefits
13.	Electric Service Reliability
14.	Electric Service Power Quality
15.	Renewables Energy Time-shift
16.	Renewables Capacity Firming
17.	Wind Generation Grid Integration
	Incidental Benefits
18.	Increased Asset Utilization
19.	Avoided Transmission and Distribution Energy Losses
20.	Avoided Transmission Access Charges
21.	Reduced Transmission and Distribution Investment Risk
22.	Dynamic Operating Benefits
23.	Power Factor Correction
24.	Reduced Generation Fossil Fuel Use
25.	Reduced Air Emissions from Generation
26.	Flexibility

Readers should note that the emphasis in this document and this section is on the financial *benefit* of storage, with very limited regard to the *cost* associated with owning and operating storage systems. Nevertheless, the benefit estimate *is* intended as a general indication of the cost at which storage is competitive.

5.1.1. Benefit Definition

5.1.1.1. Benefit Basis

In broad terms, benefits from storage can take two forms: 1) additional revenue received by the storage owner/operator or 2) cost that is avoided by the storage owner/operator (avoided cost). Examples of additional revenue include payments received for a) energy sales, b) capacity, and c) ancillary services. Examples of avoided cost associated with storage use include a) a utility's reduced or avoided need (and cost) for generation or T&D capacity and b) a utility customer's reduced cost for energy and demand charges.

Avoided cost can have at least three forms. First, if storage is the only viable alternative, then avoided cost involves the negative outcomes associated with doing nothing. Second, if storage is used *in lieu* of a conventional/standard solution, then avoided cost is the total cost that would have been incurred for the conventional/standard solution is used (where total cost includes purchase, installation, operation, and removal and disposal). Third, if there are several viable alternatives, then the avoided cost is alternative with the lowest total cost (where total cost includes cost to purchase, install, operate, and remove for disposal).

Avoided Cost for the Do Nothing Alternative

In some cases, the leading alternative is to 'do nothing.' Do nothing is a common option for needs that are relatively unlikely to materialize and/or that are expensive. Consider the example of a distribution circuit that is heavily loaded. If there is only a one-in-ten chance that overloading will occur, then the do nothing alternative may be preferable to installing an upgrade, especially if the upgrade is expensive.

Avoided Cost for the Conventional/Standard Solution

In most cases, especially those involving utilities, the benefit for storage is established based on the cost for a conventional/standard alternative. That is, if storage is to be used *in lieu* of a standard/conventional alternative then the benefit (associated with storage use) is the (avoided) cost for the standard/conventional alternative. This concept is especially important for utilities for which the conventional/standard alternative is mandated by legislation and/or regulation.

Consider the possibility that a utility would use storage to improve localized electric service reliability. The conventional/standard alternative competing with storage is whatever the utility would normally do to improve reliability. Those alternatives may range from adding equipment to manage the causes of outages to a full T&D upgrade, involving alternate circuits and transformers. Consider another example: Due to load growth, a utility needs to upgrade its T&D equipment; however, use of storage could defer or to avoid the need to make the upgrade. In that case, the storage-related benefit is the avoided cost associated with deferring or avoiding the need for the conventional/ standard alternative which is the T&D upgrade.

Avoided Cost for the Lowest Cost Viable Alternative

In some cases, the storage benefit could be based on the cost of the lowest cost alternative that is otherwise viable. Consider the possibility that a utility customer could add facility-scale storage for time-of use energy cost management and demand charge management plus electric service reliability. In that case, the lowest cost viable alternative could be energy efficiency measures plus under-desk UPSs and/or on-site backup generation.

5.1.1.2. Gross versus Net Benefit

For most benefit types, the *gross* benefit value is calculated. That is, benefits are estimated without regard to the cost. The benefit estimate is intended to provide a general indication of the price point required for storage to be financially viable. So, if storage can be owned and operated for an amount less than the estimated benefit, then the value proposition may be financially viable.

The one notable exception is electric energy time-shift. For that application, the financial merits of each possible hourly 'buy-low/sell-high' transaction must be calculated before the transaction is made, based on the difference between the benefit for the energy that is discharged *versus* the marginal cost to get that energy. Storage marginal cost includes variable operating cost, charging energy cost, and the cost for energy losses. So, the estimated benefit for electric energy time-shift is net of storage marginal cost.

5.1.1.3. Benefit Financials

For this guide, the financial benefit is defined as the total lifecycle financial benefit associated with use of storage. Although, arguably, some benefits cannot be quantified, only benefits that can be expressed in financial terms are included. For this document, storage is assumed to be in use for 10 years, the assumed price escalation is 2.5%, and the discount rate is 10%. (See Section 1.6.1 for more details about the approach used to address storage financials.)

5.1.2. Benefits Summary

Table 11 summarizes the benefit values characterized later in this section.

Table 11. Application-specific Benefit Estimates

#	Type	Benefit ($/kW)* Low	Benefit ($/kW)* High	Note
1	Electric Energy Time-shift	400	700	Low: 80% efficiency, 2¢/kWh VOC, 4 hours. High: 80% efficiency, 1¢/kWh VOC, 5.5 hours.
2	Electric Supply Capacity	359	710	Low: mid/peak duty cycle combustion turbine, cost $50/kW-year. High: combined cycle combustion turbine, cost $99/kW-year.
3	Load Following	600	1,000	Low: simple cycle combustion turbine, price $20/MW per service hour. High: combined cycle combustion turbine, price $50/MW per service hour.
4	Area Regulation	785	2,010	Low: $25/MW per hour, 50% capacity factor. High $40/MW per hour, 80% capacity factor. For up regulation and down regulation.
5	Electric Supply Reserve Capacity	57	225	Low: $3/MW per hour, 30% capacity factor. High $6/MW per hour, 60% capacity factor.
6	Voltage Support	400	800	Low: prevent 1 outage lasting 1 hour over 10 years. High: prevent 2 outages lasting 1 hour over 10 years. Storage = 5% of load.
7	Transmission Support	192		Based on DOE/EPRI storage report[14].
8	Transmission Congestion Relief	31	141	Based on CAISO congestion prices in 2007.
9.1	T&D Upgrade Deferral 50th percentile	481	687	Low: upgrade factor = 0.25. High: upgrade factor = 0.33.
9.2	T&D Upgrade Deferral 90th percentile	759	1,079	Same as above.
10	Substation On-site Power	1,800	3,000	Based on cost for standard storage solution.

Table 11. (Continued)

#	Type	Benefit ($/kW)* Low	Benefit ($/kW)* High	Note
11	Time-of-use Energy Cost Management	1,226		Based on PG&E's A6 time-of-use tariff. Six hours of storage discharge duration.
12	Demand Charge Management	582		Based on PG&E's A6 time-of-use tariff. Six hours of storage discharge duration.
13	Electric Service Reliability	359	978	Low: $20/kWh * 2.5 hours/year of avoided outages for 10 years. High: 10 Years of UPS Cost-of-ownership (present value).
14	Electric Service Power Quality	359	978	Low: avoided power quality related cost, 10 years. High: UPS cost-of-ownership, 10 years (present value).
15	Renewables Energy Time-shift	233	389	Low: bulk wind generation. High: baseload RE generation.
16	Renewables Capacity Firming	709	915	Low: fixed orientation distributed PV. High: bulk wind generation.
17.1	Wind Generation Grid Integration, Short Duration	500	1,000	Though the estimated benefit is relatively high, a modest amount of storage (<0.1 kW) is needed per kW of wind generation.
17.2	Wind Generation Grid Integration, Long Duration	100	782	Low: avoid 1 outage in 10 years from wind generation shortfall. High: high estimate of benefit for reduced transmisison congestion.

*Based on potential (kW, 10 years) times the average of low and high benefit estimates ($/kW, 10 years).

5.1.3. Economic Impact Summary

Table 12 summarizes the estimated economic impact from storage used for specific applications, given the estimated application-specific benefit and maximum market potential.

Table 12. Application-specific Potential Economic Impact Estimates

#	Type	Economic Potential ($Million)* CA	Economic Potential ($Million)* U.S.
1	Electric Energy Time-shift	795	10,129
2	Electric Supply Capacity	772	9,838
3	Load Following	2,312	29,467
4	Area Regulation	112	1,415
5	Electric Supply Reserve Capacity	90	844
6	Voltage Support	433	5,525
7	Transmission Support	208	2,646

Energy Storage for the Electricity Grid 85

#	Type	Economic Potential ($Million)* CA	U.S.
8	Transmission Congestion Relief	248	3,168
9.1	T&D Upgrade Deferral 50th percentile	226	2,912
9.2	T&D Upgrade Deferral 90th percentile	71	916
10	Substation On-site Power	47	600
11	Time-of-use Energy Cost Management	6,177	78,743
12	Demand Charge Management	1,466	18,695
13	Electric Service Reliability	483	6,154
14	Electric Service Power Quality	483	6,154
15	Renewables Energy Time-shift	899	11,455
16	Renewables Capacity Firming	2,346	29,909
17.1	Wind Generation Grid Integration, Short Duration	135	1,727
17.2	Wind Generation Grid Integration, Long Duration	637	8,122

*Based on potential (kW, 10 years) times the average of low and high benefit estimates ($/kW, 10 years).

5.2. Application-specific Benefits

5.2.1. Benefit #1 — Electric Energy Time-shift

5.2.1.1. Description

The annual financial benefit for electric energy time-shift (time-shift) is derived by using storage to make many electric energy buy-low/sell-high transactions. For a utility, the benefit may take the form of either lower energy cost or profit (if the energy is sold in the energy marketplace). For other stakeholders, the benefit is internalized as profit.[40]

To estimate the time-shift benefit, a simple storage dispatch algorithm is used. The algorithm contains the logic needed to determine when to charge and when to discharge storage in order to optimize the financial benefit. Specifically, it determines when to buy and when to sell electric energy, based on price. In simplest terms, the dispatch algorithm evaluates a time series of prices to find all possible transactions in a given year that yield a net benefit (*i.e.,* benefit exceeds cost). The algorithm keeps track of net benefits from all such transactions for the entire year to estimate an *annual* time-shift benefit. One key point regarding the approach used for this guide is worth noting: the results reflect 'perfect knowledge'. That is, a predetermined series of projected prices was used. In effect, at any given hour in the year, the algorithm 'knows' what prices will be at any other hour of the year.

Three data items are used in conjunction with the dispatch algorithm:

- Chronological hourly price data for one year (8,760 hours)
- Energy storage round-trip efficiency

- Storage system discharge duration

The chronological hourly price data used are the projected hourly electric energy prices in California for 2009.[41] Figure 16 shows prices for the entire year. Based on this data, there are about 900 hours per year when the price is above $100/MWh (10¢/kWh). During off-peak periods (when storage plants are charged), the price is frequently at about $50/MWh to $60/MWh (5¢/kWh to 6¢/kWh). (See Appendix F for more details about energy prices used.)

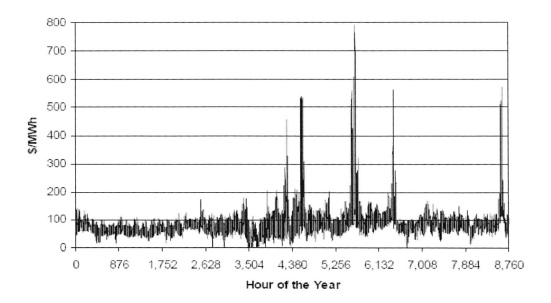

Figure 16. Chronological electricity price data for California, 2009 (projected).

Unlike the other benefits estimated in this report, the benefit for electric energy time-shift is expressed in terms of benefit net of *variable* cost. That is, before a decision is made to make any specific buy-low/sell-high transaction, the financial merits of that transaction are determined based on the cost (to purchase, store, and discharge the energy) *versus* the expected benefit (revenue or cost reduction).

If the cost for wear on the storage system, plus the cost for charging energy, plus the cost to make up for storage losses exceeds the expected benefit, then the transaction is not made. For example, 3 ¢/kWh energy could be used to charge an 80% efficient storage plant whose variable operating cost is also about 3¢ for each kWh of storage output. After accounting for storage energy losses, the total cost to charge and then to discharge is about 6.6 ¢/kWh. So, if the energy is worth more than 6.6 ¢/kWh, then the transaction is a good one.

One other consideration regarding the electric energy time-shift benefit is worth noting. The benefit for electric energy time-shift is based, in large part, on the differential between on-peak and off-peak energy prices. Even somewhat modest deployment of storage for PEVs or PHEVs and/or for utility applications could lead to a non-trivial decrease in that all-important difference between on-peak and off-peak energy prices. That would affect the potential benefit for energy time-shift.

5.2.1.2. Estimate

The storage dispatch algorithm is used to estimate the electric energy time-shift benefit for a given year. Figure 17 shows the estimated net electric energy time-shift benefit for storage systems. The three plots in that figure are for storage with the following (non-energy) variable operating costs (maintenance and replacement cost per $_{kWh_{out}}$: 1) nothing, 2) 1¢/kWh$_{out}$, and 3) 2¢/kWh$_{out}$. Note that if that non-energy variable operating cost (VOC) exceeds 2¢/kWh, then the number of cost-effective transactions in a given year drops precipitously.

The *spread* shown for each plot in Figure 17 reflects the net benefit for storage efficiencies ranging from 70% to 90% and for storage whose discharge duration ranges from one to eight hours. As the hours of storage discharge duration increase, initially the incremental benefit increases too, but the increase eventually levels off. That reflects the diminishing benefit per buy-low/sell-high transaction. The benefit decreases because storage with longer discharge duration requires charging during more hours per year. It also decreases because the additional energy used for charging is probably more expensive and the selling price is probably lower, yielding a diminishing benefit per kWh discharged.

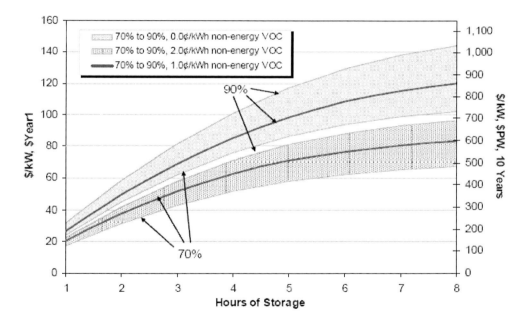

Figure 17. Annual and 10-year present worth time-shift benefit.

To estimate the lifecycle benefit for storage that provides electric energy time-shift service for 10 years, multiply the respective annual value by the 7.17 PW factor. The present worth of benefits is shown in Figure 17 on the second Y axis. The generic benefit estimate for electric energy time-shift ranges from $60/kW-year to $100/kW-year for lifecycle benefits ranging from approximately $400/kW to $700/kW.

5.2.2. Benefit #2 — *Electric Supply Capacity*

5.2.2.1. Description

In areas where electric generation capacity is limited, energy storage could be used to offset the need to purchase and install new generation and/or to offset the need to 'rent' generation capacity in the wholesale electricity marketplace. The resulting cost reduction (or avoided cost) is the benefit associated with storage used for the electric supply capacity application. Another possibility for ascribing a financial value to this benefit is price-based, where price is set by the electricity marketplace or by a designated agency, probably at the wholesale level. If applicable, electric supply capacity prices could be used to estimate this benefit.

5.2.2.2. Estimate

It is important to note that, in many wholesale electricity markets, generation capacity cost is not separated from energy costs. In those regions, the generation capacity cost is embedded in the price per unit of energy purchased. In such cases, there is no *explicit* capacity cost or charge that can be avoided, nor is there a way to sell generation capacity. Nonetheless, there is a capacity cost which is borne by electricity end users, irrespective of how the cost is recouped.

For many regions, the most likely type of new generation plant 'on the margin' is a clean, efficient natural-gas-fired combustion turbine-based power plant (state-of-the-art combined cycle or advanced simple cycle configuration) that operates for 2,000 to 6,000 hours per year.

The generic installed cost assumed for this guide is $1,000/kW. A typical annual fixed operation and maintenance (O&M) cost for such as plant is assumed to be $10/kW-year ($2007).[42] Applying the standard value of 0.11 for the utility fixed charge rate yields an annual cost of ownership of

$1,000/kW ⋅ 0.11 = $110/kW-year.

After adding the $10/kW-year fixed O&M cost, the total annual cost for the generation *capacity* is $120/kW-year. Applying the PW factor of 7.17, the lifecycle benefit (for a storage plant used for 10 years) is

$120/kW-year ⋅ 7.17 = $860/kW.

Arguably, $120/kW-year represents the maximum value for cases involving combustion-turbinebased generation, on the margin. A more conservative value would probably reflect either the cost to contract for or to own older, less efficient, higher maintenance generation – either steam-based or simple cycle combustion-based. As a lower bound, it is assumed that low-cost electric supply capacity has an equipment cost of $50/kW-year plus $10/kW-year for fixed O&M, yielding a total cost of $60/kW-year.

5.2.3. Benefit #3 — Load Following

5.2.3.1. Description

Ideally market based pricing exists for this service. For this guide, however, generic generation costs are used as proxies for market-based prices. Generation cost has two possible elements: 1) marginal cost and 2) capacity cost, described below.

Generation marginal cost consists mostly of cost for fuel and for variable maintenance. The marginal cost can be *avoided* if generation does not have to operate to provide load-following service (because storage is used instead). Generation marginal cost may be *reduced* if part load operation (of generation for load following) is reduced. (Avoiding part load operation is important because doing so reduces generation wear, fuel use and air emissions per kWh delivered.)

Generation capacity-related cost involves cost incurred to add generation capacity The need for additional generation capacity for load following is quite region-specific and year-specific, ranging from no extra load following capacity needed to a need for relatively large increments. Similarly, the type of generation preferred for new load following capacity is region-specific. That preference depends on, among other factors, the mix of existing generation, load characteristics, and regional generation fuel preferences. The type of load following capacity added ranges from hydroelectric generation capacity to simple cycle and combined cycle generation capacity.

5.2.3.2. Estimate

At the low end, the unit price for load following *service* may be based on the marginal cost for low-cost hydroelectric generation. So, the assumed low value is $20/MW per service hour. At the high end, the unit price for load following service reflects the marginal cost for combined cycle generation. Therefore, the assumed high value is $50/MW per hour of service.

The *capacity*-related benefit is estimated based on the generation capacity cost assumed for the electric supply capacity benefit (See Section 5.2.2). At the low end is a relatively clean, simple cycle combustion turbine costing $60/kW year to own or rent. At the high end of the spectrum is a new, combined cycle plant whose annual cost of $120/kW-year.

Values in Table 13 show annual and 10-year lifecycle cost calculations for generation-based load following. The table includes service-related costs and capacity-related costs. Service costs reflect a low price of $20/MW per hour, a midrange price of $35/MW per hour of service, and a high price of $50/MW per hour. Annual capacity costs include a low value of $60/kW-year and a high value of $120/kW-year. Three scenarios shown include 500, 1,000, and 2,000 hours per year of load following service.

Assuming 2,000 service hours per year and an average unit price of $30/MW per hour of service, the marginal cost is about $430/kW. Assuming that at least some capacity cost will also be incurred over 10 years, a generic load following benefit value of $800/kW is used in this guide.

5.2.4. Benefit #4 — Area Regulation

5.2.4.1. Description

At minimum, and until regulation requirements change, the internalizable benefit from storage used for area regulation will be the same amount (per kW per hour of service) as conventional generation-based regulation, with the value reflecting the prevailing price paid for the service. That price is denominated in $/MW per hour of *service*. Nonetheless, as described in Section 3.3.2 and in Appendix E, two important features may make storage the superior area regulation resource.

Table 13. Load Following Benefit Calculations

		500 Hrs./Year		1,000 Hrs./Year		2,000 Hrs./Year	
		Annual ($/kW-yr)	Ten Year ($/kW)	Annual ($/kW-yr)	Ten Year ($/kW)	Annual ($/kW-yr)	Ten Year ($/kW)
Service	$20.0/MW per hour	10.0	71.7	20.0	143.4	40.0	286.8
	$35.0/MW per hour	17.5	125.5	35.0	251.0	70.0	501.9
	$50.0/MW per hour	25.0	179.3	50.0	358.5	100.0	717.0

		Annual ($/kW-yr)	Ten Year ($/kW)
Capacity	$60/kW-year	60	430.2
	$120/kW-year	120	860.4

First, most types of storage can respond somewhat-to-very rapidly (*i.e.*, the rate of discharging and charging can change rapidly). Flywheels, capacitors, SMES, and many types of batteries have such a fast response. Even generation-like pumped hydroelectric storage and CAES can respond more quickly than many generation-based regulation resources. Because of this characteristic, regulation from such rapid-response storage may provide up to twice the benefit as regulation from generation.[43][44][45]

Second, unlike *generation* used for area regulation, *efficient* storage can provide 2 kW of service for each 1 kW of rated output. Storage can do that because it can provide regulation while discharging *and* while charging, in a fashion similar to storage used for load following.

Notably, if providing area regulation while *charging*, energy that is lost (as a function of storage efficiency) must be purchased at the prevailing price. Consider an example:

10 MW of 90% efficient storage used for area regulation; during a specific hour when storage provides regulation, it *absorbs* 4 MWh to provide down regulation, and it *injects* 4 MWh to provide up regulation. In that example, the energy losses for the hour are calculated as

$$(1 - 0.9) \cdot 4 \text{ MWh} = 0.40 \text{ MWh}.$$

Energy Storage for the Electricity Grid   91

> The price for area regulation –denominated in $/MW per hour of service – is not the same as the price for energy which is denominated in $/MWh. Rather, the price for area regulation reflects payment for one hour of service for each MW, without regard to the amount of energy involved.
>
> Although unlikely, area regulation resources could be made available during a given hour to provide regulation service without actually being used to provide the service. In that case, area regulation providers would receive a payment for one hour of service, with no energy-related implications.

5.2.4.2. Estimate

Revenue for providing up and down regulation services (regulation) for one year was estimated based on the California Independent System Operator's (CAISO's) published hourly prices for 2006. Those prices (in $/MW per hour of *service*) for up and for down regulation, are presented in Appendix E.

In 2006, in California the combined price (for up *and* down regulation) averaged about $36.70/MW per service hour (based on an annual average of $21.48/MW per service hour for up regulation and $15.33/MW per service hour for down regulation). After escalating the value for two years (at 2.5%), the price assumed is an hourly average of $38.55/MW per service hour.

Further, two storage operating scenarios for area regulation are evaluated: 1) operation 50% of the year and 2) operation 80% of the year. The single-year and 10-year lifecycle benefits for those prices and operating scenarios are shown in Table 14. The standard value for the area regulation benefit is $785/kW to $2,010/kW, for an average of $1,397/kW.

As noted above, it is possible that storage with rapid response may provide area regulation service whose benefit is twice that of the slower, generation-based regulation. If so, the benefit would be roughly double the values in Table 14.

Table 14. Area Regulation Annual and Lifecycle Benefit Summary

	Low		High	
Capacity Factor	0.50		0.80	
Annual Service Hours	4,380		7,008	
Regulation Price* ($/MW per service hour)	25.0	40.0	25.0	40.0
Annual Benefit ($/kW)	110	175	175	280
Lifecycle Value** ($/kW)	785	1,256	1,256	2,010

* For up regulation *plus* down regulation.
** For ten years, assuming PW factor = 7.17

5.2.5. Benefit #5 — Electric Supply Reserve Capacity

5.2.5.1. Description

Storage serving as electric supply reserve capacity (reserves) reduces the need and cost for those reserves that are normally supplied by generation. In many cases, the price for reserves is market-based – typically prices are a result of 'day-ahead' and 'hour-ahead' bidding.

The electric supply reserve capacity benefit is somewhat small – because generation-based reserves are inexpensive; nonetheless, it could be an important element of an attractive value proposition because providing reserves has low incremental cost. While charging, storage can provide two times its capacity as reserves (it can simultaneously cease charging and begin discharging). When charged storage can, in most cases, provide reserves merely by being ready to discharge (reserves are only used infrequently).

5.2.5.2. Estimate

The electric supply reserve capacity benefit estimate is based on the price paid for reserves and the number of hours per year during which storage provides reserves. Benefits are estimated assuming a low price of $3/MW per service hour and a high of $6/MW per service hour. Storage is assumed to provide 2,628 service hours per year at the low end and 5,256 service hours per year at the high end. The resulting annual benefit for those two scenarios is shown in Table 15.

Table 15. Electric Supply Reserve Capacity Annual Benefit

	Low	High
Capacity Factor	0.30	0.60
Annual Service Hours	2,628	5,256
Charge ($/MW per service hour)	3.0	6.0
Annual Value ($/kW-year)	7.9	31.5
Lifecycle Value* ($/kW)	57	226

*10 years, PW factor = 7.17.

5.2.6. Benefit #6 — Voltage Support

5.2.6.1. Description

Voltage support provided by storage offsets the need to use large/central generation to provide reactive power to the grid when region-wide voltage emergencies occur. Competing alternatives (to storage) may include a) do nothing and endure the cost of additional outages or the risk associated with possible outages; b) buy insurance to cover possible liabilities; c) perform load management (primarily via curtailable/interruptible loads and possibly direct load control); d) incur a forced outage; and e) add central generation capacity to provide voltage support.

5.2.6.2. Estimate

Establishing a generic benefit estimate for the voltage support application requires use of generalizations and simplifying assumptions. In general, benefit estimates should account for the limited likelihood of such an outage that may occur in any given area and the degree to which storage contributes to avoiding such an event. Furthermore, unless the utility is financially responsible for outage-related costs, it has no significant direct incentive to pay for or even to coordinate distributed resources for voltage support.

The approach used to estimate the voltage support benefit is similar to that used to estimate the benefit of storage for electric service reliability. The general concept involves segmenting the utility customer base into three groups: 1) those ascribing little or no value to avoiding outages, 2) end users for whom outages are somewhat costly, and 3) end users for whom avoiding outages has a high value. That yields a composite value for avoiding an outage of 1 kW for one hour.

The next step is to establish an assumption about how long outages may last. Finally, an assumption is needed about how many outages will be avoided over the 10-year life of the storage. These two criteria are not easy to generalize.

For the benefit estimate in this report, it is assumed that at the low end the distributed voltage support resources (including storage) would prevent one outage lasting one hour over 10 years. At the high end, distributed voltage support resources (including storage) are assumed to prevent one outage lasting two hours during its 10-year life.

The unit value assumed for this estimate is $20 per hour of unserved load. For an outage lasting one hour, that's $20/kW lifecycle (without regard to time value of money) for each kW of *system peak load*. For an outage lasting two hours, that's two hours at $20/kW or $40/kW, lifecycle (without regard to the time value of money).

The standard assumption value for market potential is based on the premise that combined voltage support resources are distributed, are located where they can provide good support, and have an aggregate rating equal to 5% of peak load. Thus, by using *distributed* storage whose power is rated at 5% of peak load to avoid a 1-hour outage, the benefit is

$20/kW$_{load}$ ÷ 0.05 = $400/kW of distributed storage. Avoiding a single *2-hour* outage over 10 years is worth

$40/kW$_{load}$ ÷ 0.05 = $800/kW of distributed storage.

5.2.7. Benefit #7 — Transmission Support

5.2.7.1. Description

To the extent that storage increases the load carrying capacity of the transmission system, a nontrivial benefit may accrue if transmission outages are avoided. Such a benefit may also accrue if additional load carrying capacity defers the need to add more transmission capacity and/or additional T&D equipment, and/or if it is rented to participants in the wholesale electric marketplace (to transmit energy) for revenue.

5.2.7.2. Estimate

When evaluating the merits of using storage for transmission support, the upper bound of the benefit value is the cost for the standard utility solution. For example, if capacitors are the proposed standard solution, then energy storage would offset the need (and cost) for those capacitors. The avoided cost (of the capacitors) is the resulting storage benefit for the transmission support application.[46]

The financial benefit values listed in Table 16 are estimated based on related research by EPRI. That research addressed SMES used for T&D support needs in Southern California during hot summer conditions when the need is greatest and when the benefits are highest.

The estimates are based on conservative assumptions.[47][48] Based on those values, the standard lifecycle benefit value assumed for transmission support is $192/kW.[49][50][51]

Table 16. Transmission Support Annual Financial Benefit

Benefit Type	Annual Benefit	Lifecycle Benefit
Transmission Enhancement	15.1	108
Voltage Control ($ capital*)	n/a	29
Subsynchronous Resonance (SSR)	n/a	16
Underfrequency load-shedding	12.8	38**
Total		192

Notes:
[1] All value are for Southern California, assuming hot summer conditions, circumstances for which benefits are highest.
[2] Based on values established in 2003 and escalated at 2.5% for six years.
*The benefit is the cost of the most likely alternative (e.g., capacitors), that would have been incurred if storage was not deployed.
**$12.8/kW, per occurrence. Assume three occurrences over the (ten year) life of the unit. This value has not been adjusted to account for time value of money.
#Based on a PV Factor of 7.17 and a ten year life.

5.2.8. Benefit #8 — Transmission Congestion Relief

5.2.8.1. Description

Alternatives that may compete with storage for transmission congestion relief include a) dumping energy upstream from congestion, b) providing load management and energy efficiency downstream from congestion, c) paying congestion charges, and d) adding transmission capacity. Note that for this application, if the generation (upstream from congestion) is already installed, then the do nothing option is the same as the dump energy option.

Given the possible shortfall of transmission capacity within and into many regions, congestion charges are possible if not likely. Currently, however, these charges cannot be generalized well – primarily because the marketplace within which transmission congestion charges will apply is in the formative stages and because congestion charges will be location-specific.

Much, if not most, of the new congestion will probably occur as more renewables (deployed in response to Renewables Portfolio Standard [RPS] targets) compete for the existing transmission capacity. Furthermore, it is assumed that the do nothing and the dump energy options are not likely. So, for this application, the benefit is based on transmission congestion charges at the low end and the cost of a transmission upgrade at the high end.

5.2.8.2. Estimate

Transmission congestion charges are becoming more common. In the parts of California's transmission system where it occurs, congestion is present for 10% to 17% of all hours during the year. Congestion charges in those areas range from about $5/MW per service hour to about $15/MW per service hour.[52] As shown in Table 17, that yields an annual benefit whose average value is on the order of $12/kW-year and a lifecycle benefit averaging about $86/kW. Although that is a small amount compared to the cost for storage, it could be an element of a value proposition that includes several benefits.

Table 17. Congestion Charges in California, $2007

	Low	High
Portion of Year	10%	15%
Hours Per Year	876	1,314
Transmission Access Charge	5	15
Annual Cost ($/kW-year)	4.38	19.71
Lifecycle Value* ($/kW)	31	141

*10 years, PW factor = 7.17
Source: derived based on data from CAISO.

More compelling are transmission corridors requiring an upgrade due to congestion. In those cases, the benefit is the cost that can be avoided by deferring or avoiding the upgrade. The cost of a transmission upgrade varies significantly depending on distance, permitting and siting challenges, and the equipment's rating.

5.2.9. Benefit #9 — Transmission and Distribution Upgrade Deferral

5.2.9.1. Description

The *single-year* T&D upgrade deferral benefit (deferral benefit) is the financial value associated with deferring a utility T&D upgrade for one year. That value reflects the utility's financial carrying charge for the new equipment involved in the upgrade. Carrying charges include the costs for financing, taxes, and insurance incurred for one year of ownership of the equipment used for the upgrade. For a utility, that amount is also known as the 'revenue requirement.'

The carrying charge (revenue requirement) *for one year* is estimated by multiplying the utility fixed charge rate times the total installed cost for the upgrade. Consider, for example, a distribution upgrade costing $1 million to purchase and install. If the utility fixed charge rate is 0.11, then the annual revenue requirement – and thus the *single year* deferral benefit – is

$1 million · 0.11 = $110,000.

Note that, for this guide, T&D operating cost avoided, if any, is assumed to be negligible. Also note that, by definition, reducing the utility revenue requirement reduces the utility's total costof-service paid by all customers as a group.

Storage power indicates the amount of storage needed for one year of deferral. It is expressed as a percentage of the existing T&D equipment nameplate rating (the equipment to be upgraded). An example: If T&D equipment to be upgraded is rated at 12 MVA, then 3% storage power is

3% · 12 MVA = 0.36 MVA or 360 kVA.

The assumed 3% storage power is intended to be representative. In practice, that value can fall within a range of as little as 1% to as much as 10%, depending on the actual peak load in the previous year plus load shape; expected load growth; load growth uncertainty; storage module sizes available; engineering philosophy and preferences, especially regarding storage oversizing to account for uncertainty; and possibly other criteria.

For more details about storage sizing for T&D upgrade deferral, readers are encouraged to refer to a report published by Sandia National Laboratories entitled *Estimating Electricity Storage Power Rating and Discharge Duration for Utility Transmission and Distribution Deferral, a Study for the DOE Energy Storage Program*.[53] Also, refer to the discussion addressing use of *modular* storage for reducing T&D investment risk in Section 5.3.

5.2.9.2. Estimate

The starting point for estimating the T&D upgrade deferral benefit is to establish the cost of the T&D upgrade to be deferred. The data used as the basis for establishing that cost is expressed in dollars per kW added – the T&D marginal cost. In California, for 50% of all locations requiring an upgrade in any given year, the *marginal* cost is $420/kW or more (*i.e.*, $420/kW *added*). For the most expensive locations requiring upgrades (90th percentile and above), the upgrade cost exceeds about $662 per kW of capacity *added*.[54][55]

As an aside, a familiar criterion for T&D planners is $/kVA *installed*. To estimate that value based on the *marginal* cost, an upgrade factor is used. The upgrade factor is the ratio of the capacity *added* to the existing capacity. Consider an example: If 4 MVA of capacity is added to a 12 MVA system, the upgrade factor is 0.34. Typical values for upgrade factor range from 0.25 to 0.50. An upgrade factor of 0.33 is assumed for this guide.

The T&D cost estimates used to estimate the T&D upgrade deferral benefit are summarized in the two tables below. Values in Table 18 indicate the single-year deferral benefit for locations whose cost is among the highest 50% of all costs for all upgrades needed. The value used, $684/kVA of storage for one year, reflects the 0.33 T&D upgrade factor, 0.11 fixed charge rate, and 3% storage power as described above.

Table 18. T&D Upgrade Cost and Benefit Summary, 50th Percentile

Upgrade Scenario Final Rating (MVA)	Capacity Added (MVA)	Upgrade Factor	Upgrade Installed Cost* $/kVA**	$	Upgrade Annual Cost*** ($)	Storage 1 Year Benefit[#] ($/kVA-year)
15	3	0.25	105.0	1,575,000	173,250	481
16	4	0.33	140.0	2,240,000	246,400	684
18	6	0.50	210.0	3,780,000	415,800	1,155

*If marginal cost per kVA of T&D capacity $/kVA added is $420/kVA. **Per kVA *installed*.
*** $Upgrade Installed Cost * 0.110 Fixed Charge Rate
# $Upgrade Annual Cost ÷ 360 kVA. (Based on 3.0% storage power)

The annual upgrade deferral value is $1,079/kVA of storage for one year for upgrades whose cost is among the highest 10% of upgrades needed, based on values shown in Table 19.

Table 19. T&D Upgrade Cost and Benefit Summary, 90th Percentile

Upgrade Scenario Final Rating (MVA)	Capacity Added (MVA)	Upgrade Factor	Upgrade Installed Cost* $/kVA**	Upgrade Installed Cost* $	Upgrade Annual Cost*** ($)	Storage 1 Year Benefit# ($/kVA-year)
15	3	0.25	165.5	2,482,500	273,075	759
16	4	0.33	220.7	3,530,667	388,373	1,079
18	6	0.50	331.0	5,958,000	655,380	1,821

*If marginal cost per kVA of T&D capacity $/kVA added is $662/kVA. **Per kVA *installed*.
*** $Upgrade Installed Cost * 0.110 Fixed Charge Rate
# $Upgrade Annual Cost ÷ 360 kVA. (Based on 3.0% storage power).

Consider this important note: The assessment described above must occur *each year* for a given deferral because, normally, the amount of load served by a given T&D node grows. So, in each year after a deferral, power engineers must reassess the merits of using storage for another year of deferral. Usually, load grows such that for each subsequent year the amount of storage needed to keep pace with load growth, and thus the amount needed to defer an upgrade for the next year, nearly doubles. In some cases, the discharge duration requirements increase too.

5.2.10. Benefit #10 — Substation On-site Power

5.2.10.1. Description

Battery storage systems (mostly lead-acid batteries) provide power at electric utility substations for switching components and for substation communication and control equipment when the grid is not energized.[56]

5.2.10.2. Estimate

Establishing a benefit value for substation on-site power is challenging. Certainly, battery systems provide critical service because the grid would be much more vulnerable to outages, and perhaps even equipment damage without an on-site, non-grid power source for times when the grid is not operational. The benefit for this application is estimated based on the price for high quality UPS systems (like those shown in Table 24 of Section 5.2.13.4).

The cost of such a state-of-the-art lead-acid battery-based system, with eight hours of discharge duration, is based on a price of $225/kW for power and $200/kWh of discharge.[57] Therefore, the presumed system (equipment) price is

$225/kW + (8 hours $200/kWh) = $225/kW + $1,600/kW = $1,825/kW.

Similarly, the presumed price for a system with 16 hours of discharge duration is

$225/kW + (16 hours $200/kWh) = $225/kW + $3,200/kW = $3,425/kW.

Given the limited discharge of these systems, variable operating costs are ignored.

5.2.11. Benefit #11 — *Time-of-use Energy Cost Management*

5.2.11.1. Description

To reduce electricity end users' time-of-use (TOU) energy cost, storage is charged with low-priced energy so that the stored energy can be used later when energy prices are high. The resulting overall electric energy cost reduction is the benefit associated with use of storage for TOU energy cost management.

TOU energy prices are specified by the applicable rate structure (tariff). Typically, those prices vary by time of day, day of the week, and season of the year. There may be two or more price points for specific days. The standard assumption value for this benefit is calculated based on PG&E's A-6 Small General TOU Service tariff. Commercial and industrial (C&I) electricity end users whose power requirements are greater than 199 kW and less than or equal to 500 kW are eligible for the A-6 tariff. TOU electricity prices for the A-6 tariff are shown in Table 20.

The summer billing period extends from May through October, and the winter billing period is November through April. Summer on-peak hours are 12:00 p.m. to 6:00 p.m. (Monday-Friday, except holidays); partial-peak hours are 8:30 a.m. to 12:00 p.m. and 6:00 p.m. to 9:30 p.m. (Monday-Friday, except holidays); and off-peak hours are 9:30 p.m. to 8:30 a.m. (Monday-Friday; all day Saturday, Sunday, and holidays). There is no winter on-peak period. Partial peak hours are 8:30 a.m. to 9:30 p.m. (Monday-Friday, except holidays); and off-peak hours are 9:30 p.m. to 8:30 a.m. (Monday-Friday; all day Saturday, Sunday, and holidays). PG&E tariffs are available at http://www.pge.com/tariffs.

Table 20. PG&E A-6 Time-of-use Energy Price Tariff

Period	Total	Generation	%	Distribution	%
Peak Summer	$0.37	$0.21	57.0%	$0.13	34.9%
Part-Peak Summer	$0.17	$0.09	53.0%	$0.05	29.8%
Off-Peak Summer	$0.11	$0.06	49.9%	$0.03	23.3%
Part-Peak Winter	$0.13	$0.06	46.0%	$0.04	31.8%
Off-Peak Winter	$0.11	$0.05	47.4%	$0.03	25.7%

Transmisison: $0.00913 for all hours.

5.2.11.2. Estimate

The A-6 tariff's on-peak energy price applies to 720 hours per year. Storage with a 6-hour discharge duration would allow the end user to avoid annual on-peak energy charges of

37¢/kWh x 720 hours/year
= $0.37/kWh x 720 hours/year = $266/kW-year.

To charge an 80% efficient energy storage system, it is necessary to use 1.25 kWh of energy in to get one kWh out. Consider a 1-MW storage plant: To discharge for 720 hours (720MWh), the storage would have to be charged with

720 x 1.25 = 900MWh.

So, the charging energy cost using low-priced, off-peak energy priced at 11¢/kWh is $0.11/kWh x 900 hours/year = $99/kW-year.
The cost *reduction* realized is

$266/kW-year − $99/kW-year = $167/kW-year.

To express that annual (cost reduction) benefit in units of $/kW lifecycle, the annual cost is multiplied by the PW factor of 7.17

$167/kW-year x 7.17 = $1,198/kW.

The storage plant could have a discharge duration that is less than the duration of the 6- hour, on-peak price period specified in the tariff. If, for example, two hours of backup are needed from a storage system with four hours of discharge, then the remaining two hours of discharge could be used for reducing energy cost. The lifecycle benefit is

(2 hours ÷ 6 hours) x $1,198/kW-year = 0.33 x $1,198/kW-year
= $395/kW.

Note that the benefit estimate illustrated above does not account for variable maintenance costs incurred as the storage plant is used (*e.g.,* overhauls and subsystem replacement, as applicable).

5.2.12. Benefit #12 — Demand Charge Management

5.2.12.1. Description

Demand charge management involves storage used to reduce an electricity end user's *power* draw on the electric grid during times when electricity use is high (*i.e.,* during peak electric demand periods). To reduce or avoid demand charges, storage is charged when low or no demand charges apply, presumably using low-priced energy. The storage is later discharged when demand charges apply. The benefit value is the overall reduction in cost due to reduced or avoided demand charges.

To one extent or another, demand charges reflect the cost for utility *equipment* needed to generate, transmit, and distribute electric energy. So, demand charges are denominated in $/kW of power draw because that criterion defines the *capacity* that the electricity *infrastructure* must have to deliver service to the customer. In most cases, the demand charge is assessed each month based on the maximum power draw within the respective month. It is important to note that tariffs with demand charges also have separate prices for energy, denominated in ¢/kWh.

Demand charges and, in most cases, energy prices are specified by the end user's electricity rate structure (tariff). Typically, demand charges vary by day of the week and by season. Demand charges may also vary by time of day.

Demand charges are assessed *each month* based on the maximum load that occurs during times when peak demand charges apply, normally 1) peak, 2) partial-peak, and 3) off-peak. Some tariffs with demand charges also include what could be called an 'anytime' demand charge. Known generically as a 'facility' demand charge, these charges are levied based on the peak demand no matter when it occurs (time or season). That is important for storage because most storage charging occurs at night when demand from utility customers' non-storage loads tends to be low. In those circumstances, charging storage at night will increase the anytime or facility demand charges incurred. Note that off-peak demand charges have a similar effect though the charges are based on maximum off-peak demand during the respective month.

The standard assumption value for this benefit is calculated based on PG&E's E-19 Medium General Demand-Metered TOU Service tariff. That tariff applies to commercial and industrial (C&I) end users with peak demand that exceeds 500 kW. PG&E tariffs are available at http://www.pge.com/tariffs.

The E-19 tariff has three monthly demand charges during six 'summer' months (May through October). Summer on-peak hours are 12:00 p.m. to 6:00 p.m. (Monday-Friday, except holidays); partial-peak hours are 8:30 a.m. to 12:00 p.m. and 6:00 p.m. to 9:30 p.m. (Monday-Friday, except holidays); and off-peak hours are 9:30 p.m. to 8:30 a.m. (Monday-Friday; all day Saturday, Sunday, and holidays). (Notably, the off-peak demand charges will apply during charging.)

During the six 'winter' months (November through April), there are only two monthly demand periods: partial-peak and off-peak. Partial peak hours are 8:30 a.m. to 9:30 p.m. (Monday-Friday, except holidays); and off-peak hours are 9:30 p.m. to 8:30 a.m. (Monday-Friday; all day Saturday, Sunday, and holidays). (As with storage use during summer months, the off-peak demand charges will apply during charging.)

Importantly, like most other tariffs with demand charges, the E-19 *energy* price (¢/kWh) paid by utility customers also depends on those time periods.

5.2.12.2. Estimate

The assumed electricity bill for a typical commercial end user using the E-19 tariff is shown in Table 21. The same end user's electric bill, after considering 80% efficient storage with 6 hours of discharge duration to eliminate peak load, is shown in Table 22. The changes due to use of storage are summarized in Table 23.

Table 21. Electricity Bill, E-19 Tariff, without Storage

	Hours Per Year*	Demand Charge ($/kW-month)	Peak Load Factor	Demand Charges ($/kW-year)	Average Load Factor	Energy Use (kWh/year)	Energy Price ($/kWh)	Energy Charges ($/kW-year)	Total Charges ($/kW-year)
Summer									
Peak	765	11.59	0.90	62.59	0.80	612	13.458	82.36	144.95
Partial-peak	893	2.65	0.80	12.72	0.60	536	9.257	49.57	62.29
Off-peak	2,723	6.89	0.60	24.80	0.55	1,497	7.541	112.92	137.72
Winter									
Partial-Peak	1,658	1.00	0.80	4.80	0.70	1,160	8.256	95.79	100.59
Off-Peak	2,723	6.89	0.55	22.74	0.50	1,361	7.286	99.18	121.92
			Total	127.65	0.590	5,166	8.513	439.82	567.47

*Approximate values.
**Average peak load during all months of the season.

Table 22. Electricity Bill, E-19 Tariff, with Storage

	Hours Per Year*	Demand Charge ($/kW-month)	Peak Load Factor**	Demand Charges ($/kW-year)	Average Load Factor	Energy Use (kWh/yea)	Energy Price (¢/kWh)	Energy Charges ($/kW-year)	Total Charges ($/kW-year)
Summer									
Peak	765	11.59					13.458		
Partial-peak	893	2.65	0.80	12.72	0.60	536	9.257	49.57	62.29
Off-peak	2,723	6.89	0.80	33.07	0.82	2,232	7.541	168.35	201.42
Winter									
Partial-Peak	1,658	1.00	0.80	4.80	0.70	1,160	8.256	95.79	100.59
Off-Peak	2,723	6.89	0.55	22.74	0.50	1,361	7.286	99.18	121.92
			Total	73.33	0.604	5,289	7.806	412.89	486.22

*Approximate values.
**Average peak load during all months of the season.
1. Storage Efficiency: 80.0%.

Table 23. Electricity Bill Comparison, E-19 Tariff, with and without Storage

	Demand Charges ($/kW-year)	Average Load Factor	Energy Use (kWh/year)	Energy Price (¢/kWh)	Energy Charges ($/kW-year)	Total Charges ($/kW-year)
With Storage ($)	73.3	0.60	5,289	7.81	412.9	486.2
w/o Storage ($)	127.6	0.590	5,166	8.51	439.8	567.5
Change, w/Storage ($)	-54.3	+0.014	+123*	-0.71	-26.9	-81.2
(%)	-42.6%	2.4%	2.4%	-8.3%	-6.1%	-14.3%

*Increase due to storage losses.

As shown in Table 23, *demand* charges are reduced by nearly 43% ($54.30), energy charges are reduced by a more modest 6.1% ($26.90), and the total annual bill is reduced by $81.20 for a total reduction of 14.3%.

5.2.13. Benefit #13 — Electric Service Reliability

5.2.13.1. Description

In simplest terms, the benefits associated with improved electric service reliability accrue if storage reduces financial losses associated with power outages. This benefit is highly end user-specific, and it applies to C&I customers, primarily those for whom power outages cause moderate to significant losses. If the utility has followed standard practices, it is usually the end user that is responsible for covering financial damages. In some cases, utilities are required to reimburse end users for financial losses due to outages.

5.2.13.2. Estimating End-user Reliability Benefit – Value-of-service Approach

For the value-of-service (VOS) approach, the benefit associated with increased electric service reliability is estimated using two criteria: 1) annual outage hours (*i.e.,* the number of hours per year during which outages occur) and 2) the value of 'unserved energy' or VOS. VOS is measured in $/kWh. The standard assumption value for annual outage hours is 2.5 hours per year. A VOS of $20/kWh is recommended as a placeholder.[58] To calculate the annual reliability benefit, the standard assumption value for annual outage hours is multiplied by the VOS:

$20/kWh ⋅ 2.5 hours per year = $50/kW-year.

To calculate lifecycle benefits over 10 years, the annual reliability benefit of $50/kW-year is multiplied by the PW factor (7.17):

$50/kW-year ⋅ 7.17 = $359/kW.

5.2.13.3. Estimating End-user Reliability Benefit – Per Event Approach

Reliability benefits may be estimated by ascribing a monetary cost to losses associated with power system events lasting one minute or more and that cause electric loads to go offline.[59] Reliability events considered are those whose effects can be avoided if storage is used.

Based on a survey of existing research and known data related to electric service reliability, a generic value of $10 per event for each kW of end user peak load is used.[60][61][62] The generic assumption for the annual number of events is 5.[63] The result is that storage used in such a way that the end user can avoid 5 electric reliability events, each worth $10 for each kW of end user peak load, yields an annual value of $50/kW-year.[64] Finally, multiplying by the PW factor of 7.17 yields a lifecycle benefit of $359/kW.

For additional information about financial considerations related to utility service reliability, please refer to a report produced by Lawrence Berkeley National Laboratory, *Evaluating the Cost of Power Interruptions and Power Quality to U.S. Electricity Consumers.*[65]

5.2.13.4. Estimating End-user Reliability Benefit – UPS Price Approach

One other possibly helpful proxy to use when estimating this benefit is the price paid for UPSs. Prices for a selection of commercially available UPSs are shown in Table 24.

Table 24. Commercially Available UPS Ratings and Prices

Item	Specifications True Power (Watts)	Apparent Power (Volt-Amps)	Power Factor	Discharge Duration* (Minutes)	Price Retail Price**	$/kW	$/kW-hour
APC Back-UPS ES 8 Outlet	200	350	0.57	2.3	44	220	5,739
Tripp Lite SMART550USB	300	550	0.55	5.3	225	748	8,472
Tripp Lite SMART1200XLHG	750	1,000	0.75	6.0	562	749	7,493
APC Back-UPS RS 1500VA	865	1,500	0.58	5.3	250	289	3,272
MGE Pulsar EX RT 3200VA	2,080	3,200	0.65	6.0	1,164	560	5,596
Tripp Lite SmartOnLine	6,000	7,500	0.80	9.0	3,493	582	3,881
Tripp Lite SmartOnLine	8,000	10,000	0.80	4.0	4,017	502	7,531
MGE Galaxy 30kVA	24,000	30,000	0.80	11.0	17,010	709	3,866
APC - Smart-UPS VT 30KVA 5 Batteries	24,000	30,000	0.80	13.7	19,410	809	3,542
	Average Power Factor	0.699			Average	574.2	5,487.9

*At full rated output.
**Based on an informal survey of retail prices.
Note: Assuming 5 year life, a rough approximation of annual cost ($/kW-year) is total cost ÷ 5.
Additional Notes:
[1] Content in Table 24 does not constitute an endorsement or recommendation of the listed products or brands.
[2] Power ratings are in units of volt-Amps (VA).
[3] Typically 1.2 to 1.3 volt-Amps are required for each Watt of load.

As shown in Table 25, a rough estimate of the 10-year lifecycle benefit is $978/kW. This estimate assumes a 5-year UPS life and one replacement of the UPS over 10 years. It is based on a 2.5%/year price escalation and 10% discount rate.

Table 25. UPS Lifecycle Cost

Year	1	2	3	4	5	6	7	8	9	10	Total
Escalator	1.00	1.03	1.05	1.08	1.10	1.13	1.16	1.19	1.22	1.25	
Cost ($Year 1)	574.2	0	0	0	0	574.2	0	0	0	0	1,148
Escalated Cost ($Current)	574.2	0	0	0	0	649.7	0	0	0	0	
Discount Factor	1.00	0.91	0.83	0.75	0.68	0.62	0.56	0.51	0.47	0.42	
Present Value ($)	574.2	0	0	0	0	403.4	0	0	0	0	978

5.2.14. Benefit #14 — *Electric Service Power Quality*

5.2.14.1. Description

The electric service power quality benefit is highly end-user-specific and, as such, is difficult to generalize. It applies primarily to those C&I customers for whom power outages may cause moderate to significant losses.

Though power quality-related technical details are not covered in depth here, they *are* summarized in Section 3.5.4. Specific types of poor power quality are well characterized in many other reports and documents.[66]

In the most general terms, power-quality-related financial benefits accrue if energy storage reduces financial losses associated with power quality anomalies. Power quality anomalies of interest are those that cause loads to go offline and/or that damage electricity-using equipment and whose negative effects can be avoided if storage is used.

As an upper bound, the power quality benefit cannot exceed the cost to add the conventional solution. An example: If the annual power quality benefit (avoided financial loss) associated with an energy storage system is $100/kW-year, and basic power conditioning equipment costing $30/kW-year would solve the same problem if installed, then the maximum benefit that could be ascribed to the energy storage plant for improved power quality is $30/kW-year.

5.2.14.2. Estimate

Power quality-related benefits may be estimated by assigning a monetary value to losses associated with power quality events that last less than one minute and cause electric loads to go offline.[67] Power quality events considered are those whose effects can be avoided if storage is used.

Based on a survey of existing research and known data related to power quality, a generic value of $5 per event for each kW of end user peak load is the standard assumption value used in this guide. Based on that same information, the generic assumption for the annual number of events is 10.[68][69][70] The result is that storage used in such a way that the C&I electricity end user can avoid 10 power quality events per year, each worth $5 per kW of end user peak load, provides an annual benefit of $50/kW-year. After multiplying by the PW factor (7.17), the lifecycle electric service power quality benefit is $359/kW. Implicit in this approach is the assumption that the power quality benefit is the same for each of 10 years.

For additional coverage of this topic, please refer to a report published by Lawrence Berkeley National Laboratory entitled *Evaluating the Cost of Power Interruptions and Power Quality to U.S. Electricity Consumers*.[71]

5.2.15. Benefit #15 — *Renewables Energy Time-shift*

5.2.15.1. Description

For the renewables energy time-shift application, storage is charged with low-value electric energy generated using renewable energy. That energy is stored so that it may be used or sold at a later time when it is more valuable.

Two cases considered in this guide are time-shift of energy from wind generation and generic *baseload* renewable energy generation. (See Section 3.6.1 for details.)

> To one extent or another, the fuelrelated cost for renewable energy is more predictable than fuel cost for conventional generation. In effect, renewable energy provides a 'hedge' against the possibility that fuel prices will be higher than expected.
>
> One simple way to quantify at least part of this effect is based on evaluations by the Lawrence Berkeley National Laboratory Electricity Market and Policy program. Based on recent work by that group, the 'forward prices' for fuel that reflect the terms of actual electricity purchase *contracts* are on the order of 10% or more higher than prices that are forecast.
>
> Indeed, a significant portion of electric energy from renewables is procured using firm prices, contracts, or power purchase agreements, rather than spot market prices. Consequently, the benefit estimated for renewable energy time-shift based on a forecast is likely to understate the energy-related benefit.[72]

5.2.15.2. End -User Time-of-Use Energy Cost Reduction Using Distributed Renewable Energy Generation

The renewables energy time-shift benefit is related to wholesale or 'spot market' electric energy for electricity supply. That is, the energy time-shift benefit described above is related to the avoided cost of purchasing electric energy from the wholesale or spot market.

An analogous opportunity exists for electricity end users to derive a renewables energy time-shift benefit. Specifically, if an end user's electric service tariff includes TOU energy prices, then the end user could use storage to time-shift energy to reduce cost for electric energy. (See Section 3.5.1 and Section 5.2.11 for more details.)

5.2.15.3. Incremental Benefit and Cost for Adding Storage for Renewables Energy Time-shift

Readers should note that the renewables energy time-shift benefit estimated in this guide accrues because it is *added* to renewable energy generation. That means that the benefit is *incremental*. Consequently, when evaluating the financial merits of adding storage to renewable energy generation, the incremental benefit is compared to incremental cost (to add storage); which means that the entire evaluation addresses the incremental benefit/cost relationship for storage.

Table 26. Wholesale Spot Energy Price Differentials, On-peak and Off-peak, Weekdays, California Forecast for 2009 (in $/MWh)

	Monthly Price "Bins"											
Hour	Month=>											
	1	2	3	4	5	6	7	8	9	10	11	12
12:00 P.M. - 5:00 P.M.	85.1	74.5	77.6	94.6	100.3	118.0	148.2	163.1	142.5	99.1	104.5	105.9
1:00 A.M. - 6:00 A.M.	-51.8	-44.4	-46.2	-61.2	-42.7	-35.2	-55.1	-69.7	-77.0	-61.3	-61.5	-72.9
Storage Losses*	-10.4	-8.9	-9.2	-12.2	-8.5	-7.0	-11.0	-13.9	-15.4	-12.3	-12.3	-14.6
Net Time-shift Benefit	23.0	21.1	22.1	21.1	49.1	75.7	82.1	79.4	50.1	25.5	30.7	18.4

	Seasonal Price "Bins"	
	May - October	November - April
12:00 P.M. - 5:00 P.M.	128.5	90.4
1:00 A.M. - 6:00 A.M.	-56.8	-56.4
Storage Losses*	-11.4	-11.3
Net Time-shift Benefit	60.3	22.8

Annual		
Hours		Value**
May - October	651.8	39,323
November - April	651.8	14,830
Total	1,304	54,152

**Net time-shift benefit * hours/year.

*Storage Efficiency = 80.0%.
Note: Values expressed in units of $/MWh.

5.2.15.4. Estimate

Although each region is different, forecast energy prices for California are used to estimate the renewables energy time-shift benefit. A summary of those prices are shown in Table 26. (See Appendix F for details about the electricity prices used.)

Although not used directly for the estimate in this guide, the range of typical variable costs for electric energy from fossil-fueled generation are shown in Figure 18. The figure is provided as context for the prices shown in Table 26. Values reflect a) fuel efficiencies ranging from 35% to 55%, b) fuel prices ranging from $3/MMBtu to $9/MMBtu, and c) a generic value of 1 ¢/kWh for non fuel variable operation cost.

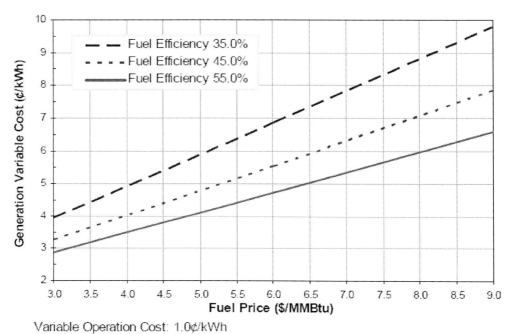

Figure 18. Generation variable cost, for various fuel prices and fuel efficiencies.

Based on the range of variable costs shown in Figure 18, variable cost for generation ranges from about 4.8 ¢/kWh for a 45% efficient combined cycle plant assuming fuel price of about $5/MMBtu to about 7 ¢/kWh for a 35% efficient simple cycle combustion turbine plant using higher priced 'on peak' fuel costing $6/MMBtu. The primary driver of those generic variable cost values is fuel price, shown on the graph's X-axis. The variable cost values in Figure 18 also reflect a generic, non-fuel-related variable operation and maintenance (O&M) cost of 1¢ /kWh. Note that the O&M cost for simple cycle combustion turbine generation is about 2.5 ¢/kWh and for combined cycle generation (a common type of new generation) is on the order of 0.43 ¢/kWh.[73]

5.2.15.5. Wind Energy Time-shift

For the wind generation case, the energy time-shift benefit is estimated based on the assumed difference between the annual average wholesale/spot value for on-peak energy and off-peak energy, net of energy storage losses. Instead of selling off-peak energy in real-time (when generated), that energy is stored and used at a later time when energy prices are high.

The off-peak *versus* on-peak price differential is estimated based on the price differential between weekday energy prices occurring during the periods of a) 12:00 p.m. to 5:00 p.m. and b) 1:00 a.m. to 6:00 a.m., as shown in Table 26. Also shown in Table 26: Time-shifting for 5 *full* hours per day (5 kWh per day per kW of wind generation), for all weekdays during the year, is worth about $54,152/MW-year or about $54.2/kW-year. Nevertheless, Figure 8 (in Section 3.6.1.2) shows that at least some of the wind generation's output occurs during the on-peak period when the energy is already most valuable. Consequently, the amount of energy from wind generation that is actually time-shifted is less than would be needed for 5 *full* hours of storage discharge (*i.e.,* is less than 5 kWh per kW of wind generation).

> Importantly, to the extent that adding energy storage for energy time-shift increases output during times when peak demand occurs, a capacity credit may also apply. Based on the benefit estimate used for the electric supply capacity application, the 10-year capacity credit could range from nothing (if generation capacity is not needed) up to $864/kW ($120/kW-year), if the need for a natural-gas-fueled combined cycle generation plant is avoided.
>
> Based on those values, the benefit for energy time-shift plus supply capacity from *baseload* renewable energy generation ranges from $389/kW over 10 years (in areas not needing additional generation capacity) up to $1,288/kW if the time-shift defers the need for combined cycle power plant capacity.

Depending on the applicable wind generation production profile(s), storage could be used to time-shift 2 to 4 kWh per day, per kW of wind generation. Assuming that storage can be used to time-shift 3 kWh per kW of wind generation during a 5-hour on-peak period, the energy time-shift benefit (based on the above values) is:

(3 kWh ÷ 5 hrs) . $54.2/kW-year = $32.5/kW-year. When multiplied by the PW factor, the benefit over 10 years is $32.5/kW-year .7.17 = $233.2/kW.

Depending on the local and regional circumstances, there may be an electric supply capacity-related benefit if the time-shift involves storage discharging as shown in Figure 8. (Also see the renewables capacity firming benefit characterization in Section 5.2.16.)

5.2.15.6. Baseload Renewables Energy Time-shift

The energy time-shift benefit for baseload renewable energy generation is based on the value achieved by storing low-value energy during off-peak periods and discharging the storage on-peak. As shown in Figure 9 (Section 3.6.1.3), the effect is to avoid using or selling the generator's energy when that energy has a low value and to increase the amount of electric energy available during times when that energy is more valuable. Based on the differential between the price for off-peak energy and the price for on-peak energy (shown in Table 26), the value related to energy time-shift for baseload renewable energy generation is approximately $54.2/kW-year or about $389/kW over 10 years (7.17 .$54.2/kW-year).

5.2.16. Benefit #16 — Renewables Capacity Firming

5.2.16.1. Description

The benefit for firming output from renewable energy generation is related to the cost that can be avoided for other electric *supply* capacity. If renewable energy generation output is constant during times when demand is high, then less conventional generation *capacity* is needed. In this guide, benefits are estimated for two cases: 1) distributed PV generation and 2) bulk wind generation. (See the benefit characterization in Section 3.6.2 for details.)

5.2.16.2. Capacity Credit

The value of a generator's capacity (capacity credit) is based on the degree to which the generator's capacity contributes to the reliability of the electric supply system, primarily during peak demand periods. It is also based on the cost for electric supply resources which may include local or regional generation plants, power purchases, or demand response. Capacity credit is an important criterion of merit used by power engineers to estimate the contribution that renewable energy-fueled generation makes toward the total amount of power required to serve load.

Perhaps the most robust way to estimate an intermittent generator's capacity credit is to calculate the effective load carrying capacity (ELCC). ELCC is a measure of a power plant's contribution to the greater electric supply system's capacity during times when the amount and reliability of capacity is important. ELCC is established using reliability and/or production cost models to estimate common reliability indices, such as loss of load probability, loss of load expectation, or expected unserved energy.

All power plants, except for the benchmark (a fully dispatchable, very reliable combustion turbine-based generator), have an ELCC that is less than the generator's rated capacity. For example, 100 MW of wind generation may have a capacity credit of 0.25; which means that the wind generation provides 0.25 · 100 MW = 25 MW of *capacity* to the electric supply system when demand is high.

5.2.16.3. Generation Capacity Cost

The cost assumed for generation capacity (which forms the basis for the capacity firming benefit) is the same as the generation cost for the electric supply capacity benefit, as described in Section 5.2.2. It is for a combustion-turbine-based generation plant whose annual cost is assumed to be $120/kW-year.

5.2.16.4. On-peak Period and Storage Operation

Renewables capacity firming is assumed to be most valuable during the hours of 12:00 p.m. to 5:00 p.m., weekdays during the summer peak demand season (May through October). Because there is also some benefit associated with energy time-shift during the winter demand season (November through April), it is also assumed that the storage is used for energy time-shift during those months, for the same five hours per day on weekdays.

5.2.16.5. Energy for Renewables Capacity Firming

Readers should note that the renewables capacity firming benefit estimated does not include benefits related to the *energy* that is discharged when storage is being discharged to firm renewable energy generation output. If storage (used for renewables capacity firming) is discharged for a small portion of the year, then the energy-related benefit may be modest. Conversely, the energy-related benefit could be more significant if storage is discharged for a larger portion of the year.

Although each circumstance is different, the total benefit for renewables capacity firming is often maximized by using low-priced, off-peak wholesale energy from the grid to charge storage. Furthermore, all energy output from the renewable energy generation is delivered directly to the grid without storage losses. Among other effects, storing low-priced energy from the grid *and* directly from renewable energy generation means that there is more energy output from the renewable energy plus storage system than could be delivered if only energy from renewable energy generation is stored.

For this report, the *wholesale* energy prices used to estimate energy benefits associated with renewables capacity firming are the same ones used for the electric energy time-shift benefit (See Section 5.2.1). Monthly and seasonal average price *differentials* for the prices used are summarized in Table 26 in the description of the renewables energy time-shift benefit (See Section 5.2.15). The price differential is the difference between on-peak energy and off-peak energy during weekdays.

5.2.16.6. Distributed PV Capacity Firming

In many parts of California, well-designed and well-operated solar generation provides a capacity credit of 0.80 or more, in part because of the good correlation between insolation and demand.[74]

For the purpose of this guide, however, the solar generation that is firmed (*i.e.,* distributed, flat-panel PV modules with a fixed orientation) is assumed to have a capacity credit of 0.40. That value is lower than the 0.80 capacity credit for a well-optimized, solar generation facility for several reasons.

First, PV systems evaluated herein have a fixed orientation; however, generation with a high capacity credit uses 'tracking' to follow the sun, so the solar collector is pointed directly at the sun for a large portion of the day. The result is more power production during peak demand periods and more energy generation during the year than a similar plant that does not employ tracking, though tracking adds complexity and cost.

Other reasons that distributed PV systems' capacity credit may be relatively low include the following: the PV modules' (fixed) orientation is suboptimal; regular dust accumulation on modules; shading of PV modules by trees, buildings, *etc.* during a portion of the peak demand period; high ambient temperatures that reduce PV's efficiency and power during the peak demand period; and the level of cloudiness over the PV's location.

Storage is used to firm PV during the five peak demand price hours in the summer months. For this report, the generic peak demand period assumed is 12:00 p.m. to 5:00 p.m., weekdays, during the summer peak demand season (May through October).

The generic storage discharge duration for storage plus PV systems ranges from 2 to 3 hours, though the discharge duration could be less in regions with good insolation and/or for well designed and maintained PV systems.

The storage plus PV system is assumed to operate as follows for PV capacity firming: low-value (and low-priced) energy *from the grid* is stored, and that energy is discharged during utility peak demand hours. Because most or all energy generated by PV has high or relatively high real-time value, all PV energy is assumed to be used or sold to the grid as it is generated.

For this analysis, adding storage to distributed fixed-orientation PV is assumed to increase the capacity credit from 0.40 to 1.0. Although a given storage plus PV system may not be reliable enough to warrant a capacity credit of 1.0, it is assumed that that unit diversity among many small storage plus PV systems leads to an effective aggregated electric supply capacity credit *approaching* 1.0.

5.2.16.7. Bulk Wind Generation Firming

Capacity firming could be applied to smaller distributed wind generation capacity; however, in this guide the wind generation that is firmed is assumed to be deployed in central/large-scale wind farms. The generic capacity credit assumed for wind generation is 0.25.[75]

Note that most *energy* production from wind generation is assumed to occur when the energy has relatively low value (*i.e.,* most energy produced is generated during evening, night, and early morning hours).

Depending on the duration of the peak demand period and the degree to which wind coincides with peak load, storage used to firm wind generation capacity is assumed to have a discharge duration of 3 to 4 hours (3.5 hours is the generic value used.)

After being firmed with storage, the wind generation is assumed to have a capacity credit approaching 1.0 (0.75 of which is attributable to the addition of storage).

5.2.16.8. Distributed Renewables Capacity Firming for Demand Charge Reduction

Note that the renewables capacity firming benefit is related to electric *supply* capacity. That is, the benefit described above is related to the avoided cost of owning a generation plant. In the previous example, the generation is a generic dispatchable resource.

An important analog for electricity *end users* allows them to derive a benefit for capacity firming based on the applicable tariff for electric service. If the end user's electric service tariff includes demand charges, then the end user could use storage to reduce those charges. Demand charges reflect the price charged by the utility for each kW of *power* draw (demand) by the end user.

(See Section 3.5.2 and Section 5.2.12 for more details about demand charge reduction using storage.)

5.2.16.9. Incremental Benefit and Cost for Adding Storage for Renewables Capacity Firming

One point worth noting is that the renewables capacity firming benefit estimated in this report is for *adding* storage to renewable energy generation, so the benefit is *incremental*. Consequently, when evaluating the financial merits of adding storage to renewables generation, the incremental benefit is compared to incremental cost (to add storage).

5.2.16.10. Estimate

The renewables capacity firming benefit is based on the avoided cost for generation capacity of $120/kW-year and on the degree to which the renewable energy generation output is firmed. As an example: For PV, the assumed capacity credit *before* firming is 0.4, whereas the assumed capacity credit *after* firming is 1.0, for an increase of

1.0 – 0.4 = 0.6 kW per kW of rated capacity.

The resulting capacity firming benefit is

0.6 · $120/kW-year = $72/kW-year.

The energy-related benefit (for the energy discharged from storage) is summarized in Table 27. The total annual benefit, including the capacity-related benefit plus the energy-related benefit, is summarized in Table 28.

Table 27. Energy Time-shift Benefit from Renewable Energy Generation During Operation for Capacity Firming

	Photovoltaics		*Wind Generation*	
	Summer	Winter	Summer	Winter
Net Unit Benefit ($/MWh)1	60.3	22.2	60.3	22.2
(¢/kWh)	6.03	2.22	6.03	2.22
Energy Time-shift (Hours/Day)2	2.5	2.5	3.5	3.5
Days/Year3	130	130	130	130
Hours/Year	326	326	456	456
Net Seasonal Benefit ($/kW-yr)	19.7	7.2	27.5	10.1
Net Annual Benefit ($/kW-yr)	**26.9**		**37.6**	

[1] On-peak energy price minus off-peak energy price minus cost for storage losses. Does *not* include consideration of storage VOC.
[2] This criterion is based on the storage discharge duration.
[3] This criterion is based on the definition of peak demand period.

Table 28. Total Annual Renewables Capacity Firming Benefit

	Storage Energy	*Renewables Effective Capacity 1*		*Storage Incremental Value ($/kW-yr)*		
	Discharge Duration	w/o Firing	Firmed	Capacity2	Energy	Total
PV	2.5	0.40	1.00	72.0	26.9	98.9
Wind	3.5	0.25	1.00	90.0	37.6	127.6

[1] During peak demand periods.
[2] Assuming $120 per kW-year for combustion turbine based generation.

The annual values are converted to 10-year lifecycle benefit by multiplying by the PW factor of 7.17. The estimated 10-year net benefit associated with firming of PV output is

$98.9/kW-year · 7.17 = $709/kW.

The estimated 10-year net benefit from firming of wind generation is

$127.6/kW-year .7.17 = $915/kW.

5.2.17. Benefit #17 — *Wind Generation Grid Integration*

5.2.17.1. Description

The wind generation grid integration (wind integration) application includes two categories and a total of six subtypes. The two categories are 1) short-duration (lasting for a few seconds to a few minutes) and 2) long-duration (lasting for many minutes to a few hours). The six subtypes are summarized in Table 29.

Table 29. Wind Generation Grid Integration Application Subtypes

Short-duration Applications
1. Reduce Output Volatility (due to momentary wind fluctuations)
2. Improve Power Quality
Long-duration Applications
3. Reduce Output Variability (lasting minutes to hours)
4. Transmission Congestion Relief
5. Backup for Unexpected Wind Generation Shortfall
6. Reduce Minimum Load Violations

The benefit associated with storage used for each subtype varies significantly. Even among the subtypes, the benefit varies from moment-to-moment, throughout the day, throughout the year and from year-to-year.

Benefit values for wind generation grid integration in this guide provide a starting point for related analyses, rather than being definitive. The rationale used to establish each benefit value is described below. Readers are left to judge the merits of that rationale for a specific region, electric supply system, or wind generation resource.

5.2.17.2. Estimate

The methodology for estimating each of the six wind generation grid integration application subtypes varies. A brief discussion of each is provided below.

Reduce Output Volatility

The leading response to grid effects from wind output volatility (characterized by variations lasting a few seconds to a few minutes) is increased use of conventional area regulation resources. For this report, the benefit for reducing aggregate wind output volatility is the avoided cost for that additional area regulation service needed to accommodate the volatility. The area regulation service is described in Section 3.3.2 and the benefit is described in Section 5.2.4.

(An alternate approach that could be used to estimate the benefit for short-duration intermittency is that used for the renewables capacity firming application in Section 5.2.16.)

Area regulation capacity needed to accommodate wind generation additions is assumed to be required during the six most productive months for wind generation (which varies depending on region). Consequently, the benefit estimate is about half that for *annual* operation. If storage can provide *rapid-response* regulation, and if the benefit from that capability can be internalized by the storage owner, then the benefit can be as high as $1,000/kW for 10 years. If the rapid-response capability does not have a specified value, then the 10-year benefit may be closer to $500/kW. In this guide, the estimated generic benefit is $750/kW for 10 years.

Improve Power Quality

The benefit for improved power quality is specific to the location, wind resource, and wind turbine type(s), and it varies from moment-to-moment, throughout a day, throughout the year, and among years. Also, newer wind turbines pose fewer and less significant power quality-related challenges than older turbines.[76]

The first option for establishing the benefit for this application is to determine the cost of the most likely existing option for addressing the specific power quality challenge and, in some cases, the 'do nothing' option. Conventional options may include replacing components of older wind turbines; upgrading circuits and/or transformers; using capacitors, static VAR compensators, or power electronics; curtailing production from wind generation; and/or using on-site/local dispatchable (*e.g.*, diesel-fueled) generation. Given the challenge of generalizing the circumstances and options for this application, estimating a generic benefit is probably not helpful, so no estimate is provided in this report.

Reduce Output Variability

Wind generation output variability involves changes that occur over periods lasting from minutes to hours. Wind variability (from minute-to-minute and throughout the day) adds to the need for load following resources that must make up the difference between load and generation that is already online. For this guide, the benefit of reducing aggregate wind output variability is the avoided cost for that additional load following service.

It is also assumed that most additional load following capacity will probably be provided by combined cycle generation plants. Furthermore, the additional load following is assumed to be needed for six hours per day (three hours during the morning when load is increasing, and three hours as load decreases at night) which is assumed to occur during the six most productive wind generation months each year.

Given that the service is provided by a combined cycle power plant, the assumed (marginal) cost for the additional service is $50/MW per service hour. As a result, the estimated annual benefit (in Year 1) for using storage with wind generation to reduce the need for additional load following resources is

6 hours/day ·7 days/week ·26 weeks/year ·$50/MW per hour of service
= 1,092 hours/year ·$50/MW per hour of service
= $54,600/MW per year of service ($54.6/kW-year).

The generic lifecycle benefit is

$54.6/kW-year · 7.17 = $391.5/kW.

Transmission Congestion Relief

The transmission congestion relief application subtype cannot be easily generalized. In some areas, there may be enough unused transmission capacity to accommodate all, or at least most, expected wind generation capacity additions. In other areas, any significant additions may overwhelm existing transmission capacity. In some cases, congestion is reflected in pricing for energy or for energy transfers.

The cost to upgrade transmission to accommodate renewables in California probably reflects relatively high costs (for new transmission capacity); however, it may still be instructive to consider the circumstances. In California, cumulative wind generation capacity additions are assumed to be 5,200 MW by 2010 and 10,600 MW by 2020. The total installed cost for new transmission capacity needed to accommodate *all* renewables in California is an estimated $2.3 billion by 2010 and $6.3 billion by 2020.[77] For this report, it is assumed that about two-thirds of the transmission cost for all renewables is attributable to *wind* generation additions (given that most new renewable generation capacity expected is wind generation).

Based on those assumptions, the estimated lifecycle cost for transmission capacity needed to accommodate wind generation capacity additions is shown in Table 30. The approach used to make that estimate is described below.

Table 30. Estimated Total Transmission Cost for Wind Capacity Additions in California

	Year	2010	2020
1	**Wind Capacity Additions (MW cum.)**	**5,200**	**10,600**
2	Transmission Total Installed Cost ($Million)	2,300	6,300
3	(Assumed) Portion of Transmission Attributable to Wind Gen. added	0.667	0.667
4	Transmission Cost Attributable to Wind Gen. added ($Million)	1,534	4,202
5	**Transmission Annual Cost for Wind Gen. Added ($Million)***	**168.8**	**462.2**
6	Transmission Cost for Wind Gen. / Wind Gen. kW ($/kW of Wind gen.)**	295	396
7	Transmission Annual Cost for Wind Gen. / Wind Gen. kW ($/kW-year of Wind gen.)	32.5	43.6
8	**Transmission Lifecycle Cost for Wind Gen. ($/kW of Wind gen. for 10 years)***	**232.7**	**312.7**
9	(Assumed) kW storage per kW of Wind gen.	0.50	0.50
10	**Lifecycle Benefit ($/kW storage, 10 years)**	**465.4**	**625.3**

* Attributable to wind generation. Based on Fixed Charge Rate = 0.11
** Transmission Annual Cost / Wind Capacity Additions
*** 10.0%/yr. discount rate, 2.5%/yr. escalation rate: PW factor = 7.17

The approach used to estimate the transmission congestion relief benefit involves assumptions about or estimates of 1) wind generation capacity to be added; 2) transmission capacity needs and the related total and annual cost attributable to increased wind generation capacity to be added (key premise: wind generation-related transmission congestion will occur if that transmission capacity is not added); 3) the value of a 10-year deferral of the upgrades needed; and 4) the lifecycle (10 year) benefit if storage is used *in lieu* of upgrades.

The following ten-step process was used to develop the generic benefit estimate shown in Table 30:

1) Determine the total amount of wind generation to be added (Line 1 in Table 30).
2) Use a current estimate of transmission *total cost* that will be incurred because all types of renewables generation will be added (Line 2 in Table 30). Total cost is defined as the *installed* cost, including land, site preparation, permits, equipment purchases, and installation.
3) Estimate the *portion* of transmission total cost that is attributable to wind generation additions (line 3 in Table 30). For the example, wind generation is assumed to account for two-thirds of the transmission needed to accommodate all renewables.
4) Calculate the *value* of transmission total cost that is attributable to wind generation additions. In the example, multiply the transmission total installed cost for renewables (Line 4 in Table 30) by two-thirds. For the example, an estimated $1.53 billion would be spent in 2010 and $4.2 billion would be spent in 2020.
5) Calculate the *annual* (financial carrying) cost for the transmission attributable to wind generation additions by multiplying the transmission total cost that is attributable to wind generation additions (Line 4 in Table 30) by the fixed charge rate of 0.11. The result (Line 5 in Table 30) is approximately $169 million in 2010 and $462 million in 2020.
6) Allocate transmission *total* cost attributable to wind generation additions to wind generation on a $/kW of wind generation basis. That is done by dividing the transmission cost attributable to wind generation added (Line 4 in Table 30) by the kW of wind generation to be added (Line 1 in Table 30). The result is $295/kW of transmission installed cost per kW of wind generation capacity added by 2010 and $396/kW of wind generation added by 2020 (Line 6 in Table 30).
7) Allocate the *annual* cost for transmission needed to serve new wind generation, on a $/kW of wind generation basis. That is done by dividing the *annual* transmission cost attributable to wind generation additions (Line 5 in Table 30) by the kW of wind generation to be added (Line 1 in Table 30). In 2010, the resulting single-year transmission cost is about $186 Million ÷ 5,200 MW = $32.5 per kW-year of wind capacity. In 2020, the annual cost for transmission added (per kW of wind generation added) is $462 Million ÷ 10,600 MW = $43.6 per kW-year (Line 7 in Table 30).
8) Estimate the *lifecycle* transmission cost attributable to wind generation additions by multiplying the annual transmission-related cost per kW of wind generation (Line 7 in Table 30) by the present worth factor of 7.17. That yields an estimated lifecycle cost for wind generation capacity added of $232.7/kW by 2010 and $312.7/kW by 2020 (Line 8 in Table 30).
9) Estimate the amount of storage needed (per kW of wind generation) to avoid the need for additional wind generation-related transmission. In the example, the

assumption is that 0.5 kW of storage (whose useful life is 10 years) is needed per kW of wind generation to offset transmission-related cost (Line 9 in Table 30). That is based on the simplifying assumption that in almost all cases wind generation output will not be more than 50% of its rated capacity during times when the transmission system is heavily loaded, overloaded, or congested.

10) Calculate the 10-year lifecycle benefit associated with each kW of storage used to provide transmission congestion relief (based on deferring transmission upgrades for 10 years). That value is derived by dividing lifecycle transmission cost attributable to wind generation additions (Line 8 in Table 30) by 0.5 (kW storage / kW wind generation). For the generic estimate, the benefit is $465.4/kW in 2010 and $625.3/kW in 2020 (Line 10 in Table 30).

This benefit estimate reflects the *average* cost for transmission. Presumably, there are some locations for which the cost to upgrade the transmission is higher. Furthermore, it is those locations for which storage may be the best alternative (given the relatively high cost).

Table 31. Transmission Cost for Wind Capacity Additions in California, High-value Locations

	Year	2010	2020
1	Wind Capacity Additions (MW cum.)	5,200	10,600
2	Applicable Portion*	0.2	0.2
3	**Wind Capacity Affected (MW cum.)**	**1,040**	**2,120**
4	Transmission Total (Installed) Cost ($Million)	2,300	6,300
5	(Assumed) Portion of Transmission Total Cost Attributable to Wind Gen. Added	0.667	0.667
6	Transmission Total Cost Attributable to Wind Gen. Added ($Million)	1,534	4,202
7	Portion (of cost for all transmission additions) In Play*	0.5	0.5
8	Transmission Cost Attributable to Wind gen. added ($Million)	767	2,101
9	**Transmission Annual Cost for Wind Gen. Added ($Million)****	**84.4**	**231.1**
10	Transmission Total Cost for Wind Gen. / Wind Gen. Added kW ($/kW of Wind Gen.)**	738	991
11	nsmission Annual Cost for Wind Gen. / Wind Gen. Added kW ($/kW-year of Wind Gen.)	81.1	109.0
12	**Transmission Lifecycle Cost for Wind Gen. Added ($/kW of Wind gen. for 10 years)*****	**582**	**782**
13	(Assumed) kW storage per kW of Wind Gen. Added	1.00	1.00
14	**Lifecycle Benefit ($/kW storage, 10 years)**	**582**	**782**

* 50% of all costs attributible to Wind gen. are incurred for 20% of Wind gen. additions.
** Attributable to wind generation. Based on Fixed Charge Rate = 0.11
*** 10% discount rate, 2.5% escalation rate: PW factor = 7.17

Consider another scenario: For the situation described above, 50% of all wind-related transmission upgrade costs are incurred to accommodate 20% of the wind capacity additions. Furthermore, those locations require 1 kW of storage per kW of wind generation to avoid the need to upgrade transmission equipment. The results of this scenario are shown in Table 31.

Based on the results shown in Table 31, the lifecycle benefit for storage used to offset need for the most expensive transmission upgrades (those needed to accommodate wind generation) would be $582/kW over 10 years in 2010 and $782/kW over 10 years in 2020 (Line 1 in Table 31).

Based on the results for the two scenarios shown in Table 30 and Table 31, the *generic* value assumed for the lifecycle benefit is $625/kW for 10 years.

Backup for Unexpected Wind Generation Shortfall

The value for this application is related to avoiding electric service outages that are caused by a sudden, unexpected drop in wind generation output. To the extent that storage allows grid operators to avoid such outages, the storage provides benefit. It is important to note that, in most cases, the ISO addresses a sudden reduction of wind generation output with one of several non-storage options, especially out-of-area energy purchases; reserve capacity; interrupting or curtailing load to reduce demand; and increasingly automated load control. Storage provides another option.

The values in Table 32 reflect a simple benefit estimate based on the value-of-service (VOS) metric described in Section 5.2.13. The assumed composite VOS for all customer classes is $10/kWh. That value reflects the cost incurred by end users per kWh of energy not delivered due to the outage. Furthermore, it reflects a composite of the value for all electricity end-user classes, ranging from residential end users at the low end, for whom the cost is close to nothing, to high-value-added manufacturing customers whose VOS may exceed $100/kWh. As shown in the table, at the lower bound, one outage is avoided over 10 years for an estimated 10-year lifecycle benefit of $100/kW or an annual benefit of about $14/kW-year. At the high end, two outages are avoided over 10 years, yielding an estimated lifecycle benefit of $200/kW and an annual benefit of $28/kW-year.

Table 32. Benefit for Avoided Service Outages Due to Sudden Drop of Wind Generation Output

	Low	High
Wind-to-Peak Load Ratio	10.0%	10.0%
Outages Avoided (10 years)	1	2
Outage Duration (hours)	1	1
Value of Unserved Energy ($/kWh)	10	10
Lifecycle Benefit ($Year1 / kW-*load*)	10	20
Lifecycle Benefit* ($ Year 1 / kW *wind gen.*)	100	200
Annual Benefit** ($/kW-year)	14	28

*Lifecycle Benefit per kW of Load / Wind/Peak Load Ratio.
**Assuming PW factor = 7.17.

For the estimate above, it is assumed that there is 1 kW of storage per kW of wind generation. To the extent that wind resources are geographically diverse, less than 1 kW of storage per kW of wind generation is conceivable. If, for example, storage of 0.5 kW per kW of wind generation capacity would suffice for a geographically diversified wind generation resource, then the benefit values in Table 32 would double.

Reduce Minimum Load Violations

Minimum load violations occur when generation capacity exceeds demand. When that occurs, some of the energy generated may not be usable. The benefit for reducing minimum load violations is assumed to be related to the value of energy that cannot be used. The generic value is estimated based on forecasted energy prices in California in 2009. A summary of those values is shown in Table 33.

Table 33. Low and High Values for Minimum Load Violations

Item Name	Low	High
Portion of the Year	1.0%	4.0%
Hours Per Year	87.6	350.4
Energy Price ($/MWh)	56.5	56.5
Benefit ($/MW-year)	4,949	19,798
($/kW-year)	4.9	19.8

Based on the values shown in Table 33, the generic value for reduced minimum load violations ranges from about $5/kW-year on the low end to about $20/kW-year on the high end. The low value reflects minimum load violations that occur during 1% of the year, or about 57 hours per year. The high value reflects minimum load violations occurring during 4% of the year, or 350 hours per year. Both values reflect an average energy price of $56.5/MWh during minimum load violations.

5.2.17.3. Wind Integration Benefits Summary

Table 34 summarizes the benefits estimated (and described above) for the wind integration application subtypes.

Table 34. Wind Integration Benefits Summary

	Benefit Estimate ($/kW)*	
Application Subtype	Low	High
Short Duration		
1. Reduce Output Volatility (due to momentary wind fluctuations)	500	1,000
2. Improve Power Quality	not estimated	
Long Duration		
3. Reduce Output Variability (lasting minutes to hours)	391	
4. Transmission Congestion Relief	465	782
5. Backup for Unexpected Wind Generation Shortfall	100	200
6. Reduce Minimum Load Violations	5	20

* 10 years, 2.5% escalation rate, 10% discount rate: Present Worth factor = 7.17.

5.3. Incidental Benefits

Some benefits are not specific to any one application, as they may accrue incidentally when storage is used for one or more applications. For example, dynamic operating benefits occur because the operation of the greater electric supply system is more optimal because storage is used. And, although avoiding transmission access charges is not an application, it may be that using storage allows stakeholders to reduce or avoid charges associated with transmitting energy through the transmission system. A discussion of nine meaningful incidental benefits which are explored in this guide is provided below.

5.3.1. Benefit #18 — *Increased Asset Utilization*

5.3.1.1. Description

In many situations, use of energy storage will increase the amount of electricity that is generated, and/or transmitted, and/or distributed using existing utility assets. The effect is commonly referred to as increased asset utilization. Two important financial implications of increased asset utilization are 1) the cost to own the equipment is amortized across more (units of) energy which reduces the unit cost/price for that energy, and 2) the payback from the investment occurs sooner, which reduces investment risk.

Consider an example: A utility installs distributed energy storage to address local electric service reliability needs and to defer an expensive T&D upgrade. Storage use increases generation asset utilization if the storage is charged using *existing* generation assets (presumably during times when demand is low). Similarly, transmission asset utilization increases assuming that *existing* transmission capacity is used to transmit the storage charging energy (presumably the transmission occurs during times when transmission asset utilization is normally low). Depending on use patterns and location, distributed energy storage may also increase distribution asset utilization.

The benefit of increased asset utilization is highly circumstance-specific. It is not estimated in this guide.

5.3.2. Benefit #19 — *Avoided Transmission and Distribution Energy Losses*

5.3.2.1. Description

As with any process involving conversion or transfer of energy, energy losses occur during electric energy transmission and distribution. These T&D energy losses (sometimes referred to as I^2R or 'I squared R' energy losses) tend to be lower at night and when loading is light and higher during the day and when loading is heavy. T&D energy losses increase as the amount of current flow in T&D equipment increases and as the ambient temperature increases. Thus, losses are greatest on days when T&D equipment is heavily loaded and the temperature is high.

If storage is charged with grid energy, then the benefit is based on the difference between the cost for losses incurred to deliver energy for charging (off-peak) and the cost that would have been incurred if the energy was delivered in real-time (on-peak). If storage is charged with energy generated locally, then the losses avoided (and benefit) may be even higher because no/limited losses are incurred to get the energy to the storage for charging.

5.3.2.2. Estimate

The generic benefit values shown in Figure 19 reflect two energy price scenarios and two scenarios for on-peak *versus* off-peak losses. The first price scenario involves an average price *difference* (labeled as Price Δ in the figure) of 6 ¢/kWh between on-peak and off-peak energy prices. For the second scenario, the average *difference* between on-peak and off-peak energy prices is 8 ¢/kWh. The values in Figure 19 also reflect a T&D energy loss *difference* (labeled as Loss Δ in the figure) between on-peak and off-peak of 3% at the low end and 5% at the higher end. An example: If on-peak T&D losses are 8% and T&D losses off-peak are 5%, then the *difference* is 3%. The estimated generic benefit for avoided T&D I^2R energy losses is $8/kW-year (net) or about $57/kW over 10 years.

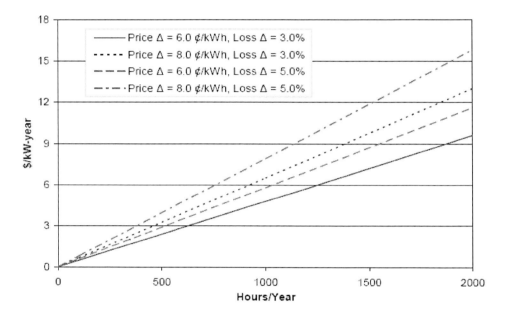

Figure 19. Benefit for T&D I^2R energy losses avoided.

5.3.3. Benefit #20 — Avoided Transmission Access Charges

5.3.3.1. Description

Typically, utilities that transmit electricity across transmission facilities that are owned by another entity must pay the owners for transmission 'service'. Similarly, utility customers must pay the cost incurred by the utility to own and to operate transmission needed to deliver the electricity. Related charges are often called transmission access charges.

Consider municipal electric utilities (munis) and electric cooperatives (co-ops). Munis and coops may own some or all of the generation capacity needed. Almost all munis and co-ops own and operate their electricity distribution system. Many, however, do not own transmission capacity. Also, most utilities transmit some power through other utilities' transmission lines. Utilities must pay transmission access charges to transmit power from their own generation plant(s) and/or from the wholesale electricity marketplace.

The benefit for avoided transmission access charges depends on, among other factors, tariff terms and pricing, location, and increasingly, time of year and time of day. In some cases, transmission access is priced based on energy used ($/kWh delivered). In other cases, the transmission charge is assessed based on capacity used, like demand charges ($/kW).

In many parts of the country, the marketplace for transmission capacity is just emerging. As the marketplace for electricity opens up, transmission access charges will be available from the various regional transmission organizations. The trend toward locational marginal pricing of energy will allow for increasingly precise, location-specific allocation of transmission costs.

5.3.3.2. Estimate

At the lower end of the spectrum, transmission access charges are estimated based on annual average transmission charges for firm point-to-point transmission service in the Midwest ISO control area. Based on an informal survey of those transmission access charges, the annual amount is approximately $25/kW-year to $30/kW-year.[78] Furthermore, the Midwest ISO's charges for off-peak transmission service are on the order of 30% less than the charge for service on-peak.

At the high end of the spectrum, consider a California-specific indication of the *retail* charge for transmission: A transmission access charge of 0.913 ¢/kWh of energy delivered is assessed for transmission under terms of PG&E's A-6 commercial TOU energy price electricity service tariff. If assuming annual energy use of 4,300 kWh per kW of peak load, the total transmission charges are about $40/kW-year.[79]

Note that the value of $40/kW-year is assumed to indicate the utility 'revenue requirement' for transmission which is the amount that the utility must collect as revenue from customers to cover cost. Furthermore, if transmission is priced based on energy delivered, rather than being based on peak demand, then storage could actually increase transmission charges for end users because for each kWh discharged from storage, transmission charges are incurred for storage charging energy *and* for storage energy losses. Finally note that, in some cases, transmission charges are lower at night than during the day.

The estimated generic benefit for avoided transmission access charges is $20/kW-year. After applying the 7.17 PW factor, the lifecycle benefit is $143.40/kW.

5.3.4. Benefit #21 — Reduced Transmission and Distribution Investment Risk

5.3.4.1. Description

Although there is no specific accounting for or price ascribed to it, there is an undetermined amount of risk associated with investments in T&D upgrades or expansion, as there is with *any* investment. While there is no formal way to account for that risk, it is an actual cost borne by electricity users.[1]

Consider a simple example: Utility power engineers decide that it is prudent to upgrade some T&D equipment. When the upgrade project is half finished, the utility receives news that a large customer load will be removed such that the in-process upgrade will not be needed for several years. Whether the project is completed or not, for several years no revenue is received to cover the cost incurred for the upgrade. As a result, utility customers at

---

[1] Although not addressed in this report, storage could also be used to reduce generation fuel price risk.

large must pay more to cover that unmet revenue requirement. The effect is the same if *aggregate* load growth is lower than expected.

Uncertainty can lead to T&D project delays, the result of which may be service outages and damage to existing equipment. Some sources of uncertainty that can cause costly project delays include a) utility staff or funding shortages, b) institutional delays such as those for permits, c) unforeseen challenges encountered during construction, and d) weather.

For most T&D upgrades, the investment risk is low to very low. A low-risk T&D investment tends to involve an upgrade that is routine, low cost, and whose cost is likely or very likely to be offset by revenues.

Storage – or any other *modular* resource that can be located downstream (electrically) from the T&D upgrade – can be used to manage risk. For example, if there is uncertainty about whether an expected block load addition will occur or staffing shortages or permitting delays will affect the upgrade, modular storage could be used to defer the upgrade for one year – enabling the utility to delay a possibly risky T&D upgrade investment until there is less uncertainty.

It is not possible to generalize this benefit given the wide range of possible circumstances that could be involved; therefore, an estimate is not provided in this guide.

5.3.5. Benefit #22 — Dynamic Operating Benefits

A dynamic operating benefit (DOB) is a *generation* operating cost that is reduced or avoided because storage is part of the electric supply system. Generation operating cost is reduced if generation equipment a) is used less frequently (*i.e.*, has fewer startups), b) operates at a more constant output when it is used (avoided part load operation), and c) operates at its rated output level most/all of the time when in use.[80]

DOBs include those for reduced generation equipment wear, reduced fuel use, and reduced emissions. Reducing equipment wear may reduce maintenance costs and/or extend equipment service life. Fuel use and emissions are reduced if a) generation output is more constant, b) generation output operates at its rated output, and c) generation is started less frequently.

Some of the DOBs reflect expenses that would otherwise be incurred by utilities and that would be reflected in utility service prices. Other DOBs reduce societal costs. DOBs that reduce actual expenses include reduced fuel cost, reduced maintenance cost, and increased equipment life. The key societal benefits include lower cost-of-service, reduced resource (fuel) use, and reduced air emissions.

This benefit is specific to the generation mix in a given region. It is not estimated in this guide.

5.3.6. Benefit #23 — Power Factor Correction

As described in Appendix C, utilities often need to compensate for reactance that causes unacceptably low power factor. The typical utility response – to improve a circuit's power factor and effectiveness – is twofold: 1) include a (low) power factor charge for commercial electricity end users' whose loads have an especially low power factor (*e.g.*, below 0.85) and 2) use capacitors to offset the effects from inductive loads (*i.e.*, to reduce the degree to which voltage and current are out of phase).

Depending on circumstances, the utility solution may involve other more expensive alternatives such as static synchronous compensators (StatComs) and static VAR compensators.

Depending on the type and characteristics of storage deployed, distributed storage could provide effective power factor correction. Battery or other storage systems whose storage media has direct current (DC) output and which include power conditioning to convert between AC and DC power are especially well-suited to power factor correction. Conventional motor-generator systems can also provide reactive power (VAR) needed for local power factor correction.

Notably, power factor correcting capacitors (the most common approach used by utilities for power factor correction) are inexpensive relative to generation capacity. Typical installed costs range from $10 to $15 per kVAR, so the avoided cost (benefit) if storage is used would be low (relative to storage system cost). Nonetheless, that benefit may still be attractive if the *incremental* cost to add power factor correction capability to storage is low enough.

5.3.7. Benefit #24 — Reduced Generation Fossil Fuel Use

One incidental benefit that may accrue if storage is used is a reduction in the use of fossil fuels used for generation. Storage use can lead to reduced fossil fuel use in at least three ways. First stored energy from more efficient fossil fueled generation and/or renewables can offset use of less efficient intermediate duty or peaking generation (energy time-shift). Second, fuel use may be reduced due to dynamic operating benefits associated with storage use (Benefit #22). Third, fossil-fueled generation tends to be more efficient when ambient temperatures are low. Coincidentally, most storage charging occurs at night, when temperatures are lower. Finally, if energy is transmitted at night when ambient temperatures and T&D loading are relatively low, then T&D energy losses are reduced (Benefit #21).

Importantly, the degree to which fuel use is reduced *or increased* (due to use of storage) depends on three key criteria: 1) the age and type of generation equipment and fuel used to generate electricity for charging storage, 2) the age and type of generation equipment and fuel that *would have been used* if storage is not deployed, and 3) storage efficiency (*i.e.,* losses).

Consider a simple example: Combined cycle combustion turbine generation (CC) whose fuel efficiency is 49% (requiring 6,965 Btu/kWh of fuel, often referred to as the generator's 'heat rate') and simple cycle combustion turbine generation (CT) whose fuel efficiency is 33% (for a heat rate of 10,342 Btu/kWh of fuel). The fuel use difference between those two generators is

10,342 Btu/kWh on-peak – 6,965 Btu/kWh off-peak = 3,377 Btu/kWh
3,377 Btu/kWh difference ÷ 10,342 Btu/kWh on-peak = 32.7%.

Then, if storage efficiency is 75%, then the *net* amount of fuel used to generate charging energy for storage is

6,965 Btu/kWh off-peak ÷ 75% efficiency = 9,292 Btu/kWh.

The result is a fuel use reduction of

10,342 Btu/kWh on-peak – 9,292 Btu/kWh charging = 1,055 Btu/kWh

1,055 Btu/kWh difference ÷ 10,342 Btu/kWh on-peak = 10.2%.

The above example and another involving charging with electric energy from coal generation are summarized in Table 35.

Notably, although the total amount of fossil fuel used for generation may be reduced if storage is used, the financial benefit associated with that reduction depends on the type and price of fuel(s) involved. Generally, the price for coal is lower than that for natural gas and petroleum-based fuels.

Given that this benefit is so circumstance-specific — being affected by on-peak and off-peak generation age and type, as well as on-peak and off-peak fuel type and price — it is not helpful to provide a generic value for fossil-fuel use reduction using storage, so no estimate is given.

5.3.8. Benefit #25 — Reduced Air Emissions from Generation

Reduction of air emissions from electricity generation is a potentially important incidental benefit of storage use. As with reduced fuel use (described above), there are at least four distinct ways that storage can reduce generation-related air emissions. The first involves using stored electric energy generated using relatively efficient and/or clean power plants (baseload and/or renewables) to offset the use of less efficient and/or dirtier on-peak generation (energy time-shift).

The remaining three ways that storage use can lead to reduced air emissions involve reduced *fuel use* (which presumably leads to reduced air emissions): 1) dynamic operating benefits (Benefit #22); 2) increased generation operation at night, for storage charging, when fuel efficiency is higher; and 3) reduced T&D energy losses that accrue if more energy is transmitted at night when T&D equipment is not heavily loaded and when ambient temperatures are lower (Benefit #21).

Importantly, storage-use-related air emission reductions are circumstance-specific. Specifically, the degree to which air emissions are reduced *or increased* (due to use of storage) depends on three key criteria: 1) the age and type of generation equipment and fuel used to generate electricity for charging storage, 2) the age and type of generation equipment and fuel that *would have been used* if storage is not deployed, and 3) storage efficiency (*i.e.,* losses).

Depending on the circumstances, storage could lead to reduced electricity generation-related emissions of carbon monoxide (CO_2), oxides of nitrogen (NO_x), oxides of sulfur (SO_x), soot/particulate, carbon monoxide (CO) and volatile organic compounds.

Consider generic emission levels shown in Table 36 for NO_x and for CO_2. Values in that table are meant to indicate two common scenarios: 1) charge storage using off-peak electricity from a natural-gas-fueled combined cycle combustion turbine to offset use of a natural-gas-fueled simple cycle combustion turbine on-peak and 2) charge storage using off-peak electricity from modern coal-fueled generation to offset use of a natural-gas-fueled simple cycle combustion turbine on-peak. (Not shown is use of renewable energy to charge storage, which would lead to a dramatic reduction or even total elimination of air emissions per kWh from storage.) Based on the values in the table, storage would lead to dramatically different results depending on the type of generation involved.

Table 35. Generation Fuel Use Implications of Energy Storage Use

Scenario	Off-peak/Charging Fuel Efficiency[1] (%)	Off-peak/Charging Heat Rate[1] (Btu/kWh)	On-peak/Avoided Fuel Efficiency[1] (%)	On-peak/Avoided Heat Rate[1] (Btu/kWh)	Difference Net Fuel Use[2,3] (Btu/kWh)	Difference Change of Fuel Use[4] (Btu/kWh)
Charge: Combined Cycle Avoid: Simple Cycle C.T.	49.0%	6,965	33.0%	10,342	9,287	-1,055 (-10.2%)
Charge: Advanced Coal Avoid: Simple Cycle C.T.	43.0%	7,937	33.0%	10,342	10,583	+241 (+2.3%)

[1] In this context "fuel" only includes fossil fuels.
[2] Off-peak generation fuel used, including additional fuel needed to make up for storage losses.
[3] Storage efficiency = 75.0%.
[4] Fuel use by on-peak resource (avoided) minus net fuel use for electrc energy used for charging. C.T. = Combustion Turbine.

Table 36. Generation CO2 and NOx Emissions Implications of Energy Storage Use

Scenario	Off-peak/Charging CO$_2$ (lbs/MWh)	NOx (lbs/MWh)	On-peak/Avoided CO$_2$ (lbs/MWh)	NOx (lbs/MWh)	Difference[1] CO$_2$ (lbs/MWh)	NOx (lbs/MWh)
Charge: Combined Cycle Avoid: Simple Cycle C.T.	922	0.260	1,131	0.320	+98.3 (+8.7%)	+0.027 (+8.3%)
Charge: Advanced Coal Avoid: Simple Cycle C.T.	2,222	3.620	1,131	0.320	+1,832 (+162%)	+4.51 (+1,408%)

Source: Hadley, S.W. VanDyke, J.W. Emissions Benefits of Distributed Generation in the Texas Market. Oak Ridge National Laboratory Report ORNL/TM-2003/100. April 2003.

1. These values reflect additional fuel used for generation required to make up for energy losses for storage whose efficiency = 75.0%

C.T. = Combustion Turbine.

Of course, it is necessary to ascribe a 'price' to (reduction of) a given type of air emission before the *internalizable* financial benefit can be estimated. That topic is beyond the scope of this study, so the financial benefit for emission reductions is not estimated.

5.3.9. Benefit #26 — Flexibility

In broad terms, flexibility can be defined as the degree to which and the rate at which adjustment to changing circumstances is possible. More specifically, flexibility may provide the means to respond adeptly to uncertainty. Flexibility allows decision makers to *manage* risk and even to take advantage of business opportunities involving risk (*i.e.,* to use 'real options'[81]).

Although it is almost impossible to generalize, in some circumstances there may be a significant financial benefit associated with flexibility, especially in a changing business environment with significant uncertainty. The benefit accrues if the flexibility allows selection and use of more optimal solutions or response to business-related needs, challenges, and opportunities. For example, modular electric resources (including storage) can be used to provide electric supply and/or T&D capacity 'on the margin,' when and where needed. In some cases that alternative could comprise a more optimal (financially) response than is possible using conventional 'lumpy' capacity additions. Indeed, depending on the circumstances, a more financially optimal solution can involve higher revenue, more profit, and/or lower cost per kW of load served.

This benefit is highly circumstance-specific and it is not estimated.

5.3.10. Incidental Energy Benefit

In some energy storage applications, energy is discharged incidentally during operation. That energy almost certainly has *some* value (benefit). For example, it may offset the need for a utility and/or a utility customer to purchase energy.

5.4. Benefits Not Addressed in this Report

As characterized in Section 3.8, the approach used in this guide does not address many storage applications explicitly. Similarly, this report does not address some benefits explicitly, especially those that are not 'utility-related'.

Consider an example provided in Section 3.8 for an application involving storage for trackside support of electrified rail transportation systems. Two possible benefits for that application are a) increased revenue related to increased ridership and b) reduced equipment wear. Clearly, those benefits are not addressed explicitly in this guide, although they may actually exist and they may be important elements of an attractive value proposition. Also not addressed are possible tax-related incentives, especially income tax credits, and to a lesser extent, income tax deductions.

5.4.1. Utility Incentives, Special Tariffs and Pricing Approaches Not Addressed

5.4.1.1. Utility Incentives Not Addressed in this Report

Although not common practice, utilities may eventually provide incentives to customers to install storage. Those incentives could be similar to those used to encourage customers to install rooftop photovoltaics, to increase energy efficiency (of loads), and to participate in demand response, smart metering, and Smart Grid programs. Those incentives are an important element of storage value propositions.

5.4.1.2. Special Electric Service Tariffs and Pricing

In addition to the reduced time-of-use energy cost and reduced demand charges described in this report, there are at least three other possible ways that utility customers can use storage to reduce their overall electricity-related cost: 1) interruptible/ curtailable tariffs, 2) critical peak pricing, and 3) load management programs.

Interruptible/curtailable tariffs provide a discount to participants who agree to allow the grid operator to 'curtail' or 'interrupt' electric service when there is a shortage of energy and/or capacity. Normally, the agreement specifies that maximum frequency and duration of curtailments/interruptions. Historically, curtailment and interruption are used during electric *supply* shortages, though in the future, they could also be used when there is *transmission* congestion and/or when *localized* T&D overloading occurs.

Critical peak pricing involves energy prices that are significantly higher than normal and that apply when there is a shortage of energy and/or capacity. Normally, critical peak prices are invoked during electric *supply* shortages. In the future, they could also be used when *transmission* congestion exists and/or when *localized* T&D overloading occurs.

Load management programs incorporate pricing and/or direct load control to 'manage' peak demand during electric supply energy and/or capacity shortfalls. The objective is to create 'dispatchable' demand reduction (*i.e.,* utility customer loads that can be remotely controlled by the ISO, when needed, to address energy or capacity shortfalls.) When needed, the power draw of the demand response 'resource' is reduced, thereby reducing the need for generation.

5.4.1.3. Electric Service Pricing Approaches Not Addressed

In addition to time-of-use energy prices that reflect predetermined price for energy used within a predetermined time period, there is a steady movement toward 'dynamic' pricing involving energy prices that reflect current conditions and that may change as frequently as several times per hour. Similarly, there is movement to location-specific electricity prices, commonly referred to as locational marginal pricing (LMP). No attempt was made to address those pricing approaches in this report.

6. STORAGE VALUE PROPOSITIONS

6.1. Introduction

This section provides an overview of the concept of storage value propositions, including coverage of important elements and considerations.

A value proposition is characterized by 1) one or more (combined) applications plus 2) attractive financial returns (*i.e.*, benefits that exceed costs by the 'hurdle rate' of return). In some cases, storage used for just one application may provide attractive returns. In other circumstances, it may be necessary to combine benefits from two or more applications so that total benefits exceed total cost. Hence, this report emphasizes the important concept of combining applications for benefits aggregation.

Of course, applications must be compatible if they are to be combined. A combination of applications is technically compatible if the same storage system can be used for all of the applications. A combination of applications has operational compatibility if there are no operational conflicts among the applications. As a general indication, the synergies matrix shown in Table 37 provides an overview of the possible compatibility among the various applications characterized in this document.

6.2. Benefits Aggregation Challenges

There are some notable challenges associated with benefits aggregation. One important theme in that regard is that much of the knowledge, perspective and experience needed for savvy and effective benefit aggregation are yet to be acquired because benefit aggregation is just becoming common practice. Given that premise, significant education and research are needed to provide important evidence to key stakeholders, especially utility regulators and utility engineers and financial decision-makers, about the merits and importance of benefits aggregation.

Table 37. Applications Synergies Matrix

● Excellent　○ Good　○ Fair　○ Poor　⊗ Incompatible

Application	Electric Energy Time-shift	Electric Supply Capacity	Load Following	Area Regulation	Electric Supply Reserve Capacity	Voltage Support[b]	Transmission Congestion Relief[1]	T&D Upgrade Deferral[1]	Time-of-Use Energy Cost Management[1]	Demand Charge Management[1]	Electric Service Reliability[1]	Electric Service Power Quality[1]	Renewables Energy Time-shift	Renewables Capacity Firming	Wind Generation Grid Integration
Electric Energy Time-shift		●	○*	○*	○*	●	●†	●†	⊗	⊗	⊗	⊗	●	●	○*
Electric Supply Capacity	●		○*	○*	○*	●	○†	●	⊗	⊗	⊗	⊗	○×*	○×*	⊗
Load Following	○	○*		○*	○*	●	○†	○×*	○*†	○*†	⊗	⊗	○	⊗	⊗
Area Regulation	○*	○*	○*		●	⊗	○×*	⊗	⊗	⊗	⊗	⊗	○	○	⊗
Electric Supply Reserve Capacity	○*	○*	○*	○*		●	○*	○*	○*†	○*†	⊗	⊗	○*	○*	○*
Voltage Support[b]	●	●	●	⊗	●		●	●	○†	○†	○†	○†	○*†	○*†	⊗
Transmission Congestion Relief[1]	●†	○†	○×	○×*	○*†	●		○×†	○†	○†	⊗	⊗	○†	○†	⊗
T&D Upgrade Deferral[1]	●†	●	○×†	⊗	○*	●	○×†		○†	○†	⊗	⊗	○†	○†	⊗
Time-of-Use Energy Cost Management[1]	⊗	⊗	○*†	⊗	○*†	○‡	○†	○†		○†	●	●	○†	○‡	⊗
Demand Charge Management[1]	⊗	⊗	○*†	⊗	○*†	○‡	○†	○†	○†		●	●	○‡	●‡	⊗
Electric Service Reliability[1]	⊗	⊗	⊗	⊗	⊗	○‡	○	○	●	●		●	○‡	○‡	⊗
Electric Service Power Quality[1]	⊗	⊗	⊗	⊗	⊗	○‡	⊗	⊗	●	●	●		⊗	⊗	⊗
Renewables Energy Time-shift	○	○×*	○	○	○*	○‡†	○‡	○†	○‡	○‡	○‡	⊗		●	○×
Renewables Capacity Firming	○	○×*	⊗	○	○*	○‡†	○†	○†	○‡	○‡	○‡	⊗	●		○×
Wind Generation Grid Integration	○	⊗	⊗	⊗	○*	⊗	⊗	⊗	⊗	⊗	⊗	⊗	○×	○×	

Notes

[a] For Area Regulation: Assume that storage cannot be connected at the distribution level.

[b] For Voltage support: Assume that a) storage is distributed and b) the storage system includes reactive power capability.

[c] For Reserve Capacity: Must have stored energy for at least one hour of discharge (i.e., so can offer useof the storage as reserve capacity on "hour-ahead"
[d] For T&D Load Following: For load following up (mornings) or down (evenings) involving charging; must pay prevailing energy price.
[e] For T&D Deferral: Annual hours of discharge range from somewhat limited to none. So storage is available for other applications during most of the year.
[f] For Time-of-use Energy Cost Management and Demand Charge Management: Assume discharge for 5 hrs./day (noon to 5:00 pm), weekdays, May to Octo
[g] Transmission Support (not shown) is assumed to be mostly or entirely incompatible with other applications.

Annotations

[1] Requires distributed storage that is located where needed.
x Somewhat to very circumstance-specific, especially regarding timing of operation and/or location.
* Most storage cannot provide power for both applications simultaneously.
† Presumably discharge is somewhat to very coincident for the two applications.
# For distributed storage: charging energy a) from onsite renewable generation and/or or b) purchased from offsite renewable generation via the grid. ‡ Requires utility dispatch of onsite storage.

The following (listed in no particular order) are some of the reasons that benefit aggregation is challenging and not common practice:

- The potential for technical and/or operational conflicts.
- Regulatory 'permission' does not exist.
- Engineering standards and tools do not exist.
- Weak or non-existent price signals make it difficult for some stakeholders to internalize some/many benefits. In other words, inefficient markets.
- Prevailing utility technological and financial biases against any untested or unfamiliar solution, and consequently, the slow pace of change in the utility industry.
- Some storage benefits have been demonstrated insufficiently or not at all.
- The benefits that do exist tend to be difficult to aggregate in practice because, for example, different benefits accruing to several stakeholders must be coordinated for a given value proposition to be financially attractive and operationally viable.

6.2.1. Technical Conflicts

In some cases, storage systems do not have the features or performance characteristics needed to serve multiple applications. One example is storage that cannot tolerate many deep discharges. Such storage systems could be well-suited for T&D deferral because storage might be used infrequently for that application, but the same storage system is not suitable for energy time-shift, which requires a lot of charging and discharging.

Another example is storage that cannot respond rapidly to changing conditions. Such systems may be suitable for energy time-shift or to reduce demand charges, but they may not be able to provide transmission support or end-user power quality benefits.

Another important criterion affecting technical compatibility is the storage's discharge duration. Storage whose discharge duration is optimized for some applications may not have enough discharge duration to serve other applications. Additionally, less reliable (though lower cost) storage systems may be suitable for energy time-shift or TOU energy cost reduction benefits; however, such systems could not be used for demand reduction, T&D support, or T&D deferral benefits because those applications require high reliability for the benefits to accrue.

6.2.2. Operational Conflicts

When estimating combined benefits for a value proposition, it is important to consider all potential operational conflicts between the applications being combined. Operational conflicts involve competing needs for a storage plant's power output and/or stored energy. For example, when storage is providing power for distribution upgrade deferral it cannot be called upon to provide backup power for electric service reliability. Another example is storage that is being used for most types of ancillary services: That same storage cannot be used for most other applications (*e.g.,* electric energy time-shift or transmission congestion relief) at the same time.

6.2.3. Aggregating Benefits among Stakeholders

One of the biggest challenges for many otherwise financially attractive value propositions is aggregating benefits that accrue to different stakeholders. Specifically, many of the benefits described in this report accrue to specific electricity end users, some to the ratepayers as a group, and others to utilities. Furthermore, various benefits accrue to different utility subsidiaries (*e.g.,* electric supply, transmission, distribution, customer service and unregulated business activities) that do not necessarily have the same incentives or biases.

Five 'beneficiary stakeholders' are worth noting because most benefits accrue to them: 1) specific electricity end users (*e.g.,* those who use storage to reduce electricity cost); 2) utility ratepayers at large; 3) the utility, especially T&D and electric supply business units; 4) 'merchant' storage project owners (*i.e.,* entities that use storage for profit only); and 5) society at large (*e.g.,* for improved environmental quality). In addition to the beneficiary stakeholders, there may other stakeholders with which aggregators must *coordinate* including regulators, ISOs, permitting agencies, and affected localities/communities.

Consider storage for T&D deferral. Utility ratepayers would be better off if the cost incurred per kWh of energy delivered is reduced, as would be the case with cost-effective T&D deferral. Nevertheless, in some circumstances ratepayers' interest may be at odds with investor-owned utilities' need to invest in *equipment* to generate dividends. (Recall that IOUs do not make any profit from mark-up on energy or fuel purchases, rather energy and fuel purchases are treated as 'pass-throughs' meaning that the utility passes the cost for energy on to end users without any mark-up or profit.)

Similarly, in some circumstances, specific electricity end users that install storage to reduce TOU energy cost and/or to reduce demand charges may actually reduce revenues needed to cover the utility's carrying cost for investments in generation and/or T&D equipment.

Consequently, when aggregating benefits into a value proposition, it is important to acknowledge and address the 'cross-cutting' nature of storage value propositions and the diversity of topics, stakeholders, motivators, and incentives that must be considered when developing or pursuing an actual project involving an electric utility-related energy storage value proposition.

Section 7.1 provides some additional details about important stakeholders and Section 7.2 provides an introduction to important challenges that may affect prospects for benefits aggregation.

6.2.4. Effect on Market Potential

As described in Section 4, it is important to consider the effect on market potential when combining applications. The market potential for specific combinations is almost certainly not the sum of the market potential for individual applications.

6.3. Notable Application Synergies

Each application characterization in Section 3 included a summary of notable synergies with other applications. A few application synergies in particular stand out within the context of developing attractive value propositions.

6.3.1. Electric Energy Time-shift and Electric Supply Capacity

Although it is important to maintain a crisp distinction between capacity-related and energy-related applications (and benefits), there are important synergies between the two. Those synergies exist if use of energy and need for capacity occur concurrently (which is fairly common). For example, storage used by an end user to reduce TOU energy charges could also reduce the same end user's demand charges; provide dispatchable load control as a system resource; or reduce loading on T&D capacity to reduce congestion or for T&D deferral. Another example is storage used for electric energy time-shift. It can provide electric supply capacity benefits because the times when energy has a high value coincide with high capacity value.

6.3.2. Electric Supply Reserve Capacity

Electric supply reserve capacity is especially compatible with other application/benefit combinations. (See Section 3.3.3 for details.) The most important reasons are 1) most times storage is used for reserves,so it may not have to discharge; 2) storage can provide two times its power as reserve capacity while charging; and 3) if there is an hour-ahead market for reserve capacity, then decisions can be made almost in real-time regarding the merits of discharging (if needed) *versus* saving the energy for later, for more benefit.

6.3.3. Load Following

Load following is somewhat compatible with storage used for other applications, primarily because storage can provide load following (up or down) while charging. (See Section 3.3 for details.) So, while storage is being charged (so that it can serve one ore more other applications), the same storage can provide load following.

6.3.4. Transmission and Distribution Upgrade Deferral

The T&D upgrade deferral application (and the closely related T&D life extension application), may be compatible with several applications. Probably the most important consideration is that storage used for T&D deferral or life extension is needed for just a few tens of hours to perhaps 200 hours per year. Consequently, storage can be used for other applications for as much as 95% of the year. And, in most cases storage discharge for T&D deferral or life extension is likely to occur when the energy and the capacity are both valuable from an electric supply perspective. Similarly, depending on the location, the same storage could also be used for transmission congestion relief.

6.3.5. Demand Charge Management and Time-of-use Energy Cost Management

Storage used to manage TOU energy cost and/or demand charges could provide other important benefits. First, the same storage used for those purposes could also be used to improve on-site electric service reliability and/or power quality. Also, if the storage is located in a part of the T&D system that is heavily loaded during peak demand times, then the same storage could also provide benefits for T&D upgrade deferral or life extension. Similarly, the same storage could be used to reduce transmission congestion, if the storage is located downstream from congested parts of the transmission system. The same storage could also provide electric service reserve capacity during much of the year.

6.3.6. Electric Service Reliability and Electric Service Power Quality

Presumably, storage used to improve electric service reliability and/or electric service power quality would have a discharge duration of a few minutes to perhaps an hour. Consequently, storage used for those applications may not be suitable for many other applications. Storage deployed mainly for *other* applications, however, may be well-suited for improving reliability and/or power quality if a modest amount of storage is added to provide additional discharge duration relative to the discharge duration needed for the other application(s).

6.4. Distributed Energy Storage

Because *distributed* energy storage can be used for more applications than larger, central storage, distributed storage may be used for a broader spectrum of value propositions.

It is important to distinguish between locational benefits and non-locational benefits. Locational benefits are those that can be realized only if distributed storage is deployed where needed. Nonlocational benefits can be realized regardless of distributed storage's location.

6.4.1. Locational Benefits
Locational benefits include transmission congestion relief, T&D upgrade deferral, TOU energy cost management, demand charge management, electric service reliability, and electric service power quality. Additionally, the way voltage support is defined in this report, storage used for voltage support should be located close to inductive loads. Depending on the circumstances, benefits for renewables energy time-shift and renewables capacity firming also may be locational, if for example, the renewable energy generation is distributed (*e.g.,* photovoltaics).

6.4.2. Non-locational Benefits
Non-locational benefits that can accrue if distributed storage is used include electric energy time-shift, electric supply capacity, load following, and electric supply reserve capacity. Depending on the circumstances, benefits for renewables energy time-shift and renewables capacity firming may be non-locational, if for example, the renewable energy generation is deployed in large wind farms or solar thermal generation that is remote to load centers.

6.5. Storage Modularity

As described in Section 2.14, to one extent or another, most storage technologies can be deployed as relatively small modules. Some storage technologies (especially batteries, capacitors and, to a lesser extent, flywheel storage) are inherently modular. Although normally considered to be suitable for large single-site storage projects, even above-ground CAES and small pumped hydroelectric storage could be modular (though above-ground CAES and pumped hydroelectric 'modules' are probably larger than those of other modular storage technologies.)

Use of modular electric resources (including electricity storage) could lead to a profoundly different electric utility capacity expansion philosophy than that which prevailed during the previous century because smaller, modular resources offer more diverse, robust, and optimizable approaches *versus* the 'limited and lumpy' options used in the past.

Furthermore, modular resources can be used for a wider array of applications than larger, less modular options.

Importantly, smaller, more modular resources tend to be more expensive (per kW, and for storage, per unit of discharge duration). Further, in many cases, more modular resources are less energy efficient.

The following notable considerations that are specific to *modular* distributed storage are described below:

- Optimal Capacity Additions
- T&D Planning Flexibility
- Unit Diversity
- Resource Aggregation
- Transportability

6.5.1. Optimal Capacity Additions

One of the most attractive aspects of modularity is that capacity can be added incrementally, where and as needed (*i.e.,* for 'just-in-time' capacity). Modularity may also enable cost-effective redeployment of storage capacity. For utilities, modularity (and redeployment) may reduce both the total cost of service for and the risk associated with larger, more 'lumpy' investments in infrastructure (*e.g.,* T&D capacity additions).

6.5.2. T&D Planning Flexibility

One important feature of any modular resource, including storage, is that it allows for more flexible responses to challenges than are possible using the limited number of conventional utility solutions. (See Section 5.3.9 for more about flexibility.)

6.5.3. Unit Diversity

One reason to use modular electricity resources is that the aggregate capacity from those resources is probably more reliable than the aggregate capacity provided by larger, less modular resources because, at any time, only one module (or at most a few modules) is likely to be unavailable for service, so the resources' aggregate capacity is only minimally affected. In contrast, the failure of a single or less diverse resource means that all or a significant portion of the resource's capacity is unavailable to serve load.

6.5.4. Resource Aggregation

For value propositions involving residential or small-to-medium commercial end users, the effort required to investigate, analyze, design, purchase, install, and operate storage and other modular electricity resources (including demand response, distributed generation, and PHEVs) is a significant and possibly expensive challenge. In those circumstances, load aggregators – or more generally, electric *resources* aggregators – may be positioned to address many of the administrative, legal, and regulatory challenges on behalf of owners of many small individual resources.

6.5.5. Transportability

Modular energy resources (including storage) that can be moved somewhat-to-very easily may be used in two (or more) locations at different times. This feature is especially attractive if the challenges addressed with the transportable resources tend to be transitory (*i.e.,* lasting just one or a few years). Thus, transportable storage used to address a challenge at one location in a given year may be relocated to address a similar or different challenge at another location, in a subsequent year. In fact, transportable storage could even be used at two different locations in the same year if the locations' challenges occur during different seasons.

Consider a realistic example: Transportable storage used 1) at one location with a sharp, but infrequent, summer peak caused by residential air conditioning loads, and 2) at another location that has demand peaks during winter driven by heating loads. Transportability is also attractive for locations where capacity or energy needs change from one year to the next.

6.6. Value Proposition Examples

This section includes a characterization of possible value propositions involving combinations of technically and operationally compatible applications. Importantly, these are just a few of the possible combinations. Not included are value propositions that are technically incompatible (*i.e.,* the application-specific storage needs are different).

6.6.1. Electric Energy Time-shift Plus Transmission and Distribution Upgrade Deferral

One notable application combination is electric energy time-shift plus T&D deferral. In many, (and perhaps most) cases, localized T&D peak demand coincides with 'system' (supply and transmission) peak demand periods. Consequently, it is likely that the energy discharged while storage is serving the T&D upgrade deferral application has a high value. Furthermore, in most cases, storage used for T&D upgrade deferral discharges for a very small portion of the year, if at all. So, storage used for T&D upgrade deferral during a small number of hours/days per year can also provide electric energy time-shift-related benefits during almost the entire year. Even if storage does not provide T&D upgrade deferral benefits in any given year, it can still be used for electric energy time-shift (and possibly other applications such as electric supply reserve capacity).

6.6.2. Time-of-use Energy Cost Management Plus Demand Charge Management

Many, and perhaps most, electricity end users who pay demand charges also pay TOU energy prices. Demand charges are most common for larger, non-residential end users, although that may be changing. An attractive scenario for this value proposition may be indicated by a combination of high on-peak demand charges, high on-peak energy prices, low or no off-peak or 'facility' or 'baseload' demand charges, and low off-peak energy prices.

6.6.3. Renewables Energy Time-Shift Plus Electric Energy Time-Shift

It is often suggested that energy storage could be used to significantly increase the value of renewables' intermittent output. In many cases, however, the incremental benefit may not be commensurate with the incremental cost of the storage plant. Another possibility is a project involving use of storage to time-shift energy from intermittent renewables *and* to time-shift wholesale electric energy from the grid. The same storage could even be physically decoupled from the generation and located where other benefits may accrue as well. For example, storage used to time-shift energy from wind generation and to time-shift energy from the grid could provide transmission support or even, conceivably, a T&D upgrade deferral benefit, depending on the storage system's location.

6.6.4. Renewables Energy Time-Shift Plus Electric Energy Time-shift Plus Electric Supply Reserve Capacity

Depending on circumstances, the same storage used for the value proposition described above (renewables energy time-shift plus electric energy time-shift) could also be used for electric supply reserve capacity. When the storage is charged and idle, it could provide reserve capacity. When it is charging, the storage could provide 2 its rated power as reserve capacity. It is even conceivable that storage could provide load following and provide reserves while charging if charging occurs during times when load is picking up (usually in the morning) and/or when load is dropping off (usually in the evening).

6.6.5. Transportable Storage for Transmission and Distribution Upgrade Deferral and Electric Service Power Quality/Reliability at Multiple Locations

For this value proposition, transportable storage is used at ten different locations for either T&D upgrade deferral or to improve electric service power quality and/or electric service reliability. The benefit for T&D upgrade deferral is assumed to be $367/kW-year of storage, and the benefit assumed for electric service power quality/reliability is $75/kW-year of storage.

Consider this hypothetical scenario: Transportable storage is used at five different locations for one year of T&D upgrade deferral at each location, in alternating years. In the other five years, when the storage is not used for T&D upgrade deferral, it provides a benefit related to improving local electric service power quality and/or electric service reliability. The benefits for that scenario are shown in Figure 20. As shown in the figure's right-side Y-axis, the present worth of the annual benefit is nearly $1,700/kW of storage. So, if storage can be owned and operated for less than $1,700/kW, for 10 years, then it would be a financially attractive option. That value would provide a helpful target for lifecycle cost for modular electric energy storage (in this case, with a 10-year life).

6.6.6. Storage to Serve Small Air Conditioning Loads

Using storage in conjunction with smaller air conditioning (A/C) units, especially residential and small commercial 'package' units, could be the basis for a compelling value position, for several reasons, most importantly 1) A/C loads comprise a significant portion of peak demand, 2) many A/C loads only operate for a few hundred hours per year, 3) small A/C motors pose an especially difficult challenge during grid-wide voltage emergencies that can exacerbate regional power outages, and 4) storage used to serve air conditioning loads could be available for most of the year for other benefits.

In many regions, A/C comprises a significant portion of peak demand. While circumstances are different in each region, based on the values shown in Figure 21, A/C accounts for 30% of summer peak demand in California. Note also that about 53% of all A/C-related demand in California is for commercial electricity users and about 47% of A/C-related demand is for residences.[82]

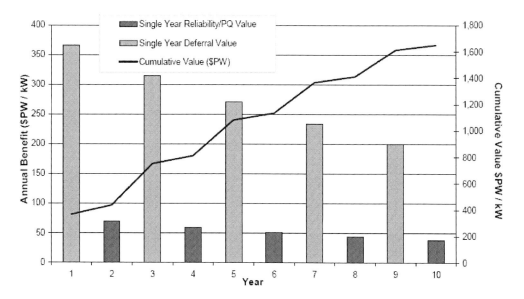

Figure 20. Value proposition for transportable storage.

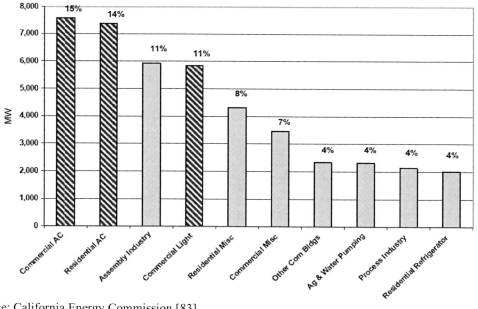

Source: California Energy Commission.[83]

Figure 21. Components of peak electric demand in California.

Given A/C's significant contribution to peak demand, utilities may incur a substantial A/C-related capacity cost – for generation, transmission, and distribution equipment to serve A/C load, but most A/C – especially small residential and commercial units – is operated for relatively few hours per year. The primary effect is that the utility receives relatively little annual revenue per kW of small A/C load served when compared to other common load types. So, smaller A/C loads cost a lot to serve (per kW) because they require so much capacity (equipment) even though limited use of small A/C equipment leads to low revenues (per kW). The consequence is very poor asset utilization.

6.6.6.1. Storage for Air Conditioning: Increased Utility Asset Utilization
The concept of poor asset utilization is illustrated graphically by the load duration curve (LDC) in Figure 22 and Figure 23. An LDC is a plot of hourly demand values, usually for one year, arranged in order of magnitude, irrespective of which hour during the year the demand occurs. Values to the left represent the highest levels of demand during the year, and values to the right represent the lowest demand values during the year.

The LDC in Figure 22 represents hourly load on a part of a distribution system during a specific year. Figure 23 includes only the highest 2% of demand values from those shown in Figure 22. The LDC shown, though real, represents a relatively extreme case (*i.e.*, the ratio of peak demand to average demand is unusually large). It was chosen because it illustrates well the concept of poor asset utilization. Specifically, as shown in Figure 23, 10% of the annual maximum demand occurs during about 0.4% of the year. Importantly, a significant portion of that demand is from A/C loads.

Storage use could increase asset utilization by reducing or eliminating the need for capacity, on the margin, and by providing charging energy for the storage during off-peak hours when generation, transmission, and distributions assets are usually underutilized.

Depending on the location and circumstances, storage serving smaller A/C loads could reduce the need for generation and T&D capacity and could lead to increased utilization of existing equipment (assets). It is likely that an energy time-shift benefit will also accrue incidentally.

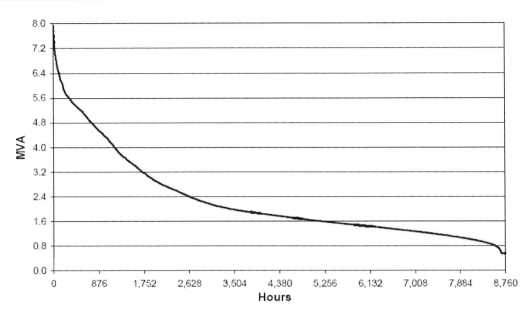

Figure 22. Load duration curve for an electricity distribution node.

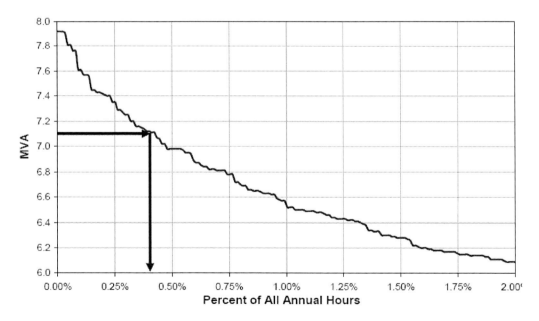

Figure 23. Portion of load duration curve with highest values.

6.6.6.2. Storage for Air Conditioning: Voltage Support

The voltage support benefit is notable because, as described in Section 3.3.4, small A/C motors pose a considerable challenge during grid emergencies by drawing additional current as voltage drops. This can pose a relatively significant challenge as the grid is re-energized after outages. Additionally, conventional capacitors used to manage localized voltage drops (due to reactance) under normal circumstances do not perform well as voltage support resources.

Consider one operational scenario: Distributed storage is used to serve small A/C equipment under normal grid conditions. If there is an 'electric supply emergency,' then the storage responds like other demand response resources by turning off the A/C equipment and providing power to the grid. If the storage's PCU has reactive power capability then the storage system could also provide reactive power as described in Appendix C.

Assuming that storage is located at or near A/C loads, the storage could provide several other important benefits, including at least two *non-locational* benefits: electric supply reserve capacity and load following. Additionally, *locational* benefits could include transmission congestion relief; improved electric service reliability and/or localized electric service power quality; and localized voltage support. Storage for smaller A/C loads could also be an important element of a robust Smart Grid and/or demand response implementation. The storage could also be used for wholesale or renewables energy time-shift on days that it is not needed for A/C loads.

One technical challenge is the amount of in-rush current needed for A/C compressor motor startup. Storage system PCUs may not be capable of providing the in-rush current needed. One way to address that issue is by using a hybrid storage system with two types of storage: one type that can provide high power for short durations, such as capacitors, and another that provides nominal power for long durations. Another possibility is to use the grid to provide some or all of the current during compressor motor startup (only during normal operating conditions for the grid). Given the diversity of compressor motor startups, presumably, providing in-rush current would not have an adverse affect on the grid.

Note that utility thermal energy storage incentives and programs are justified based on some of the same benefits described above primarily reduced demand for generation *capacity* and reduced cost for on-peak *energy* and, possibly, for reduced need for transmission capacity.

6.6.7. Distributed Storage in lieu of New Transmission Capacity

Distributed energy storage could be one important response to expected transmission capacity shortfalls. The need for new transmission capacity is driven by increasing peak demand and on-peak electric energy use; increasing interconnectedness of the grid and use of interregional generation resources; and increased deployment of renewable energy generation. Storage could help if it is located near load centers and charged during off-peak times, usually at night, when transmission systems are not heavily loaded; T&D I^2R energy losses are relatively low; and energy price tends to be low.

During on-peak times storage is used to serve load, reducing the amount of power used during peak demand periods, thus reducing loading on the transmission equipment. Four primary benefits of such use are 1) reduced need and cost for transmission capacity, 2) increased transmission asset utilization, 3) reduced T&D energy losses, and 4) energy time-shift. Of course, because the storage is distributed it could also be used for other locational benefits.

6.6.8. Distributed Storage for Bilateral Contracts with Wind Generators

In many areas, a significant portion of wind energy is produced at night when the energy's value is relatively low. Additionally, at some times of the year the supply of electric energy being generated exceeds demand for energy. One possible way to make better use of that energy is to use it to charge distributed storage.

Although several possible transactional frameworks could be used, one involves a bilateral contract between wind energy vendors and storage owners. Of course, either of those parties could use agents such as aggregators. Several benefits are possible using such a framework. The storage owner could use the storage to manage energy and demand charges or to enhance electric service reliability and/or power quality. Depending on the circumstances, distributed storage could reduce congestion of *existing* transmission capacity or delay or reduce need for *new* transmission capacity.

6.7. The Societal Storage Value Proposition

Although many benefits can be partially or totally internalized by the storage owner/user; an important factor that affects prospects for increased storage use is that some notable benefits accrue – in part or in whole – to utility customers as a group and/or to society at large. That leads to the compelling concept of a societal value proposition for storage.

The storage-related societal value proposition may include, but is not limited to, the following benefits (presented in no particular order):

- Reduced need for equipment and land for on-peak generation and transmission capacity.
- Increased asset utilization of existing utility generation, transmission, and distribution.
- Enabling superior operation of the existing generation fleet (*i.e.,* dynamic operating benefits) and transmission capacity.
- Reduced reliance on fossil fuel and increase energy security.
- Reduced air emissions.
- Reduced transmission and distribution energy losses.
- Enabling superior renewables integration to optimize benefits and to reduce integration cost and challenges.
- Enabling superior value from Smart Grid.
- Reduced cost-of-service (*e.g.,* by energy time-shift).
- Improved business productivity due to improved electric service reliability and power quality.
- Reduced need and cost for and extraction and refining of key commodities that would be needed to build conventional electric utility capacity; primarily, steel, aluminum, and copper.

The societal value proposition is an important consideration given the significant role that storage could and should play in the electricity marketplace of the future. Stakeholders that may need to understand and to consider the societal value for storage include existing and prospective storage beneficiaries, such as electric utilities and their customers; electric utility regulators; energy and electricity policymakers and policy analysts; and storage advocates.

Robust consideration of the societal value proposition for storage is important for reasons similar to those that drive the need to consider the societal value proposition for energy efficiency, demand response, distributed resources, and renewables. Perhaps the most important reason is that although the cost for storage may exceed the *internalizable* benefits, the cost may be lower than the combined value of internalizable benefits plus societal benefits. (See Section 1.4.2 which addresses the concept of internalizable benefits.)

It is important for lawmakers, regulators, and policymakers to be inclusive as they develop, consider, and promulgate regulations and policies whose outcomes/results could be improved if storage is used. For example, relevant decision-makers should consider the ways that storage could improve prospects for success regarding environment, energy, and electricity-related policy objectives such as increased use of renewables and reduced need for transmission infrastructure.

Similarly, it is important to consider incidental/unintended negative effects that laws, regulations, and policies may have on prospects for increased storage deployment. Consider an example: Many utilities do not have 'regulatory permission' to own distributed/modular resources (especially storage and generation) even though those alternatives may afford a superior means to serve load on the margin, *vis-à-vis* conventional 'lumpy' capacity additions, especially T&D capacity. (See Section 3.4.3 for more details.)

Finally, the societal value proposition may overlap with, and may be somewhat or even very coincidental to, an owner/user storage value proposition that involves direct/internalizable benefits. Consider a storage owner that uses storage to reduce on-peak TOU energy cost and peak demand charges. In that example, some societal benefits could include reduced land use impacts associated with reduced construction of new generation and transmission capacity; improved utility asset utilization; reduced air emissions; and improved business cost competitiveness.

7. ELECTRICITY STORAGE OPPORTUNITY STAKEHOLDERS, CHALLENGES, AND DRIVERS

This section presents potentially important topics and factors to consider when evaluating prospects for storage. Included are lists of the following: possibly important stakeholders, important challenges facing prospective storage users and developers, and notable storage opportunity drivers. Also included are brief characterizations of several important developments that could be significant drivers of many attractive electric utility-related storage opportunities:

- Increasing recognition by lawmakers, regulators, and policymakers of the important role that storage should play in the electricity marketplace of the future
- Increasing sophistication and savvy of load and distributed resource aggregators
- Increasingly rich price signals for electric utility-related services
- Tax and regulatory incentives
- Growing transmission capacity constraints
- Expected proliferation of PEVs and PHEVs
- Increased use of intermittent renewables
- Increasing focus on distributed resources
- Need to reduce generation fuel use and air emissions

- Innovation that drives improvements to storage technology and storage subsystem technologies
- An increasingly 'smart' grid that enables effective integration of some renewables and integration and dispatch of distributed resources including demand response, generation and storage

7.1. Stakeholders

There is a wide range of possible stakeholders in the electric-utility-related electricity storage opportunity. Of course, not all possible storage uses or projects must accommodate all of the stakeholders. The importance of particular stakeholders varies depending on factors such as the application(s), storage size and type, region, the utility or utilities involved. So, it is important to be familiar with the spectrum of possible stakeholders when formulating or evaluating value propositions.

Key 'beneficiary stakeholders' (i.e., parties that derive benefit from storage) include the following:

- Specific ratepayers that use storage to reduce electricity cost
- Utility ratepayers at large (if storage reduces the *utility's* overall cost-of-service which leads to reduced electricity price)
- Utilities
- 'Merchant' storage project owners (entities that use storage for profit only)
- Aggregators
- Storage equipment and services providers
- Society (*e.g.,* for improved environmental quality and economy)

Several important institutional or 'gatekeeper' stakeholders include the following:

- Engineering and standards community (*e.g.,* the American Society of Mechanical Engineers, the IEEE, the National Electrical Code, *etc.*)
- Federal and state energy/utility regulatory agencies
- Regional ISOs
- Local safety, siting, planning, and land use agencies
- Host communities

Other possibly important stakeholders include the following (presented in no particular order):

- Bill payers (often end users and bill payers are not the same people/entity)
- Utility functional entities (*e.g.,* electric supply, transmission, distribution, customer services, unregulated subsidiaries)
- Storage system integrators, project developers, architecture and engineering firms
- Politicians
- Electricity and environmental regulators
- Electricity, energy, and environment policymakers

- Electricity, energy, and environment researchers and research programs
- Smart Grid
- Independent power and energy services providers
- City and community planners and zoning officials
- Permitting agencies (*e.g.,* fire and health and safety)
- Landlords and property managers
- Storage advocates and advocacy organizations (*e.g.,* the Electricity Storage Association)
- Ratepayer and energy user advocacy groups
- Trade groups for specific industries and/or large commercial energy users

7.2. Challenges

To be sure, there are challenges that will affect efforts to site or deploy storage for many potentially attractive value propositions. Readers should be aware of those challenges when considering prospects for storage to be used for specific value propositions.

What follows is a summary list including some of the most important challenges that could face storage users and project developers as the storage opportunity unfolds. (See Appendix G for a more detailed list.)

- Storage's relatively high cost per kW installed
- Lack of storage-related regulatory rules and 'permission,' especially regarding distributed storage
- Prevailing electric energy and services pricing that are not economically efficient (though this is changing)
- Limited risk/reward sharing mechanisms
- Permitting and siting rules and regulations
- Limited familiarity, knowledge, and experience base (for storage)
- Existing utility technology biases
- Limited storage-related engineering standards and evaluation methodologies and tools
- Financing of *any* 'new' technology is challenging
- Investor-owned utility preference for investments in equipment and aversion to expense-based alternatives
- Inadequate infrastructure features and 'hooks' needed to accommodate or to optimize benefits from storage, especially distributed storage
- Competition among many technologies, concepts, and programs (*e.g.,* demand response, Smart Grid, distributed generation, renewables, *etc.*)
- Coordinating among numerous stakeholders, for 'permission' to use grid-connected storage and./or to aggregate benefits

7.3. Opportunity Drivers

The following is a list of possibly important drivers of the energy storage opportunity in the emerging electricity marketplace. Note that some of these drivers are also included in the list of challenges. The opportunity drivers identified by the authors include the following (in no particular order):

- Increasing interest in storage by politicians, regulators, and policymakers:
  o Battery development that is driven by automotive/transportation
  o For renewables integration
  o For transmission congestion relief and to reduce need for new transmission
- The emerging electricity marketplace:
  o Competition
  o Richer electricity-related price signals:
    - ƒ A general trend toward disaggregation of prices for energy and services
    - Locational prices
    - TOU prices
  o A broad range of new electric, control, and information technologies
- Increasing emphasis on intermittent renewable energy-fueled generation
- Generation and transmission capacity constraints and transmission congestion
- Existing and prospective incentives to install storage:
  o Tax-related issues
  o Regulatory/utility issues
  o Storage provides similar or even superior benefits to non-storage resources that are currently eligible for incentives (*e.g.,* end-use efficiency, demand response and distributed generation).
- Surging interest in electric vehicles, PEVs, and PHEVs:
  o Will affect grid cost and operations
  o Key impetus for battery technology improvements
- Growing use of demand response:
  o Especially *in lieu* of upgrading generation and transmission capacity
  o When energy is too expensive or not available
- Smart Grid
- Load aggregation
- The important role of independent power providers and energy services providers
- Growing emphasis on modular DER:
  o Distributed generation
  o Geographically targeted demand response and energy efficiency
  o Distributed energy storage
- Increasing emphasis on reducing air emissions from the electric supply
- NIMBY (not in my backyard) and BANANA (build absolutely nothing anywhere near the area):
  o Large-scale generation (conventional and renewables)
  o Transmission issues
- Growing preference for reduced fuel use

- Accelerating energy storage technology innovation (especially batteries, and to a lesser extent, capacitors and CAES)

7.4. Notable Developments Affecting Prospects for Storage

This section includes brief characterizations of ten important developments – mostly in the electricity marketplace – that could be especially important drivers of many attractive electricutility-related opportunities for storage.

7.4.1. Smart Grid and Electricity Stor Age

In broad terms, the vision for the Smart Grid is to increase operational efficiencies; improve electric service reliability; increase utility customer retention; and optimize capacity expansion (generation, transmission, and distribution) asset utilization.

Smart Grid acts as a controlling mechanism for the Advanced Metering Infrastructure (AMI) and smart meters. AMI and smart meters, in turn, enable two-way communication between a utility and its customers. Consider one concrete example: Smart Grid is expected to reduce energy use and peak demand by providing rich price signals using real-time data about energy cost and generation, transmission, and distribution capacity constraints.

Among other characteristics, Smart Grid is expected to be 'continuously upgradeable'. Also, Smart Grid will be an important element of a 'self-healing' electricity T&D network. It will add flexibility as utilities accommodate load and energy use growth. Smart Grid will also provide improved means to manage electricity transmission and distribution. Smart Grid could also be used for reactive power compensation and voltage control which, among other benefits, increases the throughput of T&D equipment. In 2008, the U.S. Department of Energy Smart Grid Task Force established the following seven 'characteristics of Smart Grid':

1) Enable active participation by consumers.
2) Accommodate all generation and storage options.
3) Enable new products, services, and markets.
4) Provide power quality for the range of needs in a digital economy.
5) Optimize asset utilization and operating efficiency.
6) Anticipate and respond to system disturbances in a self-healing manner.
7) Operate resiliently against physical and cyber attacks and natural disasters.

In the future, distributed energy storage deployed as part of, or in coordination with, Smart Grid should enable many rich value propositions that could include a wide array of benefits, possibly including the following:

- Aggregation, integration, optimization and coordination of all types of DER
- Electricity price hedging
- Ancillary services (*e.g.,* electric supply capacity reserves, voltage support provided locally, load following, area regulation)
- Reduced transmission congestion
- T&D upgrade deferral and equipment life extension

- Electric supply fleet performance and operation optimization (*i.e.,* DOBs)

Learn more about Smart Grid by visiting the U.S. DOE's Smart Grid website: http://www.oe.energy.gov/smartgrid.htm.

7.4.2. Increasing use of Demand Response Resources

Demand response is becoming an important resource, especially as an alternative to adding peak generation capacity and, to a lesser extent, to reduce need for or congestion of transmission systems. A summary of the value of demand response from the Peak Load Management Alliance includes the following primary elements:

- Reducing supplier and customer risk in the market
- Providing better reliability for the electricity system
- Reducing the costs associated with generation, transmission, and distribution
- Creating efficient markets
- Reducing environmental impact by reducing or delaying new power plant developments

7.4.3. Load Aggregators

The CAISO defines load aggregators as "..., a municipality or other governmental entity, an energy services provider, a scheduling coordinator, a utility distribution company, or any other entity representing single or multiple loads for the purpose of providing demand reduction service to the ISO."[84]

So, a load aggregator is any entity that combines loads into what is, in effect, a 'block' that can be controlled in response to requests by the ISO. Specifically, the ISO can rely on those blocks almost as if they are dispatchable generation capacity. That is, when there is not enough electric supply capacity available to serve all demand or to provide all necessary ancillary services, the ISO can request that the demand associated with load blocks be reduced or turned off.

A few points are worth considering. First, presumably, the scope of load aggregation could increase to include distributed generation and distributed storage. Although load aggregation tends to be done in response to electric-supply-related challenges, it seems likely that load aggregation could also be used to address more location-specific challenges such as overloaded T&D equipment or power-quality-related needs. It also seems likely that there could be some or perhaps significant convergence of Smart Grid, demand response, and load aggregation. Some of the advantages load aggregators have relative to individual end users, or perhaps even energy storage project developers, include the following (in no particular order):

- General business savvy regarding electricity value, pricing and markets
- Existing infrastructure
- Market familiarity
- Unit diversity
- The means to finance storage
- Opportunities to internalize more benefits

7.4.4. Increasingly Rich Electricity Price Signals

Another important development is the use of price signals for an increasing array of electric capacity, energy, and services that provide storage owners with the means to internalize more benefits. At least three important conventional pricing programs have existed for many years. As described in Section 3.5.1, some residential and many commercial electricity end users are eligible or even required to pay TOU-based prices for electric *energy*. Also, as described in Section 3.5.2, some electricity end users with somewhat large demand (>50 kW to 100 kW) often pay demand charges based on peak *load* and TOU charges for *energy*.

Many end users with medium demand or higher (>100 kW) are eligible for interruptible or curtailable rates. Under those rates, participating end users pay a discounted price for energy, and in return, the utility or the ISO may interrupt or curtail service, during grid emergencies, for a specified number of times, for specified durations. The interruptible or curtailable load is usually treated and used like reserve capacity for the electric supply system.

A more recent development is the establishment of critical peak pricing (CPP) for retail end users. Under terms of critical-peak-pricing tariffs, the utility can charge 'very high' prices for each kWh of energy used during critical peak periods. CPP tariffs allow the utility to impose the high prices a specified maximum number of times per year and for specified durations. In the U.S., the ISOs have implemented open markets for several ancillary services, including public posting of prices.

An emerging trend is the use of locational pricing or locational marginal pricing to better reflect the cost associated with delivery to specific parts of the grid. Among other factors, locational marginal prices could reflect area-specific energy cost/price, transmission capacity cost or charges, transmission congestion charges, and transmission I^2R energy losses. Importantly, load aggregators, Smart Grid, and demand response programs could be important enablers of a significant market for storage benefits when coupled with rich price signals.

7.4.5. Tax and Regulatory Incentives for Storage

One possibly important development for prospective energy storage purchasers and users is increased interest in providing related tax and regulatory/utility incentives. Tax incentives are most likely to include accelerated depreciation and possibly tax credits. Regulatory/utility (regulatory) incentives are most likely to include rebates that offset a portion of the purchase price. Although the analogy is not perfect, there is a lot of emphasis on providing tax and regulatory incentives for energy conservation and efficiency, peak demand reduction, and renewable energy systems.

Such incentives are currently offered for the following: purchasing and installing equipment for thermal energy storage; A/C efficiency improvements and/or downsizing; improving commercial lighting efficiency; installing distributed generation (*e.g.,* the Self-Generation Incentive Program in California); and/or installing renewable energy generation.

All of these programs are deemed to be important, at least in part, because they reduce peak demand, which reduces the need for electricity supply and T&D infrastructure. They also reduce on-peak energy use, which reduces fuel and operation cost for inefficient and expensive-to-run generation. It seems logical to at least consider incentives for using energy storage to the extent that it provides similar benefits.

7.4.6. Transmission Capacity Constraints

The need for additional transmission capacity is driven by several factors, including increasing deployment of bulk renewables generation that is located away from load centers; increasing the interconnectedness of the grid; increasing the use of non-utility-owned generation; increasing the use of generation located away from load centers, including increasing reliance on inter-regional energy transactions; increasing peak demand for electricity; and a heavily loaded and aging transmission infrastructure.

Importantly, storage could be used to reduce or to avoid the need for new, high-voltage, bulk transmission upgrades. That is important because one of the emerging challenges facing the new utility marketplace is the need for additional transmission capacity. Not only is *existing* transmission capacity getting older and less adequate, but siting *new* transmission is increasingly contentious.

While not addressed explicitly in this report, an approach similar to the ones used to estimate the T&D upgrade deferral benefit or T&D congestion relief benefit could also be used to estimate the benefit associated with avoided need for transmission. In simple terms, the benefit is related to the avoided cost for constructing new transmission capacity and/or upgrading existing equipment or regional transmission congestion charges.

7.4.7. Expected Proliferation of Electric Vehicles

Although the implications for energy storage generally are somewhat unclear, the expected proliferation of plug-in electric vehicles (PEVs) and plug-in hybrid electric vehicles (PHEVs) could have a significant impact on the potential for utility-related storage.[85] One possibility is that purchases of off-peak energy to charge storage will increase off-peak energy prices enough to reduce the benefit for some uses of utility-related storage, especially energy time-shift and TOU energy cost reduction.

Consider also that PEVs and PHEVs could provide some or perhaps most of the benefits that utility-related storage provides. Specifically, it may be cost-effective to charge electric vehicles when demand and energy prices are low or relatively low and then to dispatch aggregated power from those vehicles (using stored energy and/or the hybrid's fuel-driven power plant) to support the grid, especially during grid emergencies.

On the positive side, the proliferation of PEVs and PHEVs could lead to economies of scale and lower prices for advanced batteries and battery systems, including system management and grid integration (interconnection, control, and communications).

7.4.8. Increasing Use of Intermittent Renewables

Storage seems poised to be important as a complement to the expected increase of intermittent renewables. If nothing else, some output from intermittent renewables occurs when energy is not valuable and/or can change rapidly, making grid operations challenging and reducing the renewables' capacity credit. Three key facets of renewables-storage value propositions are notable: 1) capacity firming, 2) energy time-shift, and 3) grid integration.

7.4.9. Increasing Use of Modular Distributed Energy Resources

An emerging theme in the electricity marketplace is the use of modular electricity resources that are located near loads and downstream from overloaded T&D facilities. Distributed energy resources (DER) include generation, storage, and geographically-targeted load management and conservation.

On important reason for the increased interest in DER is that resources located near loads can provide more benefits than more remote resources. Other key drivers of interest in modular distributed resources include increasing congestion of regional transmission systems; challenges associated with paying for and siting large generation and transmission infrastructure; improvements in DER technologies; Smart Grid, and proliferating of rooftop/distributed photovoltaics.

7.4.10. Reducing Generation Fuel Use and Air Emissions

It is important to consider the fuel-use-related and air-emissions-related implications of storage because of trends toward reducing resource extraction, transportation and use, and policies that emphasize reducing air emissions due to generation. Depending on the circumstances, storage may be an important element of an overall strategy to reduce generation-related fuel use and air emissions.

As summarized in Section 5.3.7 and Section 5.3.8, storage can lead to reduced fuel use and air emissions in at least three ways: 1) time-shift energy from relatively efficient and/or clean baseload generation (*e.g.,* combined cycle, geothermal or wind generation) to offset use of less efficient, dirtier on-peak generation (*e.g.,* older, simple cycle combustion turbines), 2) reduce I^2R energy losses if energy is transmitted during off-peak times, and 3) dynamic operating benefits.

7.4.11. Storage Technology Innovation

Innovation by storage technology and storage system developers is accelerating, especially regarding batteries and, to a lesser extent, capacitors and CAES. Key drivers seem to be transportation-related uses, the expected increased use of intermittent renewables and a growing need for operational flexibility for the electricity grid.

8. CONCLUSIONS, OBSERVATIONS, AND NEXT STEPS

8.1. Summary Conclusions and Observations

8.1.1. The Storage Opportunity

Electric energy storage is poised to become an important element of the electricity grid and marketplace of the future. Storage has unique features and characteristics that make it useful for significant existing and emerging electric-utility-related opportunities and challenges.

Notable opportunities and challenges that storage can address include, but are not limited to, the following (presented in no particular order):

- Storage offsets the need for *additional* peaking generation capacity.
- Storage enables more optimal operation of the *existing* generation fleet, thereby reducing generation ramping and part load operation which, in turn, reduces equipment wear, fuel use, and air emissions.
- Storage is well-positioned to enable effective, optimal integration of intermittent renewables and possibly baseload renewables.

- Storage is well-suited to provide ancillary services, especially load following, area regulation, and electric supply reserve capacity. Distributed storage would be especially valuable for voltage support.
- Properly located storage can reduce congestion of existing transmission, reduce the need for additional transmission capacity, and defer the need for expensive subtransmission and distribution upgrades. Similarly, storage use can increase utilization of existing T&D assets, and in some cases it could be used to extend the life of existing T&D equipment – especially aging underground cables.
- Distributed storage will probably become a crucial element of the Smart Grid, and it can facilitate/enable increasingly important 'demand response' resources.
- Modular storage provides utility planners and engineers with flexible, reliable, and possibly less-risky alternatives to investments in conventional, inflexible, 'lumpy' T&D capacity additions.
- Distributed storage is well-suited to addressing growing electric service power quality and electric service reliability challenges, possibly by enabling utilities to provide differentiated electric service with higher quality and/or reliability (for a premium price).
- Utility customer-owned storage can be used to manage increasing electricity-related costs by time-shifting low-priced energy and by using storage to provide grid 'services', probably in conjunction with electric resources aggregators.

8.1.2. Storage Opportunity Drivers

Several current and emerging storage opportunity drivers have been recognized. The following are especially notable (presented in no particular order):

- Increasing recognition by lawmakers, regulators, and policymakers of the important role that storage should play in the electricity marketplace of the future.
- Increasing sophistication and savvy of load and distributed resource aggregators.
- Increasingly rich electricity price signals (*i.e.,* for energy, capacity, and ancillary services).
- Tax and regulatory incentives for storage.
- Expected proliferation of plug-in electric vehicles and plug-in hybrid electric vehicles.
- Increasing use of modular distributed energy resources for on-peak electric supply, ancillary services, and transmission congestion relief.
- Increasing use of intermittent renewables.
- Growing need for improved electric service power quality and reliability.
- Storage technology innovation, including improved subsystems (especially power conditioning) and storage system integration; battery innovation will accelerate, perhaps dramatically, due to development related to electric vehicles.
- An increasingly 'smart' electricity grid will enable effective integration of some renewables and integration and dispatch of distributed resources, including demand response, generation, and storage.

8.1.3. Notable Stakeholders

The storage opportunity involves numerous stakeholders. Understanding stakeholder interests and relationships is crucial for several reasons. Perhaps the most important reason is that not all benefits accrue to the same stakeholder. In fact, some benefits may involve conflicting interests. Consider a utility customer that uses storage to reduce its electricity-related costs. To the utility, the resulting 'revenue loss' increases the average price that customers at large must pay (because the utility receives less revenue without a commensurate reduction of fixed cost.)

Also, the existence of numerous stakeholders is important in that storage value propositions and storage projects may require a significant amount of coordination and cooperation among diverse stakeholders, possibly with conflicting interests.

Below are eight notable 'beneficiary stakeholders' (*i.e.*, parties that derive benefit from storage):

- Specific electricity end users who use storage to reduce electricity cost
- Utility ratepayers at large
- Utilities
- 'Merchant' storage project owners (entities that use storage for profit only)
- Aggregators
- Storage equipment and services providers
- Society at large (*e.g.*, for improved environmental quality and economy)

Several 'institutional stakeholders' or 'gatekeeper stakeholders' are also notable:

- Legislators and policymakers
- Utility engineers and capacity planners
- Engineering standards organizations
- Federal and state energy/utility regulatory agencies
- Regional independent system operators
- Local safety, siting, planning and land use agencies
- 'Host' communities

8.1.4. Notable Challenges

The storage opportunity involves some important challenges. It is prudent to be familiar with those challenges when evaluating prospects for storage in general, and for specific storage-related applications/benefits, value propositions, projects, locations, and regions/jurisdictions.

Several notable challenges include the following (in no particular order):

- Storage's relatively high cost per kW installed, compared to the benefit associated with most *technically* viable value propositions
- Lack of storage-related regulatory rules and 'permission,' especially regarding distributed storage
- Prevailing electric energy and services pricing that are not economically efficient (though, this is changing)

- Limited risk/reward sharing mechanisms (especially between utilities and customers and/or aggregators)
- Permitting and siting rules and regulations are not well-established for storage
- Limited familiarity with, knowledge about, and experience with storage
- Limited storage-related engineering standards and evaluation methodologies and tools
- Investor-owned utilities' 'rate-based' (or revenue requirement) financials that lead to a strong preference for investments in equipment and aversion to expense-based alternatives
- Storage must compete with many technologies, concepts, and programs (*e.g.,* demand response, Smart Grid, distributed generation, and renewables) for its place in the electricity marketplace of the future
- Coordinating among numerous stakeholders for 'permission' to use grid-connected storage and./or to aggregate benefits

See Appendix G for a more detailed list of challenges.

8.1.5. The Importance of Benefits Aggregation

The most important topic addressed in this guide is the aggregation of benefits into attractive value propositions (*i.e.,* a value proposition for which the total benefit exceeds the total cost by an amount that yields an acceptable-or-better return on investment). That theme is so important because in many situations two or more benefits will be required so the total benefit exceeds the total cost.

The primary purpose for this guide is to provide analysts with a framework for evaluating storage prospects for specific value propositions, including guidance about identifying and ascribing value to specific benefits that serve as building blocks for value propositions. Ideally, this framework will provide the foundation, and possibly the mindset, needed to recognize and characterize attractive value propositions.

As an aside: Given the emphasis on *benefits*, an important theme in this report is the need to maintain a crisp distinction between storage *applications* and the *benefits* that accrue if storage is used for a given application. (Applications are ways that storage is *used*, whereas benefits are primarily *financial*, including increased revenue and/or reduced or avoided cost.)

8.1.6. Multi-faceted Nature of the Storage Opportunity

Given the foregoing, clearly the storage opportunity is multi-faceted. A robust understanding of the storage opportunity requires at least some familiarity with several of those facets. Consider just a few:

- Many possible application/benefit combinations
- Numerous beneficiary stakeholders and institutional/gatekeeper stakeholders, some with conflicting interests
- Myriad rules, regulations, and permitting requirements
- Applicable market rules, tariffs, and pricing significantly affect the attractiveness of storage in specific regions and locations
- Role of storage relative to the electric supply generation fleet, renewables, demand response, Smart Grid, PEVs, and PHEVs

- Most existing storage technologies continue to improve, and advances involving emerging storage technologies are accelerating

8.2. Next Steps – Research Needs and Opportunities

Although utility-related storage opportunities are receiving increasing emphasis, more extensive research, development, and demonstration are needed. The elements of a robust storage-related research and development agenda are briefly characterized in this section.

8.2.1. Establish Consensus About Priorities and Actions
A key challenge for storage is the combination of diverse benefits and diverse stakeholders. Although that situation seems likely to persist, an important next step is to work toward a common understanding among stakeholders about several key topics, including the following: a) existence and magnitude of benefits; b) important value propositions, including the societal value proposition; c) key challenges and solutions; d) standards and rules needed (interconnection, permitting, *etc.*); e) market potential; f) the role of storage relative to and/or in conjunction with Smart Grid and demand response programs; g) storage technology and system cost and performance criteria, including definitions and values; and h) storage technology and value proposition demonstrations.

8.2.2. Identify and Characterize Attractive Value Propositions
This guide emphasizes the concept of value propositions and includes a few examples of possibly attractive value propositions. A helpful next step would be to establish a menu of model/generic value propositions that are a) generally accepted/recognized, b) financially attractive, and c) technically viable. Furthermore, value propositions targeted should be those involving somewhat-to-very significant market potential. Those value propositions would be used by storage advocates, project developers, technology and systems developers, regulators, policymakers, researchers, and prospective end users to focus their respective efforts.

8.2.3. Identify and Characterize Important Challenges and Possible Solutions
A crucial initial step towards consensus-building is to identify the most important challenges that could significantly delay and/or limit deployment of storage. First, the challenges should be characterized and then prioritized. Possible criteria to use in establishing priorities could include 1) potential showstoppers, especially those that are most likely to occur; 2) challenges whose solutions require a long lead time; 3) challenges that affect early adopters and/or users which could purchase significant amounts of storage in the near term; and 4) challenges that are most likely to create or to reinforce unhelpful misperceptions. After priorities are established, the next step would be to identify and develop an approach to address those challenges.

8.2.4. Identify, Characterize and Develop Financial and Engineering Standards, Models, and Tools
If storage is to reach its potential, one key priority is to identify, characterize, and develop the engineering and financial/accounting standards needed to evaluate important technical and financial criteria. Once those standards are established, analysts will need models and tools to

apply them. Presently, those standards, tools, and models are largely undeveloped and/or they require adaptation and evolution of existing tools.

8.2.5. Ensure Robust Integration of Distributed/Modular Storage with Smart Grid and Demand Response Programs

Smart Grid and demand response programs are poised to be important elements and enablers of the modern electricity grid and the electricity marketplace of the future. It seems likely that storage will be an important part of Smart Grid and demand response programs.

It is important to ensure robust and appropriate consideration of storage's roles and benefits as Smart Grid infrastructure and demand response, protocols, functionality, hardware, communications, and controls are developed, and as the Smart Grid and demand response programs are deployed.

8.2.6. Develop More Refined Market Potential Estimates

While the transparent, auditable, simplistic, maximum market potential estimates provided in this guide may provide a helpful point of departure, more robust methods are needed to refine those estimates. Such estimates are important metrics for politicians, policymakers, regulators, storage advocates, potential storage users, and storage vendors as they seek to gauge the potential implications and attractiveness of storage.

8.2.7. Develop Model Risk and Reward Sharing Mechanisms

As mentioned elsewhere in this guide, important discontinuities between some key stakeholders' interests – especially between utilities and customers – make risk and reward sharing difficult or impossible. Nevertheless, many otherwise attractive value propositions are not possible without risk and reward sharing, especially value propositions involving locational benefits and distributed/modular storage.

Perhaps the best example is the benefit for T&D upgrade deferral or T&D equipment life extension. Consider the example of a T&D upgrade deferral or life extension that would reduce the utility's total cost-of-service (an avoided cost) by $100,000 for one year.

Ideally, the utility would have the flexibility to share the avoided cost with customers that are willing and able reduce load, when needed, to enable the deferral. When called upon to reduce load, those customers could turn off loads (demand response) and/or use on-site generation and/or on-site storage. Peak load reduction could also be accomplished using energy efficiency.

Unfortunately, most utilities do not have the regulatory 'permission' or the transactional framework for such risk and reward sharing. If nothing else, the utility should be allowed to concentrate conventional demand response and energy efficiency incentives toward the part of the grid where T&D upgrade deferral or life extension is needed.

8.2.8. Develop Model Rules for Utility Ownership of Distributed/Modular Storage

For a variety of reasons, most utilities do not have regulatory permission to use storage *in lieu* of T&D equipment. One of the more important near terms objectives for the storage community is to advocate for utility permission to own and operate distributed/modular storage, just like any other equipment. Model rules for such utility ownership could spur the development of formalized rules at the state level.

8.2.9. Characterize, Understand, and Communicate the Societal Value Proposition for Storage

It is important to characterize, understand, and communicate the societal value proposition for storage (as described in Section 6.7) for at least two key reasons. First, society at large has a significant stake in the storage opportunity because some of the key benefits accrue, in part or in whole, to society at large (*e.g.,* reduced air emissions and reduced land use impacts from reduced need for new infrastructure). Second, some significant storage benefits may accrue to more than one stakeholder, including utility ratepayers as a group and/or to society as a whole, making 'stakeholder integration' and risk and reward sharing mechanisms especially important for societal benefits.

8.2.10. Storage Technology and Value Proposition Demonstrations

New storage technologies, subsystems, and storage system configurations must establish a record and reputation as a reliable, cost effective alternative before wide-scale acceptance. That same challenge applies to undemonstrated storage benefits and value propositions.

Establishing a track record and reputation often requires several demonstrations. Therefore, numerous demonstrations may be necessary (especially for modular/distributed storage) before wide-scale deployment of additional storage will occur.

REFERENCES

[1] Electricity Storage Association website: http://www.electricitystorage.org.
[2] Shoenung, Dr. Susan M. Hassenzahl, William M. *Long - versus Short-Term Energy Storage Technologies Analysis, A Lifecycle Cost Study*. Sandia National Laboratories, Energy Storage Program, Office of Electric Transmission and Distribution, U.S. Department of Energy. Sandia National Laboratories Report #SAND2003-2783. August 2003.
[3] Shoenung, Dr. Susan M. Eyer, Jim. *Benefit/Cost Framework for Evaluating Modular Energy Storage*. Sandia National Laboratories, Energy Storage Program, Office of Electric Transmission and Distribution, U.S. Department of Energy. Sandia National Laboratories Report #SAND2008-0978. February 2008.
[4] Eyer, Jim. Iannucci, Joe. Estimating Electricity Storage Power Rating and Discharge Duration for Utility Transmission and Distribution Deferral: A Study for the DOE Energy Storage Program. Sandia National Laboratories, Energy Storage Program, Office of Electric Transmission and Distribution, U.S. Department of Energy. Sandia National Laboratories Report #SAND2005-7069. November 2005.
[5] Mears, D. Gotschall, H. *EPRI-DOE Handbook of Energy Storage for Transmission and Distribution Applications*. Electric Power Research Institute Report #1001834. December 2003.
[6] ibid. [2].
[7] ibid. [5].
[8] *IEEE 1547 Standard for Interconnecting Distributed Resources with Electric Power Systems*. Approved by the IEEE Standards Board in June 2003. Approved as an American National Standard in October 2003. Available at:
[9] http://grouper.ieee.org/groups/scc21/1547/1547_index.html.

[10] Hirst, Eric. Kirby, Brendan. Separating and Measuring the Regulation and Load Following Ancillary Services. Oak Ridge National Laboratory. March 1999.

[11] Hirst, Eric. Kirby, Brendan. *What is the Correct Time-Averaging Period for the Regulation Ancillary Service?* Oak Ridge National Laboratory. April 2000. Available at: http://www.ornl.gov/sci/btc/apps/Restructuring/regtime.pdf.

[12] ibid. [10].

[13] Kirby, B. J. S*pinning Reserve from Responsive Loads*. Oak Ridge National Laboratory. Report #ORNL/TM-2003/19. March 2003.

[14] *2007-2016 Regional and National Peak Demand and Energy Projection Bandwidths*. Load Forecasting Working Group Of the Reliability Assessment Subcommittee North American Electric Reliability Corporation. July 2007. Available at: http://www.nerc.com.

[15] ibid. [12]

[16] Li, F. Fran. Kueck, John. Rizy, Tom. King, Tom. *Evaluation of Distributed Energy Resources for Reactive Power Supply, First Quarterly Report for Fiscal Year 2006*. Prepared for the U.S. Department of Energy by Oak Ridge National Laboratory and Energetics, Inc. November 2005.

[17] Kirby, Brendan. Hirst, Eric. *Ancillary Service Details: Voltage Control*. Oak Ridge National Laboratory, Energy Division. Sponsored by The National Regulatory Research Institute. Oak Ridge National Laboratory Report #ORNL/CON-453. December 1997.

[18] ibid. [15].

[19] ibid. [16].

[20] Electric Power Research Institute. *Reassessment of Superconducting Magnetic Energy Storage (SMES) Transmission System Benefits*. Electric Power Research Institute Report #1006795. March 2002.

[21] Torre, William V. DeSteese, J.G. Dagle, J.E. *Evaluation of Superconducting Magnetic Energy Storage for San Diego Gas and Electric*. Electric Power Research Institute Report #106286 2572-14. August 1997.

[22] ibid. [7].

[23] ibid. [8].

[24] Eyer, James M. Electric Utility Transmission and Distribution Upgrade Deferral Benefits from Modular Electricity Storage: A Study for the DOE Energy Storage Systems Program. Sandia National Laboratories. Sandia National Laboratories Report #SAND2009-4070. June 2009.

[25] Eckroad, Steve. Key, Tom. Kamath, Haresh. *Assessment of Alternatives to Lead-acid Batteries for Substations*. Proceedings of the Battcon 2004 Conference. Fort Lauderdale, Florida. Available at:

[26] http://www.battcon.com/PapersFinal2004/KamathPaper2004.pdf.

[27] ibid. [24].

[28] Bill Erdman, President, BEW Engineering. Discussion with Jim Eyer, Distributed Utility Associates regarding the incremental cost to add storage-related capabilities to PCUs and inverters. San Ramon, California. December 12, 2008.

[29] Parsons, Brian. National Wind Technology Center Presentation: *Grid Operational Impacts of Wind Power*. Presented during webcast sponsored by the National Rural Electric Cooperative Association. December 8, 2005.

[30] ibid. [27].

[31] Hawkins, David. Loutan, Clyde. *California ISO. Integration of Renewable Resources*. Presentation to Power Systems Engineering Research Center. October 2, 2007.

[32] Behnke, Michael. Erdman, William. BEW Engineering, Inc. *Impact of Past, Present and Future Wind Turbine Technologies on Transmission System Operation and Performance*. Prepared for the California Energy Commission, Public Interest Energy Research Program. Prepared by the California Wind Energy Collaborative. California Energy Commission Report #CEC-500-2006-050. May 2006.

[33] ibid. [30].

[34] ibid. [27].

[35] Renewable Integration Work Plan. R*enewables and Demand Response and Their Impact on Operational Requirements*. External Affairs, California Independent System Operator. Presentation to the California Public Utilities Commission. August 27, 2007.

[36] ibid. [29].

[37] O'Grady, Eileen. *Loss of Wind Causes Texas Power Grid Emergency*. Reuters News Service. February 27, 2008.

[38] North American Electric Reliability Council. *Electricity Supply & Demand (ES&D) Report, 2008 – 2017*. Available at: http://www.nerc.com/page.php?cid=4|38.

[39] Marshall, Lynn. Gorin, Tom. *California Demand 2008-2018, Staff Report, Revised Forecas*t. California Energy Commission Report #CEC-200-2007-015-SF2. November 2007. http://www.energy.ca.gov/electricity.

[40] Brown, Denny. *California Energy Commission Summer 2008 Electricity Supply and Demand Outlook Workshop*. California Energy Commission, Electricity Analysis Office. January 16, 2008

[41] Pew Center for Global Climate Change. *States with Renewable Portfolio Standards*. March 2008. Available at: http://www.pewclimate.org.

[42] Taylor, R. E. Hoagland, J.J. *Using Energy Storage with Wind Energy for Arbitrage*. Tennessee Valley Authority. Proceedings of the EESAT 2002 Conference. San Francisco, California. April 2002.

[43] Derived from preliminary Wholesale Electricity Price Forecast data provided by Joel Klein, California Energy Commission. April 2008.

[44] Klein, Joel. *Comparative Costs of California Central Station Electricity Generation Technologies (Cost of Generation Model)*. Presentation to ISO Stakeholders Meeting addressing California's Interim Capacity Procurement Mechanism. October 15, 2007. Available at: http://www.caiso.com/1c75/1c75c8ff34640.pdf.

[45] David Hawkins, California ISO; Mike Gravely, California Energy Commission; Bill Capp and Chet Lyons, Beacon Power. Discussion with Jim Eyer, Distributed Utility Associates at the California ISO offices. Folsom, California. April 12, 2007.

[46] California Energy Commission Press Release. California Energy Commission Applauds Beacon Power Upon Reaching Research Goal. January 10, 2007.

[47] Makarov, Dr. Yuri. Pacific Northwest National Laboratory in conjunction with the California Independent System Operator. *Relative Regulation Capacity Value of the Flywheel Energy Storage Resource*. The research was initially commissioned and funded by the CAISO. While the work was in progress, Dr. Makarov left the CAISO and published the report in November 2005.

[48] Eckroad, Steven. Electric Power Research Institute. Personal communication with Joe Iannucci, Distributed Utility Associates. June 2003.

[49] ibid. [46].
[50] Electric Power Research Institute. *Reassessment of Superconducting Magnetic Energy Storage (SMES) Transmission System Benefits*. Electric Power Research Institute Report #1006795. March 2002.
[51] ibid. [46].
[52] ibid. [48].
[53] Torre, William V. DeSteese, J.G. Dagle, J.E. *Evaluation of Superconducting Magnetic Energy Storage for San Diego Gas and Electric*. Electric Power Research Institute Report #106286 2572-14. August 1997.
[54] Annual Report on Market Issues and Performance, Section 5. Inter-Zonal Congestion Management Market. California Independent System Operator, Department of Market Monitoring. April 2007.
[55] ibid. [4].
[56] Pupp, Roger. *Distributed Utility Penetration Study*. Pacific Gas and Electric Company and the Electric Power Research Institute. 1991.
[57] Pupp, Roger. *Distribution Cost Percentiles*. Communication by e-mail message with Jim Eyer, Distributed Utility Associates. March 24, 2003.
[58] ibid. [2].
[59] ibid. [3].
[60] ibid. [3].
[61] ibid. [12].
[62] Eto, Joseph, *et al*. Lawrence Berkeley National Laboratory. *Scoping Study on Trends in the Economic Value of Electricity Reliability to the U.S. Economy*. Prepared for the Electric Power Research Institute and the U.S. Department of Energy. Coordinated by the Consortium for Electric Reliability Technology Solutions. Lawrence Berkeley National Laboratory Report #47911. June 2001; Private communications between Joseph Eto and Joseph Iannucci, Distributed Utility Associates. March and April 2003.
[63] Sullivan, Michael J., Vardell, Terry, Johnson, Mark. Power Interruption Costs to Industrial and Commercial Consumers of Electricity. IEEE Transactions on Industry Applications. November/December 1997.
[64] Sullivan, Michael J., Vardell, Terry. Suddeth, Noland B. Vojdani, Ali. Interruption Costs, Customer Satisfaction and Expectations for Service Reliability. IEEE Transactions on Power Systems. Vol. 11, No. 2. May 1996.
[65] ibid. [12].
[66] ibid. [60].
[67] LaCommare, Kristina Hamachi. Eto, Joseph H. *Evaluating the Cost of Power Interruptions and Power Quality to U.S. Electricity Consumers*. Lawrence Berkeley National Laboratory. Energy Storage Program, Office of Electric Transmission and Distribution and Office of Planning, Budget, and Analysis, Assistant Secretary for Energy Efficiency and Renewable Energy, U.S. Department of Energy. Lawrence Berkeley National Laboratory Report #LBNL-55718. September 2004.
[68] American Public Power Association. Power Quality Reference Guide – U.S. Edition. 1991.
[69] ibid. [60].
[70] ibid. [60].
[71] ibid. [61].

[72] ibid. [62].
[73] ibid. [65].
[74] Wiser, Ryan H. *Managing Natural Gas Price Volatility and Escalation: The Value of Renewable Energy*. Lawrence Berkeley National Laboratory. Proceedings of the NEMS/AEO 2004 Conference. Washington, D.C. March 23, 2004.
[75] Available at: http://www.eia.doe.gov/oiaf/archive/aeo04/conf/pdf/wiser.pdf.
[76] ibid. [42].
[77] California Renewables Portfolio Standard, Renewable Generation Integration Cost Analysis: Multi-Year Analysis Results and Recommendations Final Results. California Energy Commission Report #CEC-500-2006-064. June 2006.
[78] ibid. [74].
[79] ibid. [30].
[80] Porter, Kevin. The California Energy Commission's Intermittency Analysis Project Team. *Intermittency Analysis Project: Final Report*. Prepared for the California Energy Commission Public Interest Energy Research Program. California Energy Commission Report #CEC-500-2007-081. July 2007.
[81] Midwest Regional Transmission Organization Transmission Access and Ancillary Services Charges. Available at
[82] http://oasis.midwestiso.org/documents/miso/historical_pricing.html.
[83] Region-specific prices are available from the respective regional transmission organization or possibly the Federal Energy Regulatory Commission at http://www.ferc.gov.*ibid*.
[84] Fancher, R.B., et al. *Dynamic Operating Benefits of Energy Storage*. Decision Focus, Inc. Electric Power Research Institute Report #EPRI AP-4875. October 1986.
[85] Banerjee, Prithviraj. deWeck, Olivier L. *Flexibility Strategy – Valuing Flexible Product Options*. Massachusetts Institute of Technology. September 2004.
[86] *California's Electricity Situation: Summer 2005*. Presentation prepared by the staff of the California Energy Commission, California Public Utilities Commission, and California Independent System Operator. February 22, 2005.
[87] ibid. [82].
[88] California Independent System Operator. *Process for Participating Load Program (Ancillary Services / Supplemental Energy)*. 2008. Available at: http://caiso.com/17e5/17e5997039720.pdf.
[89] Wall, Edward. Program Manager, Vehicle Technologies Program, U. S. Department of Energy. Presentation to the Clean Energy Outlook Meeting of the Association of State Energy Research and Technology Transfer Institutions. Washington, D.C. February 2008.
[90]
[91]

APPENDIX A –
ANCILLARY SERVICES OVERVIEW

In broad terms, ancillary services are necessary services that must be provided in the generation and delivery of electricity. As defined by the Federal Energy Regulatory Commission (FERC), they include: coordination and scheduling services (load following, energy imbalance service, control of transmission congestion); automatic generation control (load frequency control and the economic dispatch of plants); contractual agreements (loss compensation service); and support of system integrity and security (reactive power, or spinning and operating reserves).

Introduction

The two primary functions of the electricity grid are 1) providing a supply of electric energy, primarily using generation that converts fuel to electricity in real-time and 2) delivering that energy to customers via the transmission and distribution (T&D) system. In addition to resources that provide the electric energy; additional resources – collectively known as ancillary services – support the overall operation of the grid. Ancillary services are defined by FERC as those services necessary to support the delivery of electricity from seller to purchaser while maintaining the integrity and reliability of the interconnected transmission system ('the network'). The specific definitions used by FERC for various ancillary services are listed in Table A-1.

To one extent or another, energy storage can provide many of those ancillary services. Storage used to provide some of the ancillary services may also be used for other applications, including power quality, reliability, and others.

Regulation *Versus* Load Following

Two ancillary services – regulation and load following – are somewhat similar; however, to understand implications for storage value propositions, it is important to distinguish between them:

> Together, regulation and load following address the temporal variations in load (and generation that does not accurately follow control signals). The key distinction between load following and regulation is the time period over which these fluctuations occur. Regulation responds to rapid load fluctuations (on the order of one minute) and load following responds to slower changes (on the order of five to thirty minutes). Load following is defined as the 30-minute rolling average of system load; regulation is then the difference between actual load for each 30-second interval and the rolling average. Hourly load following is defined as the difference between the highest and lowest values of the rolling average within the hour. Regulation is defined as the standard deviation of the 120 regulation values for the hour. Finally, the implications of the current block-scheduling conventions on load following and regulation are discussed, as is the need for a new scheduling convention.[A1]

Table A-1. Ancillary Services and Their Common Definitions

System Control	Scheduling generation and transactions ahead of time, and controlling some generation in real time to maintain generation/load balance.
Reactive Supply & Voltage Control	The generation or absorption of reactive power from generators to maintain transmission system voltages within required ranges.
Regulation	Minute-by-minute generation/load balance within a control area to meet NERC standards.
Spinning Reserve	Generation capacity that is online but unloaded and that can respond within 10 minutes to compensate for generation or transmission outages.
Supplemental Reserve	Generation capacity that may be offline or curtailable load that can respond within 10 minutes to compensate for generation or transmission outages.
Energy Imbalance	Correcting for mismatches between actual and scheduled transactions on an hourly basis.
Load Following	Meeting hour-to-hour and daily load variations.
Backup Supply	Generation available within an hour for backing up reserves or for commercial transactions.
Real Power Loss Replacement	Generation that compensates for losses in the T&D system.
Dynamic Scheduling	Real-time control to electronically transfer either a generator's output or a customer's load from one control area to another.
Black Start	Ability to energize part of a grid without outside assistance after a blackout occurs.
Network Stability	Real-time response to system disturbances to maintain system stability or security.

Please see Appendix D for more about storage for Load Following and Appendix E for more about storage for Area Regulation.

REFERENCE

[A1] Hirst, Eric. Kirby, Brendan. *Separating and Measuring the Regulation and Load Following Ancillary Services*. Oak Ridge National Laboratory. March 1999.

APPENDIX B – STORAGE REPLACEMENT COST ESTIMATION WORKSHEET

The worksheet shown below is an example of a simple methodology that can be used to estimate the cost incurred during battery operation due to battery wear (*i.e.,* damage). It spreads the damage-related cost across each unit of energy discharged to establish a value that represents the cost for battery replacement that is incurred per unit of energy output from the battery.

	Life	10																			
Annual Capacity Factor	0.07			Annual Operation Hours	613																
Discount Rate	10.0%																				
Annual Use Cycles	250			Operation Hours Per Use Cycle	2.45																

Standard Refurbishment

Use Cycles Per Refurbishment	1,000			Years per Replacement	4.00																
				Replacement Frequency	1.50																
Refurbishment Cost ($/kWh, $Year 1)	300			Total Refurbishment Cost ($/kW, $Year 1)	450																
Refurbishment Cost Escalation	2.5%			Annual Refurbishment Charge ($/kW, $Year 1)	45																

			$Year 1	$Current	$PW
Refurbishment Cost ($/kW)			450	504	321
(¢/kWh)			7.3	8.2	5.2

Caveats

1. Treats refurbishment like an expense, not investment...
2. ...but does not include tax deduction for the expense.
3. Includes fractional refurbishments if "Replacement Frequency" is not an integer.
4. This is a somewhat simplistic treatment of refurbishment cost annualization. It allocates all refurbishment costs (in the form of the annual average) across all years although it could be allocated in the years before/until the last refurbishment; though annual allocations in those years would be higher. Also, the cost escalation is applied to the annual average each year. It could be allocated only in years when allocation occurs.

	Total	Year => 1	2	3	4	5	6	7	8	9	10	11	12	13	14	15	16	17	18	19	20
Refurbishment Annual Cost ($/kW, $Year 1)	450	45	45	45	45	45	45	45	45	45	45	0	0	0	0	0	0	0	0	0	0
($Current)	504	45	46	47	48	50	51	52	53	55	56	0	0	0	0	0	0	0	0	0	0
($PW)	321	45	42	38	35	33	30	28	26	24	22	0	0	0	0	0	0	0	0	0	0
Refurbishment Unit Cost (¢/kWh)	7.3	7.3	7.3	7.3	7.3	7.3	7.3	7.3	7.3	7.3	7.3	0.0	0.0	0.0	0.0	0.0	0.0	0.0	0.0	0.0	0.0
($PW)	5.2	7.3	6.8	6.2	5.8	5.3	4.9	4.5	4.2	3.8	3.6	0.0	0.0	0.0	0.0	0.0	0.0	0.0	0.0	0.0	0.0

APPENDIX C – DISTRIBUTED ENERGY STORAGE FOR VOLTAGE SUPPORT AND REACTIVE POWER

Introduction to Reactance in AC Circuits

An important technical challenge for electric grid operators is managing the effects from a phenomenon called *reactance* in an alternating current (AC) electrical circuit.[7] Reactance is caused by elements within an AC circuit (*i.e.,* inductors and capacitors). The effects from reactance are related to an accumulation of electric or magnetic fields in the circuit elements when current is flowing. The electric and magnetic fields, in turn, produce an opposing electromotive force that is proportional to either the rate of change (time derivative) or accumulation (time integral) of the current.

Perhaps the most important manifestation of reactance in an AC circuit is that capacitors and inductors cause voltage and current to be 'out of phase' (*i.e.,* to not be synchronized). Specifically, rather than the ideal situation involving voltage and current which are synchronized, capacitors cause current to *lead* the voltage and inductors cause current to *lag* the voltage. Figure C-1 provides a graphical representation of the phenomenon.

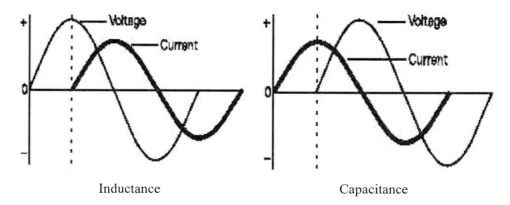

Figure C-1. Leading and lagging current due to inductance and capacitance (reactance) in an AC circuit.

In the left graph of Figure C-1, the two plots of voltage and current show capacitive reactance (current leads voltage). The two plots in the graph on the right show effects from inductive reactance (current lags voltage). The degree to which current leads or lags depends on the alternating current circuit's operating frequency (*e.g.,* electric grids operate at 50 or 60 Hz) and the capacitance and inductance in the circuit.

Importantly, to the extent that current leads or lags voltage, the effective voltage is reduced, in turn reducing the amount of usable power that can be delivered (*i.e.,* reactance reduces the effective load carrying capacity of the grid). Note that, normally, reactance in the electricity grid is dominated by reactance from *inductive* loads (causing current to lag the voltage), especially motors.

---

[7] AC power involves current flow (and voltage) that varies between a positive and a negative level. Electricity power systems use AC power that oscillates between negative and positive values 60 times per second (i.e., 60 Hertz AC). Among other advantages, AC power enables transmission over longer distances than systems using direct current (DC) power (power that has a constant current and a constant voltage).

Power Factor

The power factor of an AC electric system is defined as the ratio of *real power* to *apparent power*.

Real power (also known as 'true power') can be defined as the amount of usable power that can be delivered to loads in an AC circuit. More specifically, real power indicates the amount of work that can be accomplished over a given amount of time based on the rate at which the circuit can deliver electric energy. Real power could also be thought of as the 'effective' power or the useable power. The most common units used to express real power are watts (W) or kilowatts (kW).

Apparent power is simply the product of the voltage and current within a circuit, irrespective of whether voltage and current are synchronized and how much work can be accomplished using the electric energy that the circuit can deliver to loads. The most common unit of apparent power is volt-Ampere (volt-Amp). Note that most power equipment – such as power supplies, wires and transformers – are rated based on their apparent power (volt-Amps).

In any given circuit, the apparent power can be somewhat or significantly greater than the real power because 1) during each alternating current cycle, energy is stored within loads and then returned to the circuit; and/or 2) 'non-linear' loads distort the current's (sine) wave form within the circuit. Common non-linear loads include most electronic equipment, which have non-linear power supplies, and electronic ballasts used for lighting.

Of particular interest are effects from reactive loads that lead to the presence of reactive power in the circuit. Units of reactive power are volt-Amps reactive (VAR). VAR reduces real power because the associated reactance changes the temporal relationship between voltage and current in the AC circuit as described above. (Note that apparent power is the combination of real power and reactive power.)

The concept of power factor is important in part because – to the extent that the real (useable) power is less than apparent power – the amount of power that can be delivered to loads by a circuit with power factor that is less than one (unity) circuit is reduced. Consider the example of a circuit rated to deliver 10 MVA (apparent power) with a power factor of 0.9. That circuit could serve

0.9 × 10 MVA = 9 MW of load.

One implication is that a larger circuit (capacity) is needed to deliver a given amount of useful energy. Because more current flows within the circuit (for a given amount of energy delivered), there are more I^2R energy losses within the circuit.

(For more detail about true, reactive, and apparent power, readers could refer to the *All About Circuits* website: http://www.allaboutcircuits.com/.)

Utility Responses, Overview

Utilities use two important means to compensate for the presence of reactance (*i.e.,* to restore voltage to and/or to maintain voltage at the desired level). Generic terms for managing effects from VAR are 'VAR support' and 'VAR compensation'.

One such technique – involving an ancillary service known as 'voltage support' – is to produce reactive power (power that has lagging or leading current). The reactive power is meant to cancel out the effects of reactance in the power system.

Another more localized approach – called 'power factor correction' – involves using equipment within the T&D system to offset effects from localized reactance. In most cases, power factor correction involves use of power factor correcting capacitors that offset effects from localized inductance.

Distributed Storage for Voltage Support

The balance of this appendix is section is based largely on the research of scientists at Oak Ridge National Laboratory (ORNL). Their objective was to evaluate the potential for distributed generation as a resource for VAR compensation. In most cases, storage systems can or could be designed to provide the same service. The ORNL work tested the hypothesis that "[distributed generation] can play a larger and more significant role than at present in relieving voltage stability problems due to both a) suboptimal dispatch of reactive power supplies and b) reactive power supply shortages."[C1]

Reactive power for voltage compensation is compelling for several reasons. Among the reasons given by authors of the ORNL report, "past power blackouts have been attributed to problems with reactive power transport to load centers."[C2] Although reactive power for voltage compensation is a relatively small *portion* of total cost to generate and transmit electricity, it does account for billions of dollars in *total* cost. Another compelling reason is that most central generation technologies, especially newer ones, are not well-suited to reactive power generation because generation is usually optimized for real (*i.e.,* true) power generation at a constant output.

Importantly, unlike real power, reactive power cannot be transmitted over long distances. Consequently, central generation may not be the best source of reactive power. Conversely, a growing array of smaller, modular power technologies (*e.g.,* any type of power system with an inverter that has VAR support capability, distributed generation, energy storage, and possibly even demand response) could provide other sources of VAR support, and provide such support closer to the loads that pose the biggest challenges.

Voltage Support using Reactive Power

In simple terms, voltage control for an AC power system is accomplished primarily by managing reactive power. This is done by injecting and/or absorbing reactive power, when needed, as close as possible to the location where reactance is a problem. The amount of reactive power needed normally varies as a function of the transmission line loading. Heavily loaded lines require more reactive power than lightly loaded lines. As reactive power needs in the transmission system vary, the Independent System Operator (ISO) and regional transmission organizations (RTOs) adjust the supply of reactive power.

The Federal Energy Regulatory Commission (FERC) separates voltage control into two categories: generation-based and transmission-based.

Generation-based voltage control is an ancillary service, and transmission-based voltage control is included as an element of transmission service agreements or tariffs. Generation-based VAR support is needed to operate regional power systems and electricity markets. (Other common ancillary services include spinning reserve, contingency, emergency, or supplemental reserve, and regulation.) According to authors of the ORNL report, "It is variously estimated that providing this bundle of ancillary services costs the equivalent of 10-20% of the delivered cost of electric energy."[C3][C4]

The process of managing reactive power in transmission systems is well understood technically. The three primary objectives of reactive power management are as follows: 1) maintain adequate voltages throughout the transmission system under normal and contingency conditions, 2) minimize congestion that affects flow of real power, and 3) minimize real-power losses.

Voltage control is usually centralized because coordinated control is needed among the various entities and equipment in the electric grid to ensure effective operation of the system (*i.e.,* to keep voltage levels within necessary parameters). System operators and planners use sophisticated computer models to design and operate the power system reliably and economically. These functions are not readily distributed to individual sub-regions or to separate market participants.

An important responsibility of power system planners is to address what is generically called 'grid security'. It involves planning whose goal is to ensure adequate operation of the power system (generation and transmission) during a range of conditions and contingencies. It involves, in part, modeling the electric grid system under a broad range of conditions to ensure that the electric grid has adequate reserves when transmission lines or generators fail, as well as during peak demand periods. (Normally, power systems maintain sufficient reserves to serve load should a major generation plant or transmission line fail, commonly called an N-1 contingency).

Reactive power resource technologies differ significantly with respect to the amount of reactive power that can be produced under given conditions, response speed, and capital cost. Reactive power sources can be categorized as either *static* or *dynamic*.

Common static reactive power sources include transmission and distribution (T&D) equipment such as substation capacitors. Notably, these T&D-based options are considered to be part of the utility's capital investment portfolio (of infrastructure equipment). The equipment cost is added to the utility 'revenue requirement' – the amount of revenue required, from users, to cover all costs.

Dynamic reactive power sources include generation facilities, which are capable of producing both real *and* reactive power, and synchronous condensers, which produce *only* reactive power. Generation equipment may be owned either by utilities or independent entities. Often, reactive power is bought and sold so that the cost is covered by market-based or market-like prices.

Providing Reactive Power Locally

A key difference between VAR support (or reactive power supply) and other ancillary services is that reactive power cannot be transmitted over long distances. Reactive power needs occur in direct proportion to the distribution of load across a system and the proximity between generators and load centers.

Reactive power from distributed energy resources (DER), including distributed generation and distributed energy storage, could provide distributed dynamic voltage control in response to variations of reactive power needs within distribution systems. To serve as a reactive power resource, the DER must be able to inject and absorb reactive power. Conversely, distributed generation and distributed energy storage that do not have the ability to generate or absorb reactive power can *degrade* voltage. Notably, many DER are connected to loads and/or to the grid via equipment that incorporates solid-state power electronics that may be designed to provide reactive power compensation.

The implications and possibilities for reactive power compensation using DER capacity are not well understood. Nevertheless, reactive power is currently provided, in part, by what could be called modular/distributed sources (*e.g.*, static VAR compensators and capacitor banks). So, intuitively, it seems likely that there are exploitable synergies between the localized need for reactive power (usually near loads) and increasing emphasis on DER. Perhaps more importantly, aggregated DER capacity (if dispatched in a coordinated way) could be part of a robust approach to region-wide grid stability during major power interruptions involving declining area-wide or system-wide voltage.

As previously noted, reactive power needed to stabilize voltage cannot be transmitted very far. So, in general, *local* sources of VAR support are most helpful, especially if interruptions involve transmission corridors. Additionally, many DER types can respond rapidly to reduce the chances of a total loss of power.

Storage may be best suited to this application if rapid response is important. Some storage types reach their full discharge rate within seconds to just a few milliseconds, these include capacitors, flywheels, and superconducting magnetic energy storage. (Note that, although conventional capacitors are good for managing reactance under normal operating conditions, they do not perform well as a voltage support resource because they draw more current as voltage drops, possibly adding to cascading overloads.) In contrast, most types of generation take a few to many minutes to respond fully (*e.g.*, pumped hydroelectric and compressed air energy storage).

Aggregated modular storage deployed at or near loads, for reasons other than voltage support, could provide very helpful voltage support when and where needed. Finally, by picking up or turning off specific types of load when grid anomalies occur, DER can reduce voltage degradation, thereby reducing the possibility of cascading outages.

The most challenging loads during such an event include small motors, especially those used in smaller air conditioning equipment to operate the compressor. Figure C-2 shows that, in California, such loads account for a significant portion of peak demand. Those motors pose such a significant challenge because as grid voltage drops during local or region-wide grid emergencies, the motors draw more current to maintain power which exacerbates the voltage problem. The same motors can also pose a relatively significant challenge as the grid is re-energized after outages.

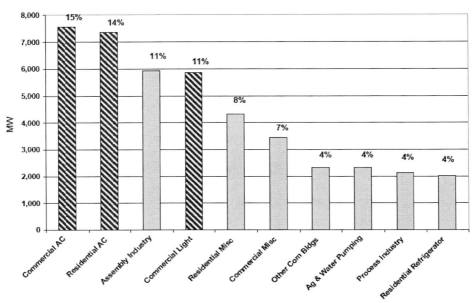

Source: California Energy Commission.[C5]

Figure C-2. Peak demand (in MW) by end use in California.

REFERENCES

[C1] Li, F. Fran. Kueck, John. Rizy, Tom. King, Tom. *Evaluation of Distributed Energy Resources for Reactive Power Supply, First Quarterly Report for Fiscal Year 2006.* Prepared for the U.S. Department of Energy by Oak Ridge National Laboratory and Energetics Incorporated. November 2005.

[C2] *ibid.* [C1].

[C3] *ibid.* [C1].

[C4] Kirby, Brendan. Hirst, Eric. *Ancillary Service Details: Voltage Control.* Oak Ridge National Laboratory, Energy Division. Sponsored by The National Regulatory Research Institute. Oak Ridge National Laboratory Report #ORNL/CON-453. December 1997.

[C5] *California's Electricity Situation: Summer 2005.* Presentation prepared by the staff of the California Energy Commission, California Public Utilities Commission, and California Independent System Operator. February 22, 2005

APPENDIX D – STORAGE FOR LOAD FOLLOWING

Storage can provide load following *up* by increasing the rate of discharge and/or decreasing the rate of charging, as described below.

Consider the example depicted in Figure D-1 which shows how charged storage with one hour of discharge duration can provide two hours of load following up by *discharging*.

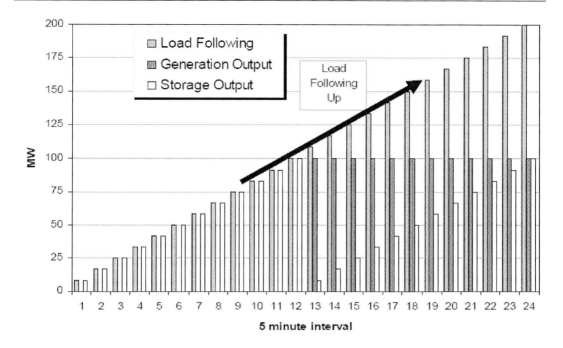

Figure D-1. Two hours of load following up with one hour of storage discharge.

In Figure D-1, the time-specific aggregated load following capacity needed is indicated by the blue bars labeled Load Following. The rate of storage *discharge* increases as load increases (shown by the yellow bars labeled Storage Output). After the first hour of load following with storage, a full 100-MW block of generation is dispatched (shown by the red bars) while storage discharge is curtailed (at interval #13). Throughout the second hour of load following, the storage output is increased every five minutes (as it was during the first hour) as load increases. At the beginning of the next hour (not shown), another 100-MW block of generation is dispatched and storage output is halted.

Storage *charging* can also be used to provide load following up by reducing the rate of charging throughout an hour, commensurate with increasing load. Consider the example shown inFigure D-2. At the beginning of the first hour of load following, a 100-MW generator is dispatched to full output (see the red bars labeled Generation Output). At the same time, storage begins charging at a rate equal to the 100-MW rating of the generator that was just dispatched. Every five minutes, the rate of storage charging is reduced to the extent that load has increased (note the yellow bars labeled Storage Charging). The resulting load following up is shown by the blue bars. At the beginning of the second hour of load following, the second 100 MW of generation is dispatched (at full output), and storage charging commences again at a rate (100 MW) equal to the output of the second generator. Finally, at the beginning of the next hour (not shown), more generation is dispatched (ideally, at full output) as storage operation (in this case, charging) ceases.

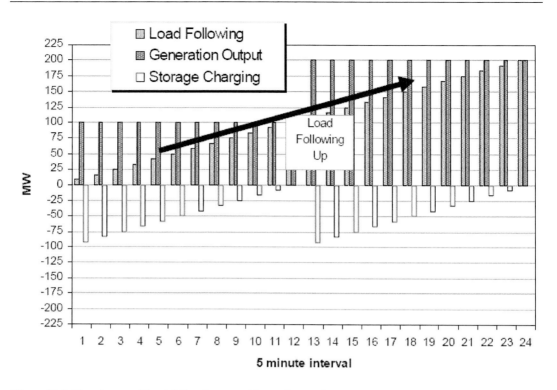

Figure D-2. Two hours of load following up with one hour of storage charging.

Storage provides load following *down* by decreasing the rate of discharge and/or by increasing the rate of charging, as described below.

For load following down involving decreasing storage discharge, the storage is cycled from full output to very low (or no) output twice in a two-hour period, providing two service hours of load following down as shown in Figure D-3. In that figure, at the end of the previous hour (not shown), a 100-MW generator is taken offline as 100 MW of storage comes online (as shown by the yellow bars labeled Storage Discharge). Another 100 MW of generation is still online (shown by the red bars labeled Generation Output). The rate of storage discharge is reduced every five minutes during the first hour as load drops. The resulting load following capacity is shown by the blue bars labeled Load Following. At the beginning of the next hour, the 100-MW generator is taken offline and the storage begins discharging again at 100 MW. Storage discharging decreases throughout the second hour as load decreases until discharging ceases at the end of the second hour.

Figure D-4 shows how storage can be used to provide load following down while charging. The example shown in Figure D-4 involves storage with one hour of discharge duration that is used to provide two hours of load following down.

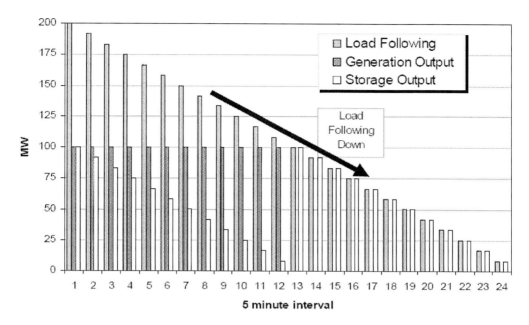

Figure D-3. Two hours of load following down with one hour of storage discharge.

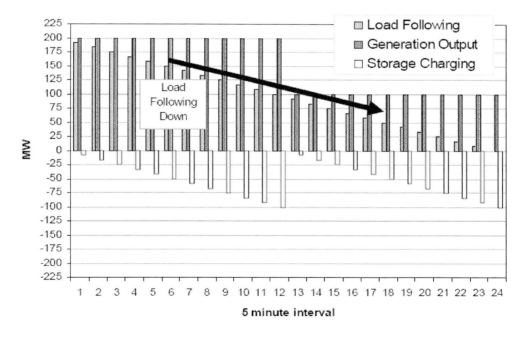

Figure D-4. Two hours of load following down with one hour of storage charging.

At the beginning of the hour, two 100-MW generators are on line for a total of 200 MW (shown by the red bars labeled Generation Output). As load decreases, there is a commensurate increase of storage charging (shown by the yellow bars labeled Storage Charging). The resulting load following capacity is shown by the blue bars labeled Load

Following. At the beginning of the second hour, 100 MW of generation is taken offline, and storage charging begins again at low power. As load continues to diminish, storage charging is increased until the beginning of the next hour (not shown) when storage charging and generator operation both cease.

Energy Associated with Load Following

When using storage *charging* for load following, the energy stored must be purchased at the prevailing wholesale price. This is an important consideration – especially for storage with lower efficiency and/or if the energy used for charging is relatively expensive – because the cost of energy used to charge storage (to provide load following) may exceed the value of the load following service.

Conversely, the value of energy *discharged* from storage to provide load following is determined by the prevailing price for wholesale energy. Depending on circumstances (*i.e.,* if the price for the load following service does not include the value of the wholesale energy involved), when discharging for load following, two benefits accrue – one for the load following service and another for the energy.

APPENDIX E – AREA REGULATION

Introduction

This appendix documents a high-level analysis of the benefit from and cost for flywheel energy storage used to provide area regulation for the electricity supply and transmission system in California. The analysis is based on results from a demonstration, in California, of flywheel energy storage developed and manufactured by Beacon Power Corporation. Demonstrated was flywheel storage systems' ability to provide rapid-response regulation. (Flywheel storage output can be varied much more rapidly than the output from conventional regulation sources, making flywheels more attractive than conventional regulation resources.)

The work was sponsored by the U.S. Department of Energy (DOE) and the Sandia National Laboratories (SNL) Energy Storage Systems Program. The demonstration was supported by the California Energy Commission (CEC) Public Interest Energy Research Program. It was located at the Distributed Utility Integration Testing facility managed by Distributed Utility Associates (DUA) and located at the Pacific Gas and Electric Company (PG&E) Technological and Ecological Services research facility in San Ramon, California.

Although the specific type of storage evaluated was flywheel storage, other types of storage that can respond rapidly when conditions change can also provide the area regulation service. Those may include some types of electrochemical batteries and capacitors. And though they respond more slowly, CAES and pumped hydroelectric storage can also be used to provide area regulation.

Another desirable storage characteristic is high efficiency, because when storage *charging* occurs during regulation, any energy that is lost must be purchased at the prevailing price.

Regulation Service

Regulation is a type of ancillary service[8] that involves managing the "interchange flows with other control areas to match closely the scheduled interchange flows" and moment-to-moment variations in demand within the control area. The primary reasons for including regulation in the power system are to maintain the grid frequency and to comply with the North American Electric Reliability Council's (NERC) Control Performance Standards 1 and 2 (NERC 1999a). Regulation also assists in recovery from disturbances, as measured by compliance with NERC's Disturbance Control Standard.[E1]

When there is a momentary shortfall of electric supply capacity, the output from regulation resources is increased to provide *up* regulation when there is a momentary shortfall of power on the grid. Conversely, regulation resources' output is reduced to provide *down* regulation when there is a momentary excess of electric supply power.

Traditionally, regulation has been provided by dispatchable thermal generation facilities. They provide up regulation by increasing output when electricity demand exceeds supply, and they provide down regulation by reducing output when electricity supply exceeds demand. Generation facilities used for up regulation and those used for down regulation are operated at levels below the facilities' maximum output and above minimum output, respectively.[E2] Generation units used for regulation must be equipped with automatic generation control (AGC) equipment and be able to change output relatively quickly (MW/minute) over an agreed upon range of power output (MW).

Flywheels for Area Regulation

Flywheel electric energy storage systems (flywheel storage or flywheels) consist of a cylinder with a shaft that can spin rapidly within a robust enclosure. A magnet levitates the cylinder to limit friction-related losses and wear. The shaft is connected to a motor/generator and stator. Kinetic energy is converted to electric power via an external power conditioning unit (PCU). High-speed flywheel electricity storage is nearing commercialization. One apparently superior application of the technology is for electric power system regulation (also known as area regulation or simply regulation). Storage provides up regulation by discharging energy into the grid and down regulation by absorbing energy from the grid.

Notably, the rate of power from (or into) flywheel storage can change quite rapidly whereas output from conventional regulation sources (primarily thermal generation plants) changes slowly. Generation plants' output (up or down) changes by percentage points per minute whereas flywheels' output can change from full *output* (discharge) to full *input* (charging) and *vice versa* within a few seconds. Additionally, thermal power plants generally are most efficient when operated at a specific and constant (power) output level. Similarly, air emissions and plant wear and tear are usually lowest when thermal generation operates at

---

[8] Ancillary services are electric resources that are used to maintain reliable and effective operation of electric supply and transmission systems. Most often, ancillary services are provided by utilities, although an increasing portion is being provided by third parties. Six key ancillary services are 1) scheduling, system control and dispatch, 2) reactive supply and voltage control from generation sources, 3) regulation and frequency response, 4) energy imbalance, 5) spinning reserve, and 6) supplemental reserve.

constant output. Unlike thermal power plants, flywheels' performance is not affected much as output varies, and the systems are virtually emissions free.

Demonstration Plant

Results described below are for a 100-kW pilot version of a Beacon Power high-speed flywheel storage system. The pilot system consisted of seven individual flywheels, a PCU, and communication and control subsystems. It can discharge at full output for 15 minutes. The response time is described by Beacon Power to be "less than 4 seconds (at full power)." The demonstration was conducted at Distributed Utility Associates' Distributed Utility Integration Test testbed located at PG&E's Technical and Ecological Services facility in San Ramon, California. Recently, Beacon has developed a 20-MW Smart Energy MatrixTM version of the flywheel system for commercial use.

Benefits

At minimum, regulation from flywheels is at least as valuable as regulation provided by slower generation capacity. Regulation from flywheels, however, may prove even more valuable. First, flywheel storage can provide both up regulation and down regulation during the same time period (although not simultaneously). Also, because of their rapid-response (*i.e.,* their ability to change power input and output rapidly), flywheels may provide regulation that is more effective than that provided by much slower generation-based resources. Because of this advantage, regulation from flywheels is assumed to provide twice the benefit to the grid as regulation from generation.[E3][E4][E5]

Revenue for providing up and down regulation services for an entire year (8,760 hours) is estimated based on California Independent System Operator (CAISO) published hourly prices for both services for the year 2006. (See the subsection 'Price for Regulation Service' in this appendix for details.) The hourly prices are multiplied by two (to reflect the higher benefit from flywheels relative to generation-based regulation) before annual revenues are estimated.

In addition to the price for regulation in specific hours of the year, another important criterion affecting the flywheel-for-regulation value proposition is flywheel plant availability. The amount of time that the flywheel is available to provide regulation affects the total profit that can be realized during the year. Because flywheel storage is modular, equipment diversity should result in high reliability. For example, a Beacon's 20-MW, commercial-scale plant is expected to comprise a few hundred flywheels.

Although not included in the financial analysis, additional benefits derived from the use of flywheels for regulation may include a reduced need for generation capacity, reduced fuel use for generation, reduced air emissions from generation, and reduced generation equipment wear-andtear.

As an indication of the prospects for reducing air emissions, consider results from a study performed by KEMA, Inc (kema.com), shown in Table E-1. Based on study results, flywheels used for regulation in California could reduce CO_2 emissions by 26% when compared to pumped hydroelectric storage, 53% if the flywheels replace baseload gas-fired generation and

59% if a natural gas-fired peaking generator is displaced. Similarly, (NO$_x$) emissions may also reduced by 20% to nearly 50%.[E6]

Table E-1. Air Emissions Reduction Potential

Flywheel Emission Savings Over 20-year Life: CA-ISO

	Coal Baseload	Coal Peaker	Natural Gas Baseload	Natural Gas Peaker	Pumped Hydro
CO2					
Flywheel	91,079	91,079	91,079	91,079	91,079
Alternate Gen.	322,009	608,354	194,534	223,997	123,577
Savings (Flywheel)	230,930	517,274	103,455	132,917	32,498
Percent Savings	72%	85%	53%	59%	26%
SO2					
Flywheel	63	63	63	63	63
Alternate Gen.	1,103	2,803	0	0	85
Savings (Flywheel)	1,041	2,741	-63	-63	23
Percent Savings	94%	98%	n/a	n/a	27%
NOx					
Flywheel	64	64	64	64	64
Alternate Gen.	499	1,269	80	118	87
Savings (Flywheel)	435	1,205	16	54	23
Percent Savings	87%	95%	20%	46%	26%

Flywheel Energy Storage Cost and Performance

The values shown in Table E-2 are flywheel storage system cost and performance assumptions plus the price for make-up energy (energy required to make up for storage losses). The cost and performance values for flywheels reflect expected values for a 20-MW commercial-scale plant. Installed cost reflects a 20% uncertainty adder. This value is used to account for the normal uncertainty associated with technology scale-up and commercial project development (*e.g.,* siting, contracts, construction delays, *etc.*).

Table E-2. Flywheel Storage Cost and Performance Assumptions

Criterion	Value
Commercial Plant Scale (MW)	20
Plant Installed Cost ($/kW)	1,566
Plant Availability	0.95
Roundtrip Efficiency	81%
Variable Operartion Cost ($/MWh$_{out}$)	3.14
Fixed Operation Cost ($/kW, Year 1)	11.60
Makeup Energy Price ($/MWh)	40

Price for Regulation Service

The key data used for estimating the regulation benefit is the hourly price for up and down regulation services. The price is denominated in $/MW per hour of service. There are two prices for the hour: up regulation and down regulation. Hourly prices for up and down regulation in California in 2006 are shown in Figure E-1 and , respectively. Annual average prices used for the valuation are $21.48/MW and $15.33/MW per service hour for up and for down regulation, respectively, for a total of $36.70/MW per service hour.

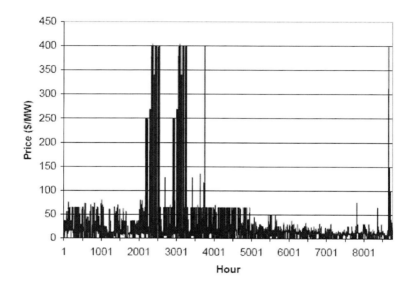

Figure E-1. Up regulation prices in California, 2006.

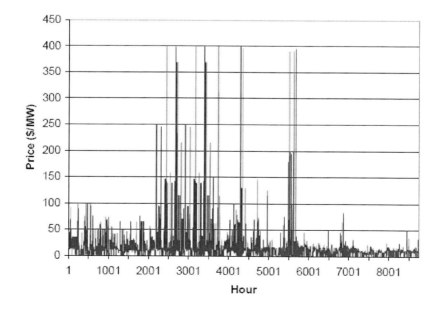

Figure E-2. Down regulation prices in California, 2006.

Value of Regulation from Flywheels

As described elsewhere in this appendix, it is assumed that flywheels used for regulation provide twice as much benefit (to the grid) as generation-based regulation. Specifically, it is assumed that regulation resources are twice as valuable if they follow the area control error (ACE) signal closely. That signal changes every several seconds to reflect the momentary difference between the amount of power that is online and the amount needed to keep supply and demand balanced and to maintain the electrical stability of the grid (especially the 60-Hz AC frequency). Based on this assumption, flywheel storage used as a regulation resource is treated as if it is eligible for payments that are twice as much as the prices shown above for conventional, generation-based regulation.

Market Potential

In addition to financials, the CEC's Public Interest Energy Research Program is interested in the market potential (in MW) for the flywheels-for-regulation value proposition. Unfortunately, the authors of this guide do not have the resources needed to establish that value rigorously or credibly. Nonetheless, the authors speculate that a conservative estimate of the market potential in California could be on the order of 50 to 60 MW of the total regulation market managed by the CAISO over the next 10 years. (The CAISO does not manage all of the regulation resources within the state. Some of that capacity could be in play as well.) This speculation has two primary bases. The first is a very cursory review of regulation capacity requirements available at the CAISO Open Access Same-time Information System website (http://oasis.caiso.com/, under the ancillary services tab). The second is a discussion with representatives from Beacon Power.[E7]

Financial Assumptions

The financial analysis used to calculate lifecycle cost and benefits include a 2.5% annual price escalation and a 10% discount rate. The annual plant carrying cost is calculated by applying an annualization factor (*i.e.*, a fixed charge rate) of 0.20 (*e.g.*, annual financial carrying charges for a $1 million plant = $200,000/year).

Results

Demonstration plant availability for three plant output levels (relative to full rating) is summarized in Table E-3. Also shown is the availability assumed for a commercial plant. As shown in the table, the demonstration unit operated 51.4% of the time at full capacity (full capacity means that all seven flywheels were operating). Similarly, the demonstration unit operated nearly 53% of the time at 85.7% of rated capacity (85.7% capacity represents six flywheels of seven). There were at least five of seven flywheels (71.4% of full rated capacity) operating almost 88% of the time.

Also shown is that the demonstration plant's availability would be somewhat higher when accounting for research-related outages. Research-related outages include downtime due to causes that would only affect operation of a research or pilot project (*e.g.,* no control signal was available, access to the demonstration facility was restricted, or the system could not be connected to the grid). Downtime to due equipment failure is *not* considered a research-related outage.

**Table E-3. Demonstration Plant Actual Availability
and Commercial Plant Expected Availability**

Capacity (% of full)	Availability (Actual)	Without "Research-related" Outages	Commercial Plant (expected)
100%	47.3%	51.4%	95.0%
85.7%	52.7%	56.9%	
71.4%	87.8%	92.0%	

The financial implications of plant availability are summarized inFigure E-3. In the figure, the left axis shows $/kW in Year 1. The axis on the right indicates the corresponding lifecycle value, over the 10-year life assumed for the plant. Results are shown for three levels of annual average power output: 71%, 86%, and 100% of plant rating (note that these values correspond to those shown in Table E-3, rounded to the nearest full percentage point). An output of 71% represents 5 of 7 flywheels in the demonstration system, 86% represents 6 of 7 flywheels, and 100% represents 7 of 7 flywheels. Results are presented, for each of those three plant output levels, for a range of plant annual availability levels. Also shown is the break-even amount, reflecting the carrying cost for a commercial plant.

The uppermost plot indicates results for plants operating at full rating. The next two plots indicate financials for a plant operating at 86% and 71% of its rating, respectively. Thicker parts (to the lower left) of the three plots reflect results from the demonstration. Endpoints on all three plots indicate financials for a plant operating at the respective portion of rated output, if the plant operates as much as a commercial plant is expected to operate (*i.e.,* 95% of the year, full-load equivalent). The box in the upper right indicates financials that would be expected for a commercial plant, based on assumptions provided in Section 3 of this guide. The financial benefit/cost ratio for such a plant ranges from $500/kW benefits . $313/kW breakeven = 1.6 up to $554/kW benefits . $313/kW breakeven = 1.77.

Note that plant designers expect a 20-year service life for a 20-MW, commercial-scale plant, although the assumed service life for this report is 10 years. To account for the difference, the present worth of additional benefits increases by about 50%.

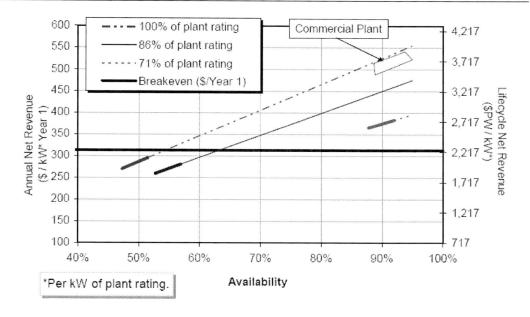

Figure E-3. First-year and lifecycle net revenue, with breakeven indicator.

Methodology Observations and Caveats

- The make-up energy price assumed was not developed rigorously. Although this value is adequate for this analysis, it should be established using a more rigorous approach when evaluating the financials for an actual project.
- Based on results from the demonstration project, flywheel systems with 15 minutes of storage can store enough energy to provide regulation during 97.5% of the time that the storage is used. For the purpose of this evaluation, the financial implications of that criterion are assumed to be modest and are ignored.
- The project was a demonstration of the flywheel's ability to respond to rapidly changing control signals without regard to the magnitude of the response (in MW) that might be needed. Consequently, the results reflect the value for regulation capacity on the margin.
- The market potential estimate used for this evaluation, although adequate for a high-level estimate of the magnitude of statewide economic impact, is imprecise. Unfortunately, little is known about the effect significant penetration of rapid-response regulation capacity will have on the need for regulation and on the price for regulation.
- The premise about how much more valuable flywheels are than generation-based regulation resources, as meritorious as it may be, may not be reflected in regulation pricing without a significant amount of confirmation, regulatory accommodation, and time.
- The 0.20 annualization factor used to estimate the annual carrying cost for the plant, though perhaps imprecise, does provide a reasonable general indication of the cost to finance the plant and equipment using non-utility capital.

- Another important assumption affecting these results is the 20% uncertainly adder (provided by Beacon Power) that increases the assumed installed cost for a commercial plant. That value is used to account for the myriad unforeseen challenges that are likely to beset *any* technology development enterprise and project development effort.
- The design service life for a commercial Beacon Power flywheel plant is 20 years; however, the assumed service life for the evaluation described in this report is 10 years. The reason for this is twofold. First, guidelines established by the CEC's Public Interest Energy Research Program for evaluating the merits of various storage demonstrations require the use of standard assumptions as bases for comparing financials for all demonstration projects sponsored. Those standard assumptions include a 10-year life, a 10% discount rate, and a 2.5 % price escalation rate. Second, while the authors do not refute the 20-year expected life assumed by Beacon Power, a more conservative 10-year life expectancy was used because both the technology and the value proposition are so new.

Conclusions

Perhaps the most important result from the Beacon flywheel demonstration is that the sponsors and vendors successfully demonstrated the ability of the flywheel to follow control signals that change very rapidly, much more rapidly than the signal used to control the output of generation-based regulation. The results indicated that the characteristics of high-speed flywheel storage are generally consistent with a possible new class of regulation resources – rapid-response energy storage-based regulation – in California. In short, it was demonstrated that high-speed flywheel storage systems are capable of following a rapidly changing (every 4 seconds) control signal (the ACE).

Based on these results and on the expected plant cost and performance, high-speed flywheel storage systems have a good chance of being a financially viable regulation resource. The results indicated a benefit/cost ratio of 1.6 to 1.8 using somewhat conservative assumptions. The results also indicated that flywheel systems with 15 minutes of storage can store enough energy to provide regulation during 97.5% of the time that the storage is used.

The market potential (in MW) is less certain. Uncertainty about technical market potential is driven in part by a lack of knowledge regarding how the use of rapid-response regulation resources on the margin will affect overall demand and prices for regulation. Regarding market share, there is always uncertainty regarding competing options (*e.g.,* other vendors/developers and other technologies or approaches).

R&D Needs and Opportunities

One compelling question for this value proposition is–*How much of this resource could be used and how much will be used?* Consistent with the hypothesis that rapid-response storage is twice as valuable as generation-based regulation capacity, another hypothesis to test is that only half as much regulation is needed if all regulation is rapid-response. Increased penetration of rapid-response regulation also means that *generation* capacity is freed to provide power or other more valuable ancillary services and less pollution will be produced and less fuel will be used per MWh delivered. Another way to broach the question is–*What are the key implications for the grid if all regulation is provided entirely by rapid-response regulation?* Those implications include impacts on: the amount of regulation needed, the total cost to ratepayers for regulation, fuel use, and air emissions from generation.

References

[E1] Hirst, Eric. Kirby, Brendan. What is the Correct Time-Averaging Period for the Regulation Ancillary Service? Oak Ridge National Laboratory. April 2000. Available at: http://www.ornl.gov/sci/btc/apps/Restructuring/regtime.pdf.

[E2] ibid. [E1].

[E3] David Hawkins, California ISO; Mike Gravely, California Energy Commission; Bill Capp and Chet Lyons, Beacon Power. Discussion with Jim Eyer, Distributed Utility Associates at the California ISO offices. Folsom, California. April 12, 2007.

[E4] California Energy Commission Press Release. California Energy Commission Applauds Beacon Power Upon Reaching Research Goal. January 10, 2007.

[E5] Makarov, Dr. Yuri. Pacific Northwest National Laboratory in conjunction with the CAISO. Relative Regulation Capacity Value of the Flywheel Energy Storage Resource. The research was initially commissioned and funded by the CAISO. While the work was in progress, Dr. Makarov left the ISO and later published the report in November 2005.

[E6] Enslin, Johan, Ph.D. Fioravanti, Richard. Emissions Comparison for a 20-MW Flywheel-based Frequency Regulation Power Plant. A study performed by KEMA, Inc. under a contract funded by the U.S. DOE via Sandia National Laboratories. Sandia National Laboratories Contract #611589 and RFQ #9058. October 2006.

[E7] Chet Lyons and Jim Arseneaux, Beacon Power. Discussion with Jim Eyer, Distributed Utility Associates. Mr. Lyons and Mr. Arseneaux indicated that discussions with representatives of various ISOs leads Beacon to assume that market penetration levels of 20% to 25% would have little to modest impact on both the need for regulation and the price paid for regulation. Beacon Power contends that that level is conservative. June 6, 2007.

APPENDIX F – ENERGY PRICES

This appendix serves two interrelated objectives: 1) provide generic electric energy costs based on a range of fuel conversion efficiencies and fuel costs and 2) provide details about projected wholesale energy prices in California. The California-specific data and figures are based on a California Energy Commission (CEC) forecast for spot electric energy prices in 2009.[F1]

Generic Electric Energy Cost

Figure F-1 and Figure F-2 show generic values for the two key components of unit energy cost: fuel and plant capital cost. Figure F-1 illustrates how fuel price and fuel conversion efficiency affect electricity price. The three plots in the figure represent three conversion efficiency values: 35%, 45%, and 55%.

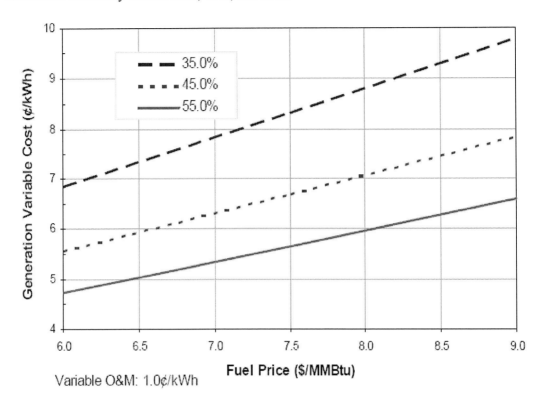

Figure F-1. Generic effect of conversion efficiency and fuel price on electricity price.

Figure F-2 shows how plant capital cost affects the price for electricity. The three plots in this figure represent three generation installed cost values: $400/kW, $1,000/kW, and $1,600/kW. These cost values reflect a generic fixed charge rate of 0.11. To adjust values to reflect a different fixed charge rate, multiply the cost values by the ratio of the actual fixed charge rate by the generic value of 0.11. For example, if the fixed charge rate is 0.13, then multiply the values in Figure F-2 by

$0.13 \div 0.11 = 1.19$.

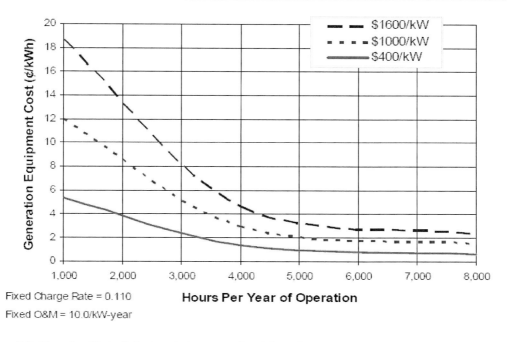

Figure F-2. Generic effect of plant capital cost on electricity price.

California Electric Energy Cost Projection

Figure F-3 shows prices in chronological order, while Figure F-4 shows hourly electric energy prices arranged in order of magnitude. In Figure F-4, two plots are shown: one is the actual price and the other is the running average value. The same data, with emphasis on the hours of the years with the highest 10% prices, are shown in Figure F-5.

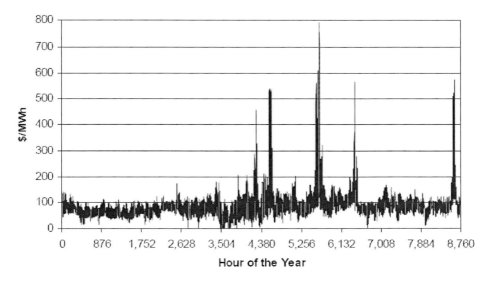

Figure F-3. Electric energy spot prices for California (2009 forecast).

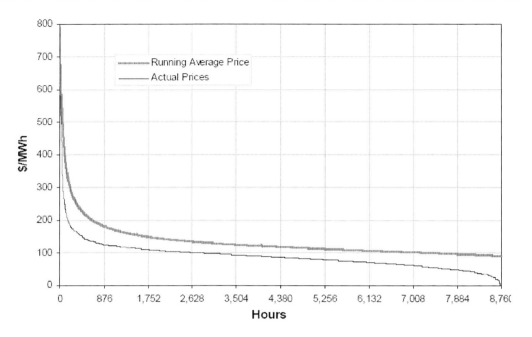

Figure F-4. Price duration curve for California (2009 forecast).

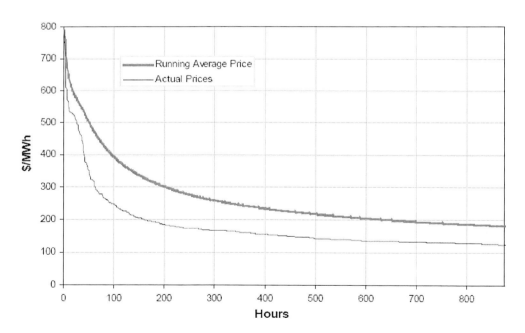

Figure F-5. Price duration curve for California (2009 forecast) 10% highest price hours.

Hourly average prices for each hour of the day for each month are listed in Table F-1. Data in Table F-2 show the net benefit for energy time-shift based on the prices in Table F-1.

Table F-1. Monthly Hourly Average Prices for California 2009 Forecast ($/MWh)

Hour	1	2	3	4	5	6	7	8	9	10	11	12
1	56.4	45.9	50.6	60.1	45.4	41.1	63.9	75.3	82.5	74.7	65.3	76.5
2	49.2	43.5	44.6	57.3	38.1	35.9	56.1	64.8	75.1	61.7	60.4	71.1
3	45.9	41.4	41.5	56.9	35.6	33.6	53.3	63.8	70.5	54.0	56.0	66.7
4	45.9	41.0	41.3	56.7	36.7	31.5	53.0	64.8	73.5	51.3	55.1	64.8
5	51.3	44.3	46.8	62.2	43.2	29.4	54.3	76.2	76.9	57.5	59.8	69.9
6	61.8	50.5	52.7	74.1	57.0	39.7	49.9	73.4	83.4	68.7	72.3	88.8
7	74.0	61.2	62.8	84.0	75.2	59.6	71.7	95.3	97.8	78.7	80.7	89.9
8	81.0	69.0	70.2	89.0	88.9	73.8	85.6	106.2	110.3	89.0	89.5	101.7
9	84.2	72.3	75.2	92.9	93.0	91.2	96.6	112.2	115.4	92.1	98.2	107.2
10	85.5	73.2	78.1	96.4	101.8	102.7	108.9	116.5	119.2	99.7	99.4	102.1
11	85.4	73.3	78.9	96.5	103.0	106.5	117.7	120.1	123.9	102.8	101.4	99.6
12	83.3	72.3	77.9	95.5	102.6	111.2	129.3	132.1	130.7	99.6	101.7	96.8
13	82.1	71.0	77.0	96.1	104.3	120.9	146.0	161.8	139.0	98.1	101.3	93.1
14	80.3	70.3	76.0	94.4	103.3	128.7	165.3	188.5	147.5	100.4	101.3	91.4
15	78.6	68.3	74.0	93.1	103.0	132.4	172.0	203.1	147.6	97.5	99.5	87.4
16	76.0	67.5	71.6	91.4	98.6	128.5	171.5	197.9	144.6	95.5	97.7	87.4
17	80.0	68.6	70.3	89.2	95.6	118.6	163.2	172.8	146.1	96.5	101.6	101.1
18	97.4	79.8	73.9	90.6	92.8	106.9	133.6	136.5	140.3	95.9	115.1	135.2
19	95.7	87.1	91.3	96.9	94.4	98.2	113.0	121.4	142.3	103.6	113.3	132.5
20	90.8	83.1	86.8	105.4	110.5	109.3	121.1	122.4	132.4	105.4	106.5	119.3
21	86.6	76.7	80.0	95.5	94.6	101.7	108.7	111.4	115.5	103.3	102.6	111.7
22	79.7	70.4	73.6	83.4	78.5	79.7	112.4	108.2	104.8	95.7	94.7	102.2
23	73.2	61.5	66.6	69.4	59.4	61.4	80.8	88.3	96.6	88.6	88.4	92.6
24	62.2	49.7	55.3	65.1	55.2	52.4	76.4	82.4	94.5	72.6	71.9	81.9

Hour	1	2	3	4	5	6	7	8	9	10	11	12
00 P.M.	85.1	74.5	77.6	94.6	100.3	118.0	148.2	163.1	142.5	99.1	104.5	105.9
00 A.M.	51.8	44.4	46.2	61.2	42.7	35.2	55.1	69.7	77.0	61.3	61.5	72.9
ference	33.3	30.0	31.4	33.4	57.7	82.8	93.1	93.3	65.5	37.8	43.0	33.0

	May - October	November - April
00 P.M.	128.5	90.4
00 A.M.	56.8	56.4
ference	71.7	34.0

Table F-2. Storage Buy-Low / Sell-High Potential for California 2009 Forecast ($/MWh)

Hour	1	2	3	4	5	6	7	8	9	10	11	12
12:00 P.M. - 5:00 P.M.	85.1	74.5	77.6	94.6	100.3	118.0	148.2	163.1	142.5	99.1	104.5	105.9
1:00 A.M. - 6:00 A.M.	51.8	44.4	46.2	61.2	42.7	35.2	55.1	69.7	77.0	61.3	61.5	72.9
Storage Losses	10.4	8.9	9.2	12.2	8.5	7.0	11.0	13.9	15.4	12.3	12.3	14.6
Net	23.0	21.1	22.1	21.1	49.1	75.7	82.1	79.4	50.1	25.5	30.7	18.4

	May - October	November - April		Hours	Value*
12:00 P.M. - 5:00 P.M.	128.5	90.4	Summer	651.8	39,323
1:00 A.M. - 6:00 A.M.	56.8	56.4	Winter	651.8	14,830
Storage Losses*	11.4	11.3	Total	1,304	54,152
Net	60.3	22.8	*Storage Efficiency = 80.0%		

*Storage Efficiency = 80.0%

References

[F1] Derived from preliminary Wholesale Electricity Price Forecast data provided by Joel Klein, California Energy Commission. April 2008.

APPENDIX G – CHALLENGES FOR STORAGE

A spectrum of challenges may affect prospects for increased use and acceptance of storage. A high-level characterization of those challenges is provided in this appendix. The purpose for this is to provide storage advocates and other interested stakeholders with a general indication of and awareness about the types of challenges that may arise for any given storage project, and more broadly, that may require attention before storage can be widely deployed. (Note that some of the items listed below are also described as opportunity drivers in Section 7.3.)

- Storage has a relatively high cost.
- Storage energy losses – 20% to 40% of energy stored is lost:
    o Storage tends to have round-trip efficiency of 60% to 80%
- 'Inefficient' electric energy and services pricing:
    o Transmission and possibly distribution
    o Demand
    o Energy
    o Reliability
- Limited risk/reward sharing mechanisms between a) utilities and utility customers and b) utilities and third parties:
    o Regulatory rules and 'permission'
    o Interconnect
    o Undetermined optimal and/or maximum storage penetration levels
        - bulk/central
        - modular/distributed
    o Operations
- Permitting and siting rules and regulations (many have yet to be developed):
    o Zoning and building codes
    o City and community planning
    o Fire, public health, and safety-related rules and codes (mostly local)
    o National Electric Code
    o Occupational safety and health (state and federal agencies)
- Limited familiarity, knowledge, and experience base:
    o Storage cost and benefits
    o Storage technology
    o Storage system integration
    o Distributed energy resources
    o Integration of storage with the grid
    o Storage benefits and value

- Existing utility technology biases (especially utilities and, to a lesser extent, regulators):
  o Utilities are technologically risk averse, for understandable reasons
  o Perceived risk for *any* new technology
- Limited engineering standards and evaluation methodologies.
- Lack of evaluation tools:
  o Electrical
  o Financial
- Financing of 'new' technology is challenging: o Unknown operational costs
  o Uncertain system life
  o Multi-year payback is difficult for commercial/residential
  o Multi-year payback is acceptable for government and utilities
- Investor-owned utilities' (IOUs') preference for *investments* in *equipment* and their aversion to expense-based alternatives (such as rentals, leases or incentives):
  o IOUs derive all profit from *investments* in *equipment*
  o IOUS will tend to avoid *expenses* related to storage involving equipment rental or leases and possibly 'risk and reward sharing'
  o IOUS will prefer to purchase storage *equipment* though financial justification will often be elusive
- Inadequate infrastructure features and 'hooks':
  o Interconnection
  o Control
  o Communication
  o Price signals

Many technologies, concepts and programs 'competing' for 'attention':

  o Renewables
    - Waste and biofuels
    - Solar thermal
    - Photovoltaics
    - Wind generation
  o Conventional fuels
    - Clean coal
    - Advanced nuclear
  o Demand response
  o Distributed resources
  o Load aggregation
  o Smart Grid
  o Conservation and efficiency
- Coordinating among numerous stakeholders, for 'permission' to use grid-connected storage and/or to aggregate benefits may be expensive and time-consuming.

In: Lightning in a Bottle: Electrical Energy Storage
Editor: Fedor Novikov

ISBN: 978-1-61470-481-2
© 2011 Nova Science Publishers, Inc.

Chapter 2

ADVANCED MATERIALS AND DEVICES FOR STATIONARY ELECTRICAL ENERGY STORAGE APPLICATIONS[*]

United States Department of Energy

ABOUT THIS REPORT

This report was supported by Sandia National Laboratories and Pacific Northwest National Laboratory on behalf of the U.S. Department of Energy's (DOE) Office of Electricity Delivery and Energy Reliability and the Advanced Research Projects Agency-Energy (ARPA-E).

This document was prepared by Sarah Lichtner, Ross Brindle, Lindsay Kishter, and Lindsay Pack of Nexight Group under the direction of Dr. Warren Hunt, Executive Director, The Minerals, Metals, and Materials Society (TMS). The cooperation of ASM International through the Energy Materials Initiative, as well as the American Ceramic Society, the Electrochemical Society, and the Materials Research Society, is gratefully acknowledged.

Notice: This report was prepared as an account of work sponsored by an agency of the United States government. Neither the United States government nor any agency thereof, nor any of their employees, makes any warranty, express or implied, or assumes any legal liability or responsibility for the accuracy, completeness, or usefulness of any information, apparatus, product, or process disclosed, or represents that its use would not infringe privately owned rights. Reference herein to any specific commercial product, process, or service by trade name, trademark, manufacturer, or otherwise does not necessarily constitute or imply its endorsement, recommendation, or favoring by the United States government or any agency thereof. The views and opinions of authors expressed herein do not necessarily state or reflect those of the United States government or any agency thereof.

---

[*] This is an edited, reformatted and augmented version of the United States Department of Energy publication, dated December 2010.

Sponsored by: U.S. Department of Energy, Office of Electricity Delivery and Energy Reliability
Advanced Research Projects Agency—Energy
Organized by: Sandia National Laboratories
Pacific Northwest National Laboratory
The Minerals, Metals & Materials Society (TMS)

EXECUTIVE SUMMARY

Reliable access to cost-effective electricity is the backbone of the U.S. economy, and electrical energy storage is an integral element in this system. Without significant investments in stationary electrical energy storage, the current electric grid infrastructure will increasingly struggle to provide reliable, affordable electricity, jeopardizing the transformational changes envisioned for a modernized grid. Investment in energy storage is essential for keeping pace with the increasing demands for electricity arising from continued growth in U.S. productivity, shifts in and continued expansion of national cultural imperatives (e.g., the distributed grid and electric vehicles), and the projected increase in renewable energy sources.

Stationary energy storage technologies promise to address the growing limitations of U.S. electricity infrastructure. A variety of near-, mid-, and long-term storage options can simultaneously provide multiple benefits that have the potential to greatly enhance the future resilience of the electric grid while preserving its reliability. These benefits include providing balancing services (e.g., regulation and load following), which enables the widespread integration of renewable energy; supplying power during brief disturbances to reduce outages and the financial losses that accompany them; and serving as substitutes for transmission and distribution upgrades to defer or eliminate them.

Significant advances in materials and devices are needed to realize the potential of energy storage technologies. Current large-scale energy storage systems are both electrochemically based (e.g., advanced lead-carbon batteries, lithium-ion batteries, sodium-based batteries, flow batteries, and electrochemical capacitors) and kinetic-energy-based (e.g., compressed-air energy storage and high-speed flywheels). Electric power industry experts and device developers have identified areas in which near-term investment could lead to substantial progress in these technologies. Deploying existing advanced energy storage technologies in the near term can further capitalize on these investments by creating the regulatory processes and market structures for ongoing growth in this sector. At the same time, a long-term focus on the research and development of advanced materials and devices will lead to new, more cost-effective, efficient, and reliable products with the potential to transform the electric grid.

Strategic Priorities for Energy Storage Device Optimization through Materials Advances

Advanced materials, device research and development, and demonstrations are required to address many of the challenges associated with energy storage system economics, technical performance, and design that must be overcome for these devices to meet the needs and performance targets of the electric power industry. The advancement of large-scale energy storage technologies will require support from the U.S. Department of Energy (DOE), industry, and academia. Figure 1 outlines the high-priority research and development activities that are necessary to overcome the limitations of today's storage technologies and to make game-changing breakthroughs in these and other technologies that are only now starting to emerge, such as metal-air batteries, liquid-metal systems, regenerative fuel cells, advanced compressed-air energy storage, and superconducting magnetic electrical storage. The priority activities outlined in this report focus on understanding and developing materials coupled with designing, developing, and demonstrating components and systems; however, there is also recognition that this work needs to be done in the context of strategic materials selection and innovative system design.

STRATEGIC MATERIALS SELECTION implies that while significant cost reduction in storage is paramount and materials make up the largest portion of system cost, it is critical that storage devices utilize materials that are both lower in cost and abundant in the United States. New materials development can expand the options available to equipment developers, potentially offering important cost and performance advantages.

INNOVATIVE DESIGNS of storage technologies can drive the development of devices that can be affordably manufactured at grid scale. Design simplifications and designs for efficient manufacturing can enable storage systems to be produced at lower costs via automated manufacturing with necessary quality control processes. Effective system design also ensures that control systems and power electronics enable efficient, secure, and reliable interoperability with the electric grid.

	NEAR TERM (< 5 years)	MID TERM (5–10 years)	LONG TERM (10–20 years)
ADVANCED LEAD-ACID AND LEAD-CARBON BATTERIES	Conduct DOE-funded validation tests of system lifetime, ramp rates, etc. Develop high-power/energy carbon electrode for lead-carbon battery	Understand poor materials utilization through diagnostics and modeling	
LITHIUM-ION BATTERIES		Develop models for ion transport through solids (inorganic solids, polymers) Conduct experiments to develop a quantitative understanding of catastrophic cell failure and degradation Design and fabricate novel electrode architectures to include electrolyte access to redox active material and short ion and electron diffusion paths (e.g., non-planar geometries) Develop a highly conductive, inorganic, solid-state conductor for solid-state Li-ion batteries	Develop new intercalation compounds with low cycling strain and fatigue; aim for 10,000 cycles at 80% depth of discharge
SODIUM-BASED BATTERIES	Develop robust planar electrolytes to reduce stack size and resistance Implement pilot-scale testing of battery systems to develop performance parameters for grid applications	Decrease operating temperature, preferably to ambient temperature	Develop a true sodium-air battery that provides the highest value in almost any category of performance Use surface-science techniques to identify species on sodium-ion anodes and cathodes

	NEAR TERM (< 5 years)	MID TERM (5–10 years)	LONG TERM (10–20 years)
FLOW BATTERIES	Establish a center for stack design and manufacturing methods, including joint and seal design Develop low-cost, formable, chemically and thermally tolerant resins for piping, stacks, and tanks Develop an inline, real-time sensor that can detect impurities in electrolyte composition for various flow battery chemistries Create a computational fluidics center at a national laboratory or university Identify low-cost hydrogen suppression materials (anti-catalysts) and redox catalysts for negative electrodes	Improve membranes to enable minimum crossover, lower system cost, increased stability, and reduced resistance Improve mass transport via a tailored catalyst layer and flow field configurations to increase operating current density and reduce system cost per kilowatt	Develop non-aqueous flow battery systems with wider cell operating voltages to improve efficiency
POWER TECHNOLOGIES	Develop a 1-megawatt flywheel motor capable of vacuum operation and superconduction Develop high-power/energy carbon electrode for electrochemical capacitors	Optimize materials utilization through diagnostics and modeling Develop hubless flywheel rotor with four times higher energy	
EMERGING TECHNOLOGIES		Improve thermal management in endothermic electrolysis reactions and exothermic fuel cell reactions in regenerative fuel cells	Develop new catalysts for metal-air batteries with low overpotentials for oxygen reduction in order to make systems more efficient, cost-effective, and bifunctional Explore the untapped potential of multivalent chemistries Develop air electrodes for metal-air batteries with high electrochemical activity and lower polarization and resistance
CROSSCUTTING ACTIVITIES	Combine technologies for synergy Conduct DOE-funded demonstrations of all energy storage technologies Specify cycle and life tests for stationary power applications	Take an integrated approach to degradation by combining microstructure/chemistry observations with mechanistic modeling (both degradation and electrochemical models) and accelerated testing	

Figure 1. Prioritized Activities to Advance Energy Storage Technologies.

The success of these activities and initiatives will require significant support from DOE. To help DOE better focus its resources over time, Figure 1 divides the solutions for each storage technology by the time frame in which they will impact the market: near term (less than 5 years), mid term (5–10 years), and long term (10–20 years). Committing to these activities will allow DOE, technology developers, and the electric power industry to pursue a coherent technology development and demonstration strategy for energy storage technologies in grid-scale applications.

INTRODUCTION AND PROCESS

Cost-effective energy storage technologies are a key enabler of grid modernization, addressing the electric grid's most pressing needs by improving its stability and resiliency. Investment in energy storage is essential for keeping pace with the increasing demands for electricity arising from continued growth in U.S. productivity, shifts in and continued expansion of national cultural imperatives (e.g., emergence of the distributed grid and electric vehicles), and the projected increase in renewable energy sources. Materials, their processing, and the devices into which they are integrated will be critical to advancing clean and competitive energy storage devices at the grid scale.

Current research and demonstration efforts by the U.S. Department of Energy (DOE), national laboratories, electric utilities and their trade organizations, storage technology providers, and academic institutions provide the foundation for the extensive effort that is needed to accelerate widespread commercial deployment of energy storage technologies. For grid-scale storage to become pervasive, the electric power industry, researchers of advanced materials and devices, equipment manufacturers, policymakers, and other stakeholders must combine their expertise and resources to develop and deploy energy storage systems that can address the specific storage needs of the electric power industry.

Seeking to accelerate the commercialization of stationary energy storage at grid scale, The Minerals, Metals & Materials Society (TMS) joined with the DOE Office of Electricity Delivery and Energy Reliability, the DOE Advanced Research Projects Agency-Energy, Pacific Northwest National Laboratory, and Sandia National Laboratories to sponsor a facilitated workshop. This workshop was designed to garner critical information from key stakeholders to develop a path forward for grid-scale energy storage.

Thirty-five stakeholders and experts from across the materials science and device communities attended the workshop on June 21–22, 2010, in Albuquerque, New Mexico. Immediately preceding the advanced materials and devices workshop, stakeholders and experts from the electric power industry, research, and government communities came together to identify targets for energy storage technologies in specific grid applications, which resulted in the workshop report, *Electric Power Industry Needs for Grid-Scale Storage Applications*. The participants of the advanced materials and devices workshop used the targets determined in the previous workshop to identify the limitations of existing energy storage technologies and the advances necessary for these devices to achieve widespread commercialization.

While all energy storage technologies and systems were within the scope of the workshop, the main focus was on technologies for which DOE involvement could accelerate progress toward commercial deployment at grid scale. The time frame under consideration was present day through 2030, with particular emphasis on the 1- to 5-year and 5- to 10-year time frames.

Based on the results of the workshop, this report provides guidance to DOE for advancing the following energy storage technologies:

- Advanced lead-acid and lead-carbon batteries
- Lithium-ion batteries
- Sodium-based batteries
- Flow batteries
- Power technologies (e.g., electrochemical capacitors and high-speed flywheels)
- Emerging technologies (e.g., metal-air batteries, liquid-metal systems, regenerative fuel cells, and advanced compressed-air energy storage

The reports from these workshops will inform future DOE program planning and ultimately help to commercialize energy storage at grid scale.

40-MEGAWATT ENERGY STORAGE FACILITY IN FAIRBANKS, ALASKA

ENERGY STORAGE: THE NEED FOR MATERIALS AND DEVICE ADVANCES AND BREAKTHROUGHS

Electricity demand in the United States is steadily rising; in 2009, electricity consumption was more than five times what it was 50 years ago.[1] This demand is projected to increase by 1% per year through 2035.[2] To meet the increased electricity demands expected by 2035 (excluding those expected from the introduction of electric vehicles), an additional 250 gigawatts of generating capacity will have to be added to the electricity generation infrastructure.[3] However, the aging electric grid does not have the ability to transmit these large amounts of electricity from the point of generation to the end user or to accommodate the proposed increases in generation from renewable sources like wind and solar.[4] Advanced storage technologies have the potential to fulfill applications across the entirety of the grid to address these growing issues. To meet increasing electricity demands while continuing to provide consumers with electricity at the level of cost and reliability they have come to expect, the U.S. electric grid requires immediate and cost-effective updates.

Stationary energy storage technologies promise to address the growing limitations of U.S. electricity infrastructure and meet the increasing demand for renewable energy use. With a variety of near-, mid-, and long-term storage options, energy storage devices can simultaneously provide multiple benefits that have the potential to greatly enhance the future resilience of the electric grid while preserving its reliability. These benefits include providing balancing services, such as regulation and load following; supplying power during brief disturbances to reduce outages and the financial losses that accompany them; and serving as substitutes to help defer or eliminate transmission and distribution upgrades.

Significant advances in energy storage materials and devices are needed to realize the potential of these technologies.

Industry members and device developers have identified areas in which short-term investment could lead to substantial progress. Deploying existing advanced energy storage technologies in the near term can further capitalize on these investments by encouraging utility operating experience and acceptance. At the same time, a long-term focus on the research and development of advanced materials and devices will lead to new, lower-cost, and more efficient and reliable products with the potential to revolutionize the electric grid.

The Current State of Energy Storage Technologies

Storage technologies currently being researched, developed, and deployed for grid applications include high-speed flywheels, electrochemical capacitors, traditional and advanced lead-acid batteries, high-temperature sodium batteries (e.g., sodium-sulfur and sodium-nickel-chloride), lithium-ion batteries, flow batteries (e.g., vanadium redox and zinc bromine), compressed-air energy storage, pumped hydro, and other advanced battery chemistries, such as metal-air, nitrogen-air, sodium-bromine, and sodium-ion.

The grid applications for these technologies can be loosely divided into power applications and energy management applications, which are differentiated based on storage discharge duration. Technologies used for power applications are typically used for short durations, ranging from fractions of a second to approximately one hour, to address faults and operational issues that cause disturbances, such as voltage sags and swells, impulses, and flickers. Technologies used for energy management applications store excess electricity during periods of low demand for use during periods of high demand. These devices are typically used for longer durations of more than one hour to serve functions that include reducing peak load and integrating renewable energy sources.

Table 1. Suitable Grid Applications and Current Status of Energy Storage Technologies[5]

ENERGY STORAGE TECHNOLOGY	SUITABLE APPLICATIONS	CURRENT DEVELOPMENT AND COMMERCIALIZATION STATUS
HIGH-SPEED FLYWHEELS (FW)	High potential for power applications	Currently, FWs are used in many uninterrupted power supply and aerospace applications, including 2 kW / 6 kWh systems used in telecommunications. FW farms are being planned and built to store megawatts of electricity for short-duration regulation services.
ELECTROCHEMICAL CAPACITORS (EC)	High potential for power applications	Small ECs are a mature technology; systems with a higher energy capacity are still in development.

Table 1. (Continued).

ENERGY STORAGE TECHNOLOGY	SUITABLE APPLICATIONS	CURRENT DEVELOPMENT AND COMMERCIALIZATION STATUS
TRADITIONAL LEAD-ACID BATTERIES (TLA)	High potential for power applications and feasible for energy applications	TLAs are the oldest and most mature energy storage technology available. The largest TLA battery system in operation has a 10 MW / 40 MWh capacity.
ADVANCED LEAD-ACID BATTERIES WITH CARBON-ENHANCED ELECTRODES (ALA-CEE)	High potential for both power and energy applications	ALA-CEEs were developed as an inexpensive battery for use in hybrid electric vehicles.
SODIUM SULFUR BATTERIES (NaS)	High potential for both power and energy applications	The battery has been demonstrated at more than 190 sites in Japan, totaling more than 270 MW in capacity. U.S. utilities have installed 9 MW for peak shaving, firming wind power, and other applications. The development of another 9 MW system is in progress.
SODIUM-NICKEL-CHLORIDE (Na-NiCl2) BATTERIES	High potential for both power and energy applications	The battery operates at lower temperatures than NaS batteries.
LITHIUM-ION (Li-ION) BATTERIES	High potential for power applications and reasonable for energy applications	Li-ion batteries currently dominate the consumer electronic market. Manufacturers are working to reduce system cost and increase safety, enabling these batteries to be used in large-scale markets.
ZINC-BROMINE BATTERIES (ZnBr)	High potential for energy applications and reasonable for power applications	ZnBr batteries with 1 MW / 3 MWh capacities have been tested on transportable trailers for utility use. Larger systems are currently being tested.
VANADIUM REDOX BATTERIES (VRB)	High potential for energy applications and reasonable for power applications	VRB batteries up to 500 kW / 5 MWh have been installed in Japan. These batteries have been tested and used for power applications, supplying up to 3 MW over 1.5 seconds.
COMPRESSED-AIR ENERGY STORAGE (CAES)	High potential for energy applications	The first commercial CAES plant was built in Germany in 1978 and has a 290 MW capacity. An additional plant with a 110 MW capacity was built in Alabama in 1991. Advanced adiabatic CAES systems are currently being developed.
PUMPED HYDRO (PH)	High potential for energy applications	PH represents approximately 3% of global generation capacity—more than 90 GW of PH storage is installed worldwide. While PH has achieved widespread deployment, all of the suitable PH locations are currently being used.

The technologies in Table 1 represent those with the greatest potential for widespread grid-scale deployment. The table indicates the applications for which the technologies are best suited and provides an overview of the current development and commercialization status of each technology. While other technologies are currently under development, they are not advanced enough for grid-scale evaluation.

The energy storage technologies in Table 1 are currently at different stages of development, demonstration, and commercialization. Increasing the amount of installed and planned technologies is critical to the widespread deployment of energy storage systems sooner rather than later; for this reason, many other installations are planned for the next five years. The energy and power capacities of some of the current and planned worldwide installations are provided in Figure 2.

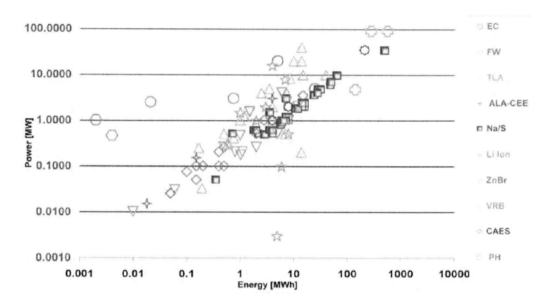

Figure 2. Installed and Planned Energy Storage Systems, April 2010[6].

Continued investment in the research and development of new and existing energy storage technologies has the long-term opportunity to revolutionize the electric power industry. Large-scale demonstrations of energy storage systems encourage the utility buy-in needed to make energy storage feasible at grid scale and provide researchers and technology developers with critical performance data. A strategic approach to the deployment of grid-scale energy storage technologies has the potential to provide a cost-effective, near-term solution.

INTEGRATING ENERGY STORAGE INTO THE ELECTRIC GRID

With the increasing penetration of variable renewable energy sources, electricity generation is no longer constant, yet must continue to meet fluctuating electricity demands. This imbalance, along with the current grid limitations and aging infrastructure, has the potential to challenge grid operators as they manage an increasingly dynamic electric grid.

Stationary energy storage technologies can be harnessed for a variety of applications to help the electric power industry provide customers with reliable and affordable electricity.

Energy storage devices provide necessary services to the electric grid, including balancing services (e.g., regulation and load following), to reduce outages and the financial losses that accompany them. By responding to the grid faster than traditional generation sources, operating efficiently at partial load, and varying discharge times depending on application need, the same storage devices can aid in the deferral of transmission and distribution infrastructure to keep electricity rates low.

Metrics for Storage Technologies and Applications

To provide the maximum benefit to electricity end users and gain acceptance from the electric power industry, storage technologies must meet certain economic, technical performance, and design targets for energy storage applications. While each energy storage application will require different specifications, these three interrelated factors must be met to ensure the widespread deployment of grid-scale energy storage.

SYSTEM ECONOMICS is the most important metric to the electric power industry. Consumers are accustomed to having electricity when they need it and at affordable prices, which makes the lifecycle cost of storage technologies critical to their widespread adoption. Some stakeholders in the electricity industry believe that an energy storage technology must be competitive with the cost of currently available technologies used for peak electricity generation (e.g., gas turbines) and must provide increased efficiency and other benefits that adequately offset capital, operating, and lifetime costs. While this view fails to recognize the full benefits of energy storage, some decision makers in the electricity industry continue to view storage as a peak generation substitute and value it accordingly. In order to achieve widespread implementation at grid scale, the cost of stationary storage devices overall must continue to decline.

The *TECHNICAL PERFORMANCE* of an energy storage device has a significant impact on overall system economics. In addition to being an affordable option, technologies must be able to meet the performance needs of a particular application. These performance needs are application-specific, but include the device's cycle life, energy density, response time, rate of charge and discharge, and efficiency. If a system does not meet the specific needs of its intended application, it is unlikely to be adopted. Advanced technologies must not only lower costs but exceed technical performance requirements when compared to currently available technologies.

SYSTEM DESIGN is also interrelated with the cost and technical performance of an energy storage system. This factor includes the storage device (e.g., battery, flywheel, regenerative fuel cell, or electrochemical capacitor), the power conditioning and control systems that allow the system to communicate with the electric grid, and any other ancillary equipment necessary for the device's operation (e.g., auxiliary cooling systems). The scalability of a system will depend on many factors, such as materials availability, the feasibility of automated manufacturing, and overall system complexity. In addition, the system design must meet the safety standards of the electric power industry and manage any health and safety risks to utility workers and the surrounding community.

Storage technologies that meet the economic, technical performance, and system design requirements of the intended application are well positioned to achieve widespread adoption in the electric power industry.

Priority Applications for Energy Storage Technologies

The wide range of chemistries and structures of energy storage devices enables them to meet the duration, capacity, and frequency demands of specific applications. Of the 15 to 20 unique storage applications that have been identified, there are five storage applications that have the greatest overall potential to benefit power system planning and operations: area and frequency regulation, renewables grid integration, transmission and distribution upgrade deferral and substitution, load following, and electric energy time shift. Area and frequency regulation and certain aspects of renewables grid integration are short-duration power management applications, while transmission and distribution upgrade deferral and substitution, load following, and electric energy time shift (including renewables) are long-duration energy management applications.

The stationary energy storage technologies used in these applications must meet certain economic, technical performance, and design targets in order to optimize grid functionality. While the metrics and targets will vary depending on the specific energy storage technology or device and the location of the application, they can serve as guidelines for researchers and the electric power industry to assess the value of individual technologies. Many storage technologies currently meet one or several of the proposed metrics. However, in order to achieve widespread commercial deployment, storage systems must meet the targets that will offer the right combination of performance and cost-effectiveness required for market acceptance. The metrics and targets for storage technologies applied to area and frequency regulation, renewables grid integration, transmission and distribution upgrade deferral and substitution, load following, and electric energy time shift are provided in Table 2. These performance targets were set at the workshop held prior to the advanced materials workshop; additional detail is available in the *Electric Power Industry Needs for Grid-Scale Storage Applications* report.

Understanding the Cost Targets

The normalized cost of energy storage systems is a key consideration for the electric power industry. Setting realistic and achievable cost targets for energy storage technologies can help guide research and development efforts from an end-user perspective while increasing the likelihood that device developers will be able to achieve them. The targets in Table 2 represent an attempt to accomplish this objective; however, such targets require several caveats. Storage system costs depend on the system location, size, and grid storage application. The complexity of cost targets emphasizes the need for both device developers and the electric power industry to recognize the imprecise nature of these targets. The actual cost of a storage technology must reflect the value of storage when used for a single grid application or for multiple simultaneous applications.

Table 2. Targets for Energy Storage Technologies Used for Priority Grid Applications

APPLICATION	PURPOSE	KEY PERFORMANCE TARGETS
Area and Frequency Regulation (Short Duration)	Reconciles momentary differences between supply and demand within a given area. Maintains grid frequency	**SERVICE COST:** $20 per MW per hour **SYSTEM LIFETIME:** 10 years with 4,500 to 7,000 cycles per year **DISCHARGE DURATION:** 15 minutes to 2 hours **RESPONSE TIME:** less than one second **ROUNDTRIP EFFICIENCY:** 75%–90%
Renewables Grid Integration (Short Duration)	Offsets fluctuations of short-duration variation of renewables generation output	**ROUNDTRIP EFFICIENCY:** 75%–90% **SYSTEM LIFETIME:** 10 years with high cycling **CAPACITY:** 1 MW–20 MW **RESPONSE TIME:** 1–2 seconds
Transmission and Distribution Upgrade Deferral and Substitution (Long Duration)	Delays or avoids the need to upgrade transmission and/or distribution infrastructure using relatively small amounts of storage. Reduces loading on existing equipment to extend equipment life	**COST:** $500 per kWh **DISCHARGE DURATION:** 2–4 hours **CAPACITY:** 1 MW–100 MW **RELIABILITY:** 99.9% **SYSTEM LIFETIME:** 10 years
Load Following (Long Duration)	Changes power output in response to the changing balance between energy supply and demand. Operates at partial output or input without compromising performance or increasing emissions. Responds quickly to load increases and decreases	**CAPITAL COST:** $1,500 per kW or $500 per kWh for 3-hour duration **OPERATIONS AND MAINTENANCE COST:** $500 per MWh **DISCHARGE DURATION:** 2–6 hours
Electric Energy Time Shift (Long Duration)	Stores inexpensive energy during low demand periods and discharges the energy during times of high demand (often referred to as arbitrage). Accommodates renewables generation at times of high grid congestion by storing energy and transmitting energy when there is no congestion	**CAPITAL COST:** $1,500 per kW or $500 per kWh **OPERATIONS AND MAINTENANCE COST:** $250–$500 per MWh **DISCHARGE DURATION:** 2–6 hours **EFFICIENCY:** 70%–80% **RESPONSE TIME:** 5–30 minutes

A MATERIALS-BASED APPROACH TO ADVANCING ENERGY STORAGE TECHNOLOGIES

The system economics, technical performance, and design of current energy storage technologies do not adequately meet the wide-ranging needs of the electric power industry. The high cost, low energy capacity, low efficiency, and current complexity of many of today's storage technologies present major obstacles to the production scale-up and integration of storage devices at grid scale. The materials composing these technologies determine the majority of their performance specifications.

Advancing materials, their processing, and the devices into which they are integrated will be critical to meeting the needs of the electric power industry and the performance targets of priority grid storage applications. Material selection will play an essential role in making storage technologies affordable, efficient, and reliable options for addressing the increasing demand for electricity and penetration of renewables-based generation.

10-MEGAWATT, 30-SECOND ENERGY STORAGE FACILITY AT MICROCHIP PLANT

Energy Storage Device Optimization through Materials Advances

Addressing energy storage system economics, technical performance, and design issues requires advanced materials research and development. While the necessary research and development activities focus on understanding and developing materials coupled with designing, developing, and demonstrating components and systems, there is also recognition that this work needs to be done in the context of strategic materials selection and innovative system design.

Strategic Materials Selection implies that while significant cost reduction in storage is paramount and materials make up the largest portion of system cost, it is critical that storage devices utilize materials that are both low in cost and abundant in the United States. New materials development can expand the options available to equipment developers, potentially offering important cost and performance advantages.

Innovative Designs of storage technologies can drive the development of devices that can be affordably manufactured at grid scale. If a storage technology design is unnecessarily complex, it will be difficult and costly to put automated manufacturing and quality control processes into place. System design also ensures that control systems and power electronics enable efficient, secure, and reliable interoperability with the electric grid.

Focus Areas of Materials Advancements

Continuous basic and applied research supporting both new and existing energy storage technologies will provide the advancements needed to deliver affordable storage devices that meet utility needs. While each storage technology has its own specific limitations and potential solutions, several key focuses in the advanced materials area could significantly encourage commercial success:

- *Basic Materials Research* – Current energy storage devices utilize only a small portion of the extensive electrochemical materials combinations available for use, so it is likely that more effective, safe, inexpensive, and robust combinations exist. Energy storage device experts need to explore the potential of lower-cost and more readily available materials such as iron, aluminum, magnesium, and copper for use in energy storage technologies.
- *Advanced Electrochemical Combinations* – New electrochemical combinations (electrolyte and electrode couples) and the more efficient utilization of current electrolytes and electrodes have the potential to increase conductivity, amplify capacity, reduce resistance, improve thermal tolerance, and extend the life of energy storage devices. Further research into non-flammable electrolytes can increase the safety of energy storage devices. Energy storage device researchers and manufacturers can also develop electrodes that can increase device conductivity while resisting overcharging and degradation.
- *Solid-State Ionics* – Electrolytes can be engineered into thin and flexible crystalline solids, which can provide storage technologies with decreased resistance, reduced cost, improved reliability, and increased efficiency in comparison to systems with liquid electrolytes.
- *Innovative Membranes and Seals* – Improved membranes and seals in storage technologies will help to limit the contamination of electrolytes, electrodes, and other contaminant-sensitive device components.
- *Nanomaterials* – Research into nanomaterials may be a promising focus that can help to develop high-power and quick-response energy storage devices.
- *Advanced Control Systems and Power Electronics* – In addition to researching materials for specific storage technologies, energy storage device experts must also advance the control systems and power electronics that enable efficient and reliable interoperability with the electric grid.
- *Novel Cell Stack Designs* – Developing novel cell and stack designs for particular stationary applications could have an impact in the long term.

The following sections discuss the technology-specific limitations of current energy storage offerings, including advanced lead-carbon batteries, lithium-ion batteries, sodium-based batteries, flow batteries, power technologies (e.g., electrochemical capacitors and high-speed flywheels), and emerging technologies (e.g., metal-air batteries, liquid-metal systems, regenerative fuel cells, and advanced compressed-air energy storage). Each technology section also includes a timeline of technology-specific activities and initiatives that are intended to explore the untapped potential of new and current materials to overcome those limitations.

The success of these activities and initiatives will require significant support from DOE. To help DOE better focus its resources over time, solutions are divided by the time frame in which they will impact the market: near term (less than 5 years), mid term (5–10 years), and long term (10–20 years).

ADVANCED LEAD-ACID AND LEAD-CARBON BATTERIES

Lead-acid batteries are the oldest type of rechargeable battery and one of the least expensive energy storage devices currently available, in terms of capital cost ($/kWh). However, the short cycle life and significant maintenance requirements of traditional lead-acid batteries leads to a high lifecycle cost ($/kWh/cycle), limiting their use for commercial and large-scale operations. To improve traditional lead-acid batteries while maintaining their low system cost, energy storage developers added carbon-enhanced electrodes to create advanced lead-carbon batteries. These batteries were originally developed as an inexpensive power option for hybrid electric vehicles; adding enhanced carbon electrodes elevates the value of these technologies for grid-scale storage by extending system life and enhancing the performance of the batteries in both power and energy management grid applications.

CURRENT PERCEIVED LIMITATIONS OF ADVANCED LEAD-ACID AND LEAD-CARBON BATTERIES

Challenges involving cycle life, maintenance requirements, specific energy, and high-voltage operation must be addressed before advanced lead-carbon batteries can realize their full potential for use in grid-scale power and energy management applications. The gaps and limitations that, if overcome, could make the most significant advances toward this end goal include the following:

- *Today's Lead-Carbon Batteries Have a Short Cycle Life.* While lead-carbon batteries have a higher cycle life than traditional lead-acid batteries, the number of lifetime cycles is still significantly lower than grid storage applications require.
- *Lead-Carbon Batteries Require Significant Maintenance.* The maintenance requirements of lead-carbon batteries increase the operational costs of the systems and limit the lifetime of the technology. The lifetime of these devices is also significantly shortened if they are not located in an air-conditioned environment.
- *The Specific Energy of Lead-Carbon Batteries Is Limited by Insufficient Materials Utilization.* The theoretical specific energy of lead-carbon batteries is 166 watt-hours per kilogram (including the weight of sulfuric acid and assuming 2 volts per cell). However, the current specific energy of these devices is only 30–55 watt-hours per kilogram, which is 67%–80% lower than the actual potential of these technologies. Since the weight of these devices can result in increased building costs in response to load-bearing issues, more advanced carbon materials or other higher-rate materials are needed to help advanced lead-carbon batteries achieve a specific energy that is closer to their theoretical potential.
- *The Battery Systems Operate at High Voltages, Increasing Design Requirements.* The large systems that are required for lead-based, grid-scale energy storage operate at high voltages, increasing the possibility of ground faults. Such faults can lead to system damage or, in extreme events, fires. Addressing this risk requires careful system design, which could potentially include bipolar designs.

Priority Activities to Advance Lead-Acid and Lead-Carbon Batteries

With targeted research and development, lead-carbon batteries have the potential to contribute to the advancement of grid-scale energy storage in both power and energy management applications. Lead-carbon batteries can serve as a promising intermediate solution that can be deployed in the near term as newer technologies are being developed and improved. There are a variety of activities and initiatives that could help overcome the current gaps and limitations of these technologies in areas such as electrolyte advances, electrode development, diagnostics and modeling, and technology demonstration and validation. These solutions aim to optimize the effectiveness and increase the value of energy storage devices by increasing the energy density and efficiency of these devices. For lead-carbon batteries, activities and initiatives can accelerate progress in the following areas:

- *Electrolyte Advances* – Exploring the use of electrolyte additives and acid mixing can help to address the performance of lead-carbon batteries related to acid stratification.
- *Electrode Development* – Developing high-energy carbon electrodes can help to increase the energy density of lead-carbon batteries to a level that is suitable for grid-scale operation. Further research of natural carbon electrodes, as opposed to nanoelectrodes, can help to keep down the cost of lead-carbon batteries, maintaining their cost-competitiveness in comparison to other energy storage technologies.
- *Diagnostics and Modeling* – Modeling could help device developers gain an understanding of why the current battery design has a specific energy well below the theoretical specific energy of lead. Once this issue is better understood, energy storage device experts may be able to increase the specific energy of lead-carbon batteries, making them more attractive options for grid-scale storage.
- *Technology Demonstration and Validation* – Testing and demonstrating lead-carbon batteries can help to validate technology lifetime, ramp rates, and other performance characteristics that need to be proven to encourage stakeholder buy-in.

The success of these activities and initiatives will require significant support from DOE. To help DOE better focus its resources over time, Figure 3 divides the solutions by the time frame in which they will impact the market: near term (less than 5 years), mid term (5–10 years), and long term (10–20 years). The bolded activities are high-priority initiatives.

Advanced Materials and Devices for Stationary Electrical Energy Storage... 209

	NEAR TERM (< 5 years)	MID TERM (5–10 years)	LONG TERM (10–20 years)	NO TIMELINE SPECIFIED
ELECTROLYTES				Investigate electrolyte additives to reduce stratification Research in situ acid mixing in flooded lead-acid batteries for grid storage
ELECTRODES	Develop high-power/energy carbon electrode for lead-carbon battery Develop natural carbon sources and materials for lead-carbon batteries	Develop bipolar battery design Develop better positive grid material and active material for lead-acid and lead-carbon batteries Develop expander materials that can accept higher charges		Study functionalized carbon for mixed electron and ionic conductivity
DIAGNOSTICS AND MODELING		Understand poor materials utilization through diagnostics and modeling		Study current collector design
DEMONSTRATION AND VALIDATION	Conduct DOE-funded validation tests of system lifetime, ramp rates, etc.			

Figure 3. Prioritized Activities to Advance Lead-Acid and Lead-Carbon Batteries.

4 X 1-MEGAWATT, 15-MINUTE LITHIUM ION SYSTEM IN PJM INTERCONNECTION

LITHIUM-ION BATTERIES

Lithium-ion (Li-ion) batteries are currently used in many electronics (e.g., laptop computers and mobile telephones) and are expected to become a major power source for electric vehicles. These batteries are commonly composed of lithium electrolytes in the form of salts or solvents and carbon and metal-oxide electrodes. This composition enables these devices to operate at high energy density, high power, and nearly 100% efficiency, making them ideal for power applications and suitable for energy applications. Despite the widespread use of Li-ion batteries in electric vehicles and electronics, these batteries face challenges for grid applications because of the differences in performance and cost requirements for such stationary applications.

Current Perceived Limitations of Lithium-Ion Batteries

Current Li-ion batteries are developed for mobile electronic and vehicle applications that require high energy, power density, and specific energy due to the volume and weight constraints of the particular applications. In comparison, grid-based applications place more emphasis on cost and cycle life, though a high energy density is still desirable. Most of the existing Li-ion chemistries have a short cycle life (<1,000 cycles) and high cost (~$1,000/kWh) when used for stationary applications. Heat management, safety, and reliability issues must also be addressed before Li-ion batteries can achieve widespread deployment at grid-scale storage levels. The gaps and limitations that, if overcome, could make the most significant advances toward the widespread deployment of Li-ion batteries for grid-scale storage include the following:

- *The High Capital Cost of the Current Li-Ion Batteries Is a Fundamental Issue for Grid Applications.* It has been shown that 80% of the high capital cost of Li-ion

batteries is due to the relatively high cost of materials for electrodes, separators, electrolytes, etc.
- *Today's Li-Ion Batteries Have a Short Life and Cycle Count.* The current nominal capacity of Li-ion batteries decreases after repeated cycling, which diminishes the efficiency of the device. The low cycle count and resulting short life of Li-ion batteries could compromise this technology's ability to provide reliable and affordable grid-scale storage.
- *Organic Electrolytes Compromise the Safety of Li-Ion Batteries.* There is a need to develop inorganic electrolytes to improve the performance and safety of Li-ion batteries. The current electrolytes used in Li-ion batteries are unstable and potentially flammable at high voltages.
- *Lithium Systems Cannot Operate at Temperature Extremes.* Lithium systems are unable to effectively operate at temperatures lower than -10°C and present a potential safety hazard at temperatures greater than 70°C. The battery systems generate significant amounts of heat during operation, which requires thermal management mechanisms to keep the device temperature within its operational limits.

Cost Targets of Lithium-Ion Batteries

Based on current knowledge of Li-ion batteries, the following installed cost targets (set for everything needed up to direct current output to the converter) reflect the push and pull of the energy storage market:

- Current: $1,000/kWh (cell: $700/kWh, rest of system: $300/kWh)
- 2015: $500/kWh (cell: $400/kWh, rest of system: $100/kWh)
- 2020: $250/kWh (cell: $210/kWh, rest of system: $40/kWh)
- 2030: $250/kWh (cell: $210/kWh, rest of system: $40/kWh

Priority Activities to Advance Lithium-Ion Batteries

With targeted research and development, Li-ion batteries have the potential to contribute to the advancement of grid-scale energy storage. Increased understanding of current Li-ion batteries and their suitability for stationary grid-scale storage can encourage the optimization and subsequent adoption of these technologies. Additionally, new Li-ion battery systems that incorporate new materials, such as cost-effective, optimized materials used for electrodes and other components, can help overcome the current gaps and limitations of Li-ion batteries, creating Li-ion systems more suited to grid-scale storage applications. For Li-ion batteries, activities and initiatives can accelerate progress in the following areas:

- *Materials Discovery and Performance Optimization* – While capable of high energy density, the materials sets for current Li-ion batteries are too expensive and may not offer sufficient performance for stationary applications. Designing and fabricating novel electrode architectures to include electrolyte access to redox active material and short ion and electron diffusion paths (e.g., non-planar geometries) is a solution with a near-term market impact. Developing a highly conductive, inorganic, solid-state conductor for solid-state Li-ion batteries presents another near-term solution. In the long term, significant reduction in cost will likely require the use of cost-effective alternative materials or the development of new Li-ion batteries, though in some cases, these alternative materials may reduce energy density. The development of new intercalation compounds with low cycling strain and fatigue for Li-ion batteries could also have a significant impact. In order to do so, these compounds should have a goal of 10,000 cycles at 80% depth of discharge. Some long-lived Li-ion chemistries, such as lithium titanate and lithium iron phosphate have already been explored; such work should continue and be expanded in pursuit of the cycle and depth of discharge goals. Aqueous electrolytes may also hold promise for reducing cost and improving the safety of Li-ion batteries.

- *Mechanisms and Modeling* – Developing models for ion transport through inorganic solids and polymers, as well as developing a quantitative understanding of cell failure (both catastrophic and degradation) through experiments, could have a market impact in the mid term. Another activity with mid-term potential is characterizing the interfaces needed to address system lifetime and performance by using predictive models of interfaces and reactions to understand performance and degradation and by developing diagnostics to probe interfaces.
- *SAFETY* – The improvement of existing solid polymer electrolytes and the development of new solid polymer electrolytes could have a market impact in the near term, while the development of non-flammable electrolytes could have a mid-term impact.

The success of these activities and initiatives will require significant support from DOE. To help DOE better focus its resources over time, Figure 4 divides the solutions by the time frame in which they will impact the market: near term (less than 5 years), mid term (5–10 years), and long term (10–20 years). The bolded activities are high-priority initiatives.

Li-ION BATTERIES	NEAR TERM (< 5 years)	MID TERM (5–10 years)	LONG TERM (10–20 years)
MATERIALS DISCOVERY AND PERFORMANCE OPTIMIZATION	Develop highly uniform manufacturing processes to increase cell uniformity (e.g., performance, life, efficiency, yield) and understanding of lifetime operation Improve battery packaging by making it more lightweight, improving safety, and ensuring long-term stability Develop unique Li-ion chemistries that hold promise of meeting stationary storage requirements for cost, cycle life, etc.	Design and fabricate novel electrode architectures to include electrolyte access to redox active material and short ion and electron diffusion paths (e.g., non-planar geometries) Develop a highly conductive, inorganic, solid-state conductor for solid-state Li-ion batteries Use silicon to develop negative materials for Li-ion because silicon is a higher-energy material than graphite Perform thermodynamic and kinetic modeling to resolve the deposition of lithium on the negative electrode Evaluate suitability of existing Li-ion vehicle batteries for grid applications	Develop new intercalation compounds with low cycling strain and fatigue for Li-ion batteries; aim for 10,000 cycles at 80% depth of discharge Develop heterogeneous hybrid electrolytes at nanoscale to optimize properties (e.g., ion transport, electrochemical stability, and mechanical integrity) Develop fast-charging Li-ion negatives other than lithium titanate Develop high-energy-density electrodes with high ionic and electric conductivity
MECHANISMS AND MODELING		Develop models for ion transport through solids (inorganic solids, polymers) Conduct experiments to develop a quantitative understanding of catastrophic cell failure and degradation Characterize interfaces using predictive models and diagnostics to address system lifetime and performance	
SAFETY	Develop new solid polymer electrolytes and improve existing electrolytes Reduce the cost and increase the energy density of lithium titanate anodes to be able to use them to improve system safety	Develop a non-flammable electrolyte Develop inexpensive ionic liquid electrolytes or additives Develop self-extinguishing fire-initiated foam to encapsulate cells/packs	Develop a self-balancing chemistry to eliminate the need for balancing electronics

Figure 4. Prioritized Activities to Advance Li-Ion Batteries.

SODIUM-BASED BATTERIES

Sodium-based batteries include those that either utilize a solid sodium-ion conducting membrane or liquid electrolyte. The use of solid electrolytes typically requires operation at elevated temperatures (around 300°C or higher) to reduce electrical resistance and deliver satisfactory performance. Of the sodium-based batteries that use solid electrolytes, sodium-sulfur and sodium-metal-halide chemistries are relatively mature; in fact, sodium-sulfur batteries are commercially available and have been deployed in significant amounts in Japan. These batteries are constructed with a beta alumina membrane, offer a high efficiency (up to 90%), and have energy densities comparable to those of Li-ion batteries. Efforts to develop sodium-ion batteries that employ liquid electrolytes and operate at room temperature are under way in order to reduce or eliminate the need to operate at elevated temperatures.

Current Perceived Limitations of Sodium-Based Batteries

The fundamental challenge for current sodium-based batteries is that their cost is still higher than the targets for broad penetration in stationary markets. Reducing the cost of sodium-based batteries requires improvements in performance, reliability, and durability. Challenges involving chemistries, materials, battery design, manufacturing and stack design, controls and monitoring, and testing and deployment must be identified and addressed before sodium batteries can achieve widespread deployment at grid-scale storage levels. The gaps and limitations that, if overcome, could make the most significant advances toward this end goal include the following:

- *Current Sodium-Sulfur Batteries Pose a Potential Safety Concern.* In the event that the beta alumina membrane were to break down, sulfur would contact molten sodium, leading to an energetic reaction that could potentially cause a fire. While this risk is successfully managed in more commercial installations today, the potential for a damaging incident is a perceived limitation to the widespread deployment of sodium-sulfur batteries.
- *Sodium Batteries Must Operate at High Temperatures.* Sodium batteries must operate at temperatures in the range of 300°C–350°C. These systems require costly thermal management systems to maintain this operating temperature regime because repeated freeze and thaw cycles dramatically reduce system cycle life.
- *Current Electrolyte Structures Limit Sodium Battery Performance and Incur High Production Cost.* Current electrolytes used in sodium batteries are made in a tubular shape with a wall thickness of about 1–2 mm to maintain structural and mechanical stability. The thick tubular electrolyte is difficult to scale up and requires high operating temperatures to have satisfactory performance. Additionally, the beta alumina membrane is sensitive to moisture and can short while operating at a high current density.
- *Corrosive Cathodes in Sodium-Sulfur and Sodium-Metal-Halide Batteries Limit Materials Selection and Reduce Durability of the Device.* Molten sulfur in the cathode chamber of sodium-sulfur batteries is corrosive, as is the second electrolyte (NaAlCl4 melt) in the cathode of sodium-metal-halide batteries. The corrosive environment prevents the use of cost-effective materials for packaging and degrades the materials and battery performance.
- *Current Systems Have Limited Portability.* The current size, weight, and high-temperature operation of sodium batteries makes them difficult to transport. Utilities will want the ability to move energy storage systems during their useful lifetimes as energy storage needs evolve with the grid. Limited portability, therefore, is a significant drawback for sodium-based systems.
- *System Manufacturing Processes Are Costly.* The manufacturing processes of high-temperature casing materials are costly and not easily automated.

- *Emerging Sodium-Ion Batteries Are Limited by the Availability of Materials.* These materials need to be able to achieve the desired system capacity and allow for facile sodium ion insertion/deinsertion.

Cost Targets of Sodium Batteries

Over time, better materials utilization can reduce the cost of sodium batteries. Based on current knowledge of sodium batteries, the following installed cost targets (set for everything needed up to direct current output to the converter) reflect the push and pull of the energy storage market:

- Current: $3,000/kW
- 2020: $2,000/kW
- 2030: $1,500/kW

The following range of lifecycle costs could also help achieve system targets:

- Current: $0.04–$0.75/kWh/cycle
- 2020: $0.01–$0.27/kWh/cycle
- 2030: $0.01–$0.08/kWh/cycle

Priority Activities to Advance Sodium Batteries

With targeted research and development, sodium batteries have the potential to contribute to the advancement of widespread grid-scale energy storage. There are a variety of activities and initiatives that could help overcome the current gaps and limitations of sodium batteries in areas such as electrochemical combinations, system construction, and pilot-scale testing. For sodium batteries, activities and initiatives can accelerate progress in the following areas:

- *Modification of Electrode Chemistries and Optimization of Interfaces* – The performance of sodium batteries is largely determined by interfaces and minor chemistries at the cathode side. The ceramic electrolyte often does not demonstrate a satisfactory wetting property to the molten sodium. Surface treatment and interfaces are required to enhance electrical contact to decrease resistance. Minor additions, such as a second electrolyte in the cathodes, have to be optimized to maximize the battery performance. Using surface-science techniques to identify and understand impurities on sodium battery anodes and cathodes can increase the cost-effectiveness and reliability of sodium batteries. Increased understanding of battery degradation modes can help to increase the tolerance of system components to impurities, extending system life.
- *New Solid Sodium-Ion Conducting Electrolyte* – Beta alumina is the only mature sodium-ion conducting membrane, and Nasicon has been investigated for potential use as a membrane. Discovery of a new solid-state electrolyte that can demonstrate satisfactory sodium-ion conductivity and other required properties can lead to the

development of more cost-effective devices that allow satisfactory operation at reduced temperatures.
- *Stack/System Construction* – Developing robust planar electrolytes can reduce stack size and resistance and provide an alternative to current cylindrical electrolyte designs. Identifying low-cost materials that can encase high-temperature cells, reducing operation temperature, and implementing U.S. manufacturing processes also have the potential to reduce system cost and increase the manufacturability and ease of integrating sodium batteries into the electric grid.
- *Operational* – Implementing pilot-scale testing of battery systems can help to develop performance parameters for grid applications.
- *New Concepts for Sodium Batteries* – Developing cost-effective sodium-air and sodium-ion batteries could expand the potential for sodium batteries by providing new technologies with electrochemical compositions that are different from sodium-beta alumina batteries.

SODIUM-BASED BATTERIES	NEAR TERM (< 5 years)	MID TERM (5–10 years)	LONG TERM (10–20 years)
ELECTRODE CHEMISTRIES AND INTERFACES	Increase understanding of degradation modes and leverage in situ characteristics and technologies Modify and optimize minor electrode chemistries and interfaces in the sodium-beta alumina batteries Develop methods for low-cost sodium purification	Develop a new second electrolyte and leverage existing molten salt and electro-plating expertise (for medium-temperature batteries) Increase impurity tolerance, toughness, moisture resistance, and conductivity of sodium conducting layers	Use surface-science techniques to identify species on sodium-ion anodes and cathodes
ELECTROLYTES		Develop new sodium-ion conducting electrolytes	
STACK/SYSTEM CONSTRUCTION	Develop robust planar electrolytes to reduce stack size and resistance Develop low-cost (<$1/m²), corrosion-resistant foils and coatings for current collectors (metal and electron-conducting) for moderate and low temperatures Optimize enclosures in U.S. batteries by enhancing portability, increasing the ambient temperature of the envelope, and implementing U.S. manufacturing Identify low-cost materials for encasing high-temperature cells		
OPERATIONAL	Implement pilot-scale testing of battery systems to develop performance parameters for grid applications Develop an accelerated durability testing protocol and identify key failure modes for a 5-year and 10-year lifetime	Decrease operating temperature, preferably to ambient temperature	
NEW SODIUM BATTERY CONCEPTS			Develop a true sodium-air battery that provides the highest value in almost any category of performance Develop low-cost anodes and cathodes for a new generation of sodium-ion batteries

Figure 5. Prioritized Activities to Advance Sodium Batteries .

The success of these activities and initiatives will require significant support from DOE. To help DOE better focus its resources over time, Figure 5 divides the solutions by the time frame in which they will impact the market: near term (less than 5 years), mid term (5–10 years), and long term (10–20 years). The bolded activities are high-priority initiatives.

FLOW BATTERIES

Flow batteries are electrochemical devices that store electricity in liquid electrolytes. During operation, the electrolytes flow through electrodes or cells to complete redox reactions and energy conversion. The electrolytes on the cathode side (catholyte) and the anode side (anolyte) are separated by a membrane or separator that allows for ion transport, completing the electrical circuit. Researchers have identified a number of potential redox flow battery chemistries, including iron-chromium, all vanadium, and zinc-bromide in varied supporting electrolytes, such as sulfuric acid or hydrochloric acid. The capability of flow batteries to store large amounts of energy or power, combined with their potential long cycle life and high efficiency, makes them promising energy storage devices for grid energy applications and reasonable options for power applications. While multi-MW/MWh systems have been demonstrated, these technologies still have challenges to overcome to meet market requirements—most notably, achieving the cost reductions necessary to gain market acceptance for grid-scale applications.

Current Perceived Limitations of Flow Batteries

The fundamental challenge currently restraining the market penetration of existing flow batteries is their inability to fully meet the performance and economic requirements of the electric power industry. To do so, flow battery developers must identify and resolve materials, cell chemistries, and stack and system design and engineering challenges, all of which factor into system cost. The gaps and limitations that, if overcome, could make the most significant advances toward the end goal of widespread commercial deployment include the following:

- *Unwanted Cross-Transport Can Lead to Efficiency Loss and Can Contaminate Electrolytes.* This issue is particularly important for flow batteries that employ different active species in the catholyte and anolyte. For example, in the iron-chromium system, cross-transport of chromium and iron cations or complexes could lead to columbic efficacy loss and contamination.
- *Flow Battery Materials May Be Unstable in Certain Conditions.* The stability and durability of membranes and electrolytes at various temperatures and in the presence of strong reduction and oxidation conditions can also threaten the performance and reliability of flow batteries.
- *The Stack Design of Flow Batteries May Cause Issues at Grid Scale.* There are trade-offs between flow rates, shunt currents, and cell performance. Conductive paths of shunt currents can short out, which creates potential scale-up problems.

- *Hydraulic Subsystems Are Needed to Ensure System Robustness.* Hydraulic subsystems, including valves, pipes, and seals, do not currently have the low cost, long life, chemical robustness, and efficiency that flow batteries require. Flow batteries also need low-cost (<$5/lb), media-compatible plastics, as well as the materials, designs, and manufacturing processes to allow less expensive (less than $0.50/gal) and more robust anolyte and catholyte tanks.
- *Real-Time Electrolyte Analysis Tools Are Limited.* Flow batteries require advanced sensors, real-time monitoring systems, and other real-time analysis tools to assess the state of charge, flow rates, balance, and state of health of vanadium redox flow batteries.
- *Flow Batteries Have Experienced Poor Industry Perception.* The electric power industry has a poor perception of flow batteries. Inconsistent and unclear rules for materials containment also make it difficult to advance these systems.

Cost Targets of Flow Batteries

Over time, better materials utilization and device design can reduce the cost of flow batteries. Based on current knowledge, the following capital cost targets can be reached through realistic and achievable technology improvements. While these cost targets reflect the goals of some developers, they may not be generally accepted by the energy storage industry:

- 2015: $200–$250/kWh capital cost
- 2020: $150–$200/kWh capital cost
- 2030: $100–$150/kWh capital cost

Priority Activities to Advance Flow Batteries

With targeted research and development, flow batteries have the potential to contribute to the advancement of grid-scale energy storage. There are a variety of activities and initiatives that could help overcome the current gaps and limitations of flow batteries in areas such as membranes, modeling and design, stack design and manufacturing, impurities, redox chemistry, and materials compatibility. For flow batteries, progress can be made in the following areas:

- *Membranes* – Improving membranes and developing layered, multi-functional membranes can reduce electrolyte crossover, lower system cost, increase stability, and lower resistance.
- *Modeling and Design* – Developing a national computational fluidics center at a national laboratory or university will enable energy storage device experts to perform multi-scale modeling to improve system performance and cost. Tailoring catalyst layer and flow field configurations will improve mass transport and reduce cost.
- *Stack Design and Manufacturing* – Funding or creating a center for stack design and manufacturing methods will help to facilitate and optimize the scale-up and integration of flow batteries in the electric grid.

- *Impurities* – Identifying which impurities to screen for and developing an inline, real-time sensor for detecting electrolyte composition can enable the lower cost and resistance of flow batteries.
- *Redox Chemistry* – Identifying low-cost anti-catalysts and redox catalysts for negative electrodes, and developing non-aqueous flow battery systems with wider cell operating voltages will improve the efficiency of flow batteries.
- *Materials Compatibility* – Developing low-cost, chemically and thermally tolerant resins for piping, stacks, and tanks, and establishing a components database will better enable integration.

The success of these activities and initiatives will require significant support from DOE. To help DOE better focus its resources over time, Figure 6 divides the solutions by the time frame in which they will impact the market: near term (less than 5 years), mid term (5–10 years), and long term (10–20 years). The bolded activities are high-priority initiatives.

FLOW BATTERIES	NEAR TERM (< 5 years)	MID TERM (5–10 years)	LONG TERM (10–20 years)
MEMBRANES	Investigate cost-effective membrane material alternatives, such as hydrocarbon-based materials	**Improve membranes to enable minimum crossover, lower system costs, increased stability, and reduced resistance** Develop layered, multi-functional membranes	
MODELING AND DESIGN	**Create a computational fluidics center at a national laboratory or university** Perform multi-scale modeling of the reaction mechanism, battery cell, and energy storage system that includes modeling, analysis, and diagnostics to improve system performance and cost	**Improve mass transport via tailored catalyst layer and flow field configurations to increase operating current density and reduce system cost per kilowatt**	
STACK AND MANUFACTURING	**Establish a center for stack design and manufacturing methods, including joint and seal design** Leverage lessons learned from the DOE hydrogen fuel cell program to manufacture robust seals Develop advanced cell and stack designs that leverage known chemistries	Jointly select membranes and electrolytes to allow high current and high isolation	
IMPURITIES	**Develop an inline, real-time sensor that can detect impurities in electrolyte composition for various flow battery chemistries** Identify driving forces for parasitic side reactions and determine which impurities should be screened Identify lower-cost electrodes ($5–$10/m²) and characterize their suitability and availability for specific flow batteries		

CONTINUED ON PAGE 32

FLOW BATTERIES	SHORT TERM (< 5 years)	MID TERM (5–10 years)	LONG TERM (10–20 years)
REDOX CHEMISTRY	Identify low-cost hydrogen suppression materials (anti-catalysts) and redox catalysts for negative electrodes Identify low-cost oxidation-resistant materials and redox catalysts for positive electrodes	Modify low-cost redox couples to improve solubility and reduce cost per kWh	Develop non-aqueous flow battery systems with wider cell operating voltages to improve efficiency
MATERIALS COMPATIBILITY	Develop low-cost, formable, chemically and thermally tolerant resins for piping, stacks, and tanks Fix leaks and develop combustible materials for media, stack, tanks, and piping to enable high-quality, low-cost manufacturing Develop a materials compatibility database for production plastics that includes tensile strength and ductility over time under various conditions (chemical, temperature, time) Establish a components database—including valves, pumps, contactors, and sensors—for integration that includes cost Develop manufacturable, corrosion-resistant plates		

Figure 6. Prioritized Activities to Advance Flow Batteries .

POWER TECHNOLOGIES

From a grid-management perspective, all energy storage devices could be classified as power technologies, as they provide power to the grid when needed for any of the applications described in Table 2. For the purposes of this report, the group of technologies considered to be power technologies are those that are designed to provide high rates of charge acceptance and injection over short time durations. Utilities and energy storage providers have successfully demonstrated two power technologies—electrochemical capacitors and high-speed flywheels—in grid power applications. Flywheels generate power by accelerating or decelerating a rotor that is coupled to an electromagnetic field. Electrochemical capacitors store energy in the electric double layer at the electrode-electrolyte interface, and, in some instances, as a fast Faradaic process referred to as a psuedocapacitance.

Power technologies must have sufficient capacity to support the requisite pulse durations demanded by the grid and must be able to do so with high efficiency. These devices are most suited for the grid application of area and frequency regulation, as described in Table 2, which requires these devices to cycle tens or hundreds of thousands of times. The renewables grid integration application area is also a possible application for these systems, as long as these systems can supply the energy necessary to support longer pulse loads (1–2 seconds). While flywheels and electrochemical capacitors are currently being demonstrated and deployed, their energy capacity must be increased through significant advances in materials technology in order to achieve widespread adoption.

Current Perceived Limitations of Power Technologies

Current electrochemical capacitors and flywheels are limited by their low energy storage capacities and their high normalized costs. These technologies need to be more inexpensive and must be able to store larger amounts of energy to increase their suitability for grid applications. The following gaps and limitations have been identified and, if overcome, could make the most significant advances toward more widespread application of these technologies:

- *The Electrolytes in Capacitors Are Not Optimized for Grid Use.* The electrolytes in current designs have high wetting with low voltages and are also potentially flammable, which poses safety concerns.
- *The Normalized Cost of Electrochemical Capacitors Is too High for Grid Applications.* Current materials used in electrochemical capacitors are too high for widespread grid-scale deployment and have low energy densities, high equivalent resistances, and limited operating temperature ranges.
- *The Energy Density of High-Speed Flywheels Is too Low for Widespread Grid-Scale Use.* Materials have not yet been sufficiently developed that provide flywheels with optimized energy densities (e.g., high-strength materials that allow for increased rotor rotation rates).
- *Flywheel Designs Are Complex.* The complicated design of flywheels can enable high cycling, but stress on the flywheel hub can increase friction and consequently reduce efficiency and cycle life.

From a grid-management perspective, all energy storage devices could be classified as power technologies, as they provide power to the grid when needed for any of the applications described in Table 2. For the purposes of this report, the group of technologies considered to be power technologies are those that are designed to provide high rates of charge acceptance and injection over short time durations. Utilities and energy storage providers have successfully demonstrated two power technologies—electrochemical capacitors and high-speed flywheels—in grid power applications. Flywheels generate power by accelerating or decelerating a rotor that is coupled to an electromagnetic field. Electrochemical capacitors store energy in the electric double layer at the electrode-electrolyte interface, and, in some instances, as a fast Faradaic process referred to as a psuedocapacitance.

Power technologies must have sufficient capacity to support the requisite pulse durations demanded by the grid and must be able to do so with high efficiency. These devices are most suited for the grid application of area and frequency regulation, as described in Table 2, which requires these devices to cycle tens or hundreds of thousands of times. The renewables grid integration application area is also a possible application for these systems, as long as these systems can supply the energy necessary to support longer pulse loads (1–2 seconds). While flywheels and electrochemical capacitors are currently being demonstrated and deployed, their energy capacity must be increased through significant advances in materials technology in order to achieve widespread adoption.

Current Perceived Limitations of Power Technologies

Current electrochemical capacitors and flywheels are limited by their low energy storage capacities and their high normalized costs. These technologies need to be more inexpensive and must be able to store larger amounts of energy to increase their suitability for grid applications. The following gaps and limitations have been identified and, if overcome, could make the most significant advances toward more widespread application of these technologies:

- *The Electrolytes in Capacitors Are Not Optimized for Grid Use.* The electrolytes in current designs have high wetting with low voltages and are also potentially flammable, which poses safety concerns.
- *The Normalized Cost of Electrochemical Capacitors Is too High for Grid Applications.* Current materials used in electrochemical capacitors are too high for widespread grid-scale deployment and have low energy densities, high equivalent resistances, and limited operating temperature ranges.
- *The Energy Density of High-Speed Flywheels Is too Low for Widespread Grid-Scale Use.* Materials have not yet been sufficiently developed that provide flywheels with optimized energy densities (e.g., high-strength materials that allow for increased rotor rotation rates).
- *Flywheel Designs Are Complex.* The complicated design of flywheels can enable high cycling, but stress on the flywheel hub can increase friction and consequently reduce efficiency and cycle life.

Priority Activities to Advance Power Technologies

There are a variety of activities and initiatives that could help overcome the current gaps and limitations of high-speed flywheels and electrochemical capacitors in areas such as technology testing and validation, diagnostics and modeling, and system design. For power technologies, activities and initiatives can accelerate progress in the following areas:

- *Electrochemical Capacitors* – Testing and demonstrating electrochemical capacitors can help to validate technology lifetime, ramp rates, and other performance characteristics that need to be proven to encourage stakeholder buy-in. Diagnostics and modeling could help provide an understanding of the limitations of current electrochemical capacitor designs and could help to drive the development of high-energy electrodes.
- *High-Speed Flywheels* – Developing a 1-megawatt motor capable of vacuum operation and superconduction, and developing a hubless flywheel rotor with four times the energy capacity of existing flywheel technologies, can significantly increase the efficiency and reduce the cost of flywheels. A magnet with a higher mechanical strength and continuous operating motors is necessary to help mitigate the challenges derived from increased friction on the flywheel hub.
- *Development of New Power Technologies* – New, transformational or complementary power devices beyond electrochemical capacitors and flywheels

could play a role in advancing grid-scale storage. Such devices include hybrid capacitors, which combine an electrochemical capacitor electrode with a battery electrode, and large-scale dielectric capacitors, which could be enabled by the development of new materials and production processes.

The success of these activities and initiatives will require significant support from DOE. To help DOE better focus its resources over time, Figure 7 divides the solutions by the time frame in which they will impact the market: near term (less than 5 years), mid term (5–10 years), and long term (10–20 years). The bolded activities are high-priority initiatives.

POWER TECHNOLOGIES	NEAR TERM (< 5 years)	MID TERM (5–10 years)	LONG TERM (10–20 years)
ELECTROCHEMICAL CAPACITORS	**Develop high-power/ energy carbon electrode** Develop natural carbon sources and materials for electrochemical capacitors	Optimize materials utilization through diagnostics and modeling	
HIGH-SPEED FLYWHEELS	**Develop a 1 MW motor capable of vacuum operation and superconduction** Build magnets with higher mechanical strength Initiate on-the-fly curing of composite flywheel rotor manufacturing	**Develop hubless flywheel rotor with four times higher energy** Increase energy capacity of flywheel with new carbon nanotube materials for rotor, translating to lower cost Develop touchdown bearing for hubless flywheel design Develop lower-cost composites for flywheels and compressed-air energy storage via nanotube-enhanced composites for above-ground pressure tanks	Push power level of electrostatic motor to 100 kW Achieve 1 million rotations per minute (up from 140,000) and overcome bearing issues Develop long-length carbon nanotube systems for rotors to increase energy capacity to about 10,000 watt-hours per kilogram

Figure 7. Prioritized Activities to Advance Power Technologies.

A NUMBER OF EMERGING ENERGY STORAGE TECHNOLOGIES HAVE THE POTENTIAL TO TRANSFORM THE ELECTRIC GRID IN THE LONG TERM.

EMERGING TECHNOLOGIES

The potential for stationary energy storage to transform the electric power industry is driving the development of many emerging storage technologies, including metal-air batteries, regenerative fuel cells, liquid-metal systems, and adiabatic compressed-air energy storage. While these technologies are in their infancy, they have the potential to improve the stability and resiliency of the electric grid in the long term, following extensive testing and demonstration.

Current Perceived Limitations of Emerging Technologies

Emerging technologies can also contribute to the commercialization of grid-scale storage. Metal-air batteries, regenerative fuel cells, liquid-metal systems, and compressed-air energy storage (CAES) are among the technologies with the highest potential. Yet, these technologies have their own set of limitations and gaps preventing them from having a significant impact:

- *Metal-Air Batteries Are Not Yet Rechargeable.* To be truly sustainable and cost-effective when implemented, metal-air batteries must be able to recharge.
- *Contaminant Control for Metal-Air Batteries Is Not Cost-Effective.* Metal-air batteries are susceptible to a variety of contaminants, such as water and carbon dioxide, that can compromise their safety and performance.
- *Regenerative Fuel Cells Have Low Round-Trip Efficiency.* At present, the low efficiency of regenerative fuel cells from slow oxygen kinetics, as well as the unknown long-term stability of these devices, is inhibiting their development.

- *The Scalability of Liquid-Metal Systems Has Not Been Demonstrated.* Modeling of liquid-metal systems is currently too immature to demonstrate the potential for these systems to achieve the cost and scale requirements of grid applications.
- *There Are No Fossil-Fuel-Free (Adiabatic) Caes Systems.* Because of this limitation, cost-effective heat storage is unavailable.

Cost Targets of Emerging Technologies

Over time, better materials utilization and device design can reduce the cost of emerging technologies. Based on current knowledge of metal-air batteries, multivalent chemistries, and regenerative fuel cells, the following installed cost targets reflect ambitious but potentially achievable technology improvements. These cost targets are not intended to be predictions of cost reductions under "business as usual" conditions. Rather, they represent the aggressive cost reductions needed to accelerate widespread deployment of energy storage at grid scales:

Metal-Air Batteries

- 2020: $2,500/kWh
  2030: $200/kWh (assuming scale is achieved)

Multivalent Chemistries

- 2020: $1,500/kWh
- 2030: $250/kWh

The potential for stationary energy storage to transform the electric power industry is driving the development of many emerging storage technologies, including metal-air batteries, regenerative fuel cells, liquid-metal systems, and adiabatic compressed-air energy storage. While these technologies are in their infancy, they have the potential to improve the stability and resiliency of the electric grid in the long term, following extensive testing and demonstration.

Current Perceived Limitations of Emerging Technologies

Emerging technologies can also contribute to the commercialization of grid-scale storage. Metal-air batteries, regenerative fuel cells, liquid-metal systems, and compressed-air energy storage (CAES) are among the technologies with the highest potential. Yet, these technologies have their own set of limitations and gaps preventing them from having a significant impact:

- *Metal-Air Batteries Are Not Yet Rechargeable.* To be truly sustainable and cost-effective when implemented, metal-air batteries must be able to recharge.

- *Contaminant Control for Metal-Air Batteries Is Not Cost-Effective.* Metal-air batteries are susceptible to a variety of contaminants, such as water and carbon dioxide, that can compromise their safety and performance.
- *Regenerative Fuel Cells Have Low Round-Trip Efficiency.* At present, the low efficiency of regenerative fuel cells from slow oxygen kinetics, as well as the unknown long-term stability of these devices, is inhibiting their development.
- *The Scalability of Liquid-Metal Systems Has Not Been Demonstrated.* Modeling of liquid-metal systems is currently too immature to demonstrate the potential for these systems to achieve the cost and scale requirements of grid applications.
- *There Are No Fossil-Fuel-Free (Adiabatic) Caes Systems.* Because of this limitation, cost-effective heat storage is unavailable.

Cost Targets of Emerging Technologies

Over time, better materials utilization and device design can reduce the cost of emerging technologies. Based on current knowledge of metal-air batteries, multivalent chemistries, and regenerative fuel cells, the following installed cost targets reflect ambitious but potentially achievable technology improvements. These cost targets are not intended to be predictions of cost reductions under "business as usual" conditions. Rather, they represent the aggressive cost reductions needed to accelerate widespread deployment of energy storage at grid scales:

Metal-Air Batteries

- 2020: $2,500/kWh
- 2030: $200/kWh (assuming scale is achieved)

Multivalent Chemistries

- 2020: $1,500/kWh
- 2030: $250/kWh

Regenerative Fuel Cells

- Current: $4,000/kW (alkaline and electrolysis and polymer fuel cell with separate module for electrolysis and for fuel cells)
- 2015: $2,000/kW (alkaline and electrolysis and polymer fuel cell with one module for electrolysis and regeneration); $800–$1,000/kW (solid oxide fuel cell with one module for electrolysis and regeneration)
- 2020: $1,500/kW (for both types of fuel cells)
- 2030: $250/kW (for both types of fuel cells)

Priority Activities to Advance Emerging Technologies

With targeted research and development, emerging technologies such as metal-air batteries, regenerative fuel cells, and multivalent chemistries have the potential to contribute to the advancement of grid-scale energy storage. There are a variety of technology-specific and crosscutting activities and initiatives that could help overcome the current gaps and limitations of these technologies in areas such as materials performance, modeling, lifecycle testing, and degradation analysis:

- *Metal-Air Batteries* – Developing new catalysts with low overpotentials for oxygen reduction could make batteries more efficient, cost-effective, and bifunctional in the long term. The development of air electrodes with high electrochemical activity for metal-air batteries to lower their polarization and resistance could also have a long-term impact.
- *Regenerative Fuel Cells* – Improving thermal management in endothermic electrolysis reactions and exothermic fuel cell reactions could have a mid-term market impact. Developing alkaline membranes, which do not require the use of precious metals, and extending nano-structured, thin-film catalysts to electrolyzers could significantly increase the potential for regenerative fuel cells to impact grid storage.
- *Multivalent Chemistries* – Exploring currently untapped multivalent chemistries, such as magnesium-ion and aluminum-ion, could have a significant impact on the efficiency and cost of energy storage technologies.

EMERGING TECHNOLOGIES	NEAR TERM (< 5 years)	MID TERM (5–10 years)	LONG TERM (10–20 years)
METAL-AIR BATTERIES	Evaluate newly developed metal-air battery concepts		Develop new catalysts with low overpotentials for oxygen reduction to make the system more efficient, cost-effective, and bifunctional Develop air electrodes with high electrochemical activity and lower polarization/resistance Develop low-cost organometallic catalysis for air electrodes
REGENERATIVE FUEL CELL	Evaluate newly developed regenerative fuel cell concepts	Improve thermal management in endothermic electrolysis reactions and exothermic fuel cell reactions Research electrolyte and electrode materials and microstructures that have increased thermal balance and can reduce polarization Develop alkaline membranes that are capable of faster kinetics at high pH Extend nano-structured thin film catalysts to electrolyzers	
MULTIVALENT CHEMISTRIES		Discover new battery chemistries or other energy conversion approaches that hold potential to store large amounts of energy or power in cost-effective ways	Explore the untapped potential of multivalent chemistries Develop a hexavalent (super oxidized) iron battery Develop new catalysts to make systems more efficient, cost-effective, and bifunctional

Figure 8. Prioritized Activities to Advance Emerging Technologies.

The success of these activities and initiatives will require significant support from DOE. To help DOE better focus its resources over time, Figure 8 divides the solutions by the time frame in which they will impact the market: near term (less than 5 years), mid term (5–10 years), and long term (10–20 years). The bolded activities are high-priority initiatives.

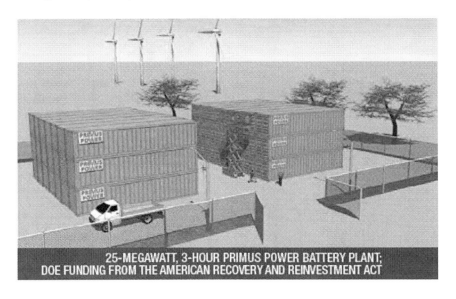

25-MEGAWATT, 3-HOUR PRIMUS POWER BATTERY PLANT; DOE FUNDING FROM THE AMERICAN RECOVERY AND REINVESTMENT ACT

THE PATH FORWARD

Research and development in advanced materials and devices has the potential to overcome many of the economic, technical performance, and design barriers that are currently preventing energy storage devices from meeting the needs of the electric power industry. Strategic materials selection and innovative system designs can reduce system cost, increase device efficiency, and ensure the reliability of storage technologies operating at grid scale. This workshop report will aid the U.S. Department of Energy in targeting investments toward the activities and initiatives that will most effectively realize the potential of energy storage materials and devices for grid-scale applications.

As the activities outlined in this workshop are put into motion, knowledge sharing within the energy storage research community will be critical to both preventing redundant efforts and developing a thorough understanding of how these technologies work together and with the grid to form a cohesive storage system. Increased understanding of various storage technologies and their ideal applications will enable energy storage device experts and DOE to explore hybrid solutions that match the complementary strengths of several storage technologies and offset the weaknesses of individual technologies. This approach will enable devices to operate across a larger range of discharge times, ultimately improving device economics and enabling grid operators to integrate energy storage in the near term.

Energy storage technologies are the solution to meeting growing electricity demands, accommodating proposed renewable energy increases, and deferring infrastructure upgrades. The development and deployment of cost-effective, widespread energy storage technologies will reduce U.S. energy dependence on foreign imports, provide electricity with fewer emissions than ever before, and enable the nation to implement the advanced and efficient grid of the future. The materials and device advances stemming from the activities and initiatives outlined in this workshop report will enable grid-scale storage and secure dependable, affordable access to electricity for nearly all U.S. citizens in the decades to come.

REFERENCES

[92] U.S. Department of Energy, Energy Information Administration, "Electricity End Use, Selected Year, 1949–2009," http://www.eia.gov/emeu/aer/pdf/pages/sec8_37.pdf (accessed August 30, 2010).

[93] U.S. Department of Energy, Energy Information Administration, *Annual Energy Outlook 2010 with Projections to 2035* (May 2010), http://www.eia.doe.gov/oiaf/aeo/electricity.html.

[94] Ibid.

[95] U.S. Department of Energy, *National Transmission Grid Study* (May 2002), http://www.ferc.gov/industries/electric/indus-act/transmission-grid.pdf.

[96] Energy Storage Association, "Technologies," (April 2010).

[97] Ibid.

WORKSHOP REPORT CONTRIBUTORS

Terry Aselage	Sandia National Laboratories
Davorin Babic	Johnson Research & Development
Nitash Balsara	University Of California, Berkeley
Scott Barnett	Northwestern University
Jacquelyn Bean	U.S. Department Of Energy
John Boyes	Sandia National Laboratories
Ed Buiel	Axion Power International
Yet-Ming Chiang	Massachusetts Institute Of Technology
Lynn Coles	U.S. Department Of Energy
Grover Coors	Coorstek/Ceramatec
Kevin Dennis	Zbb Energy
Bruce Dunn	University Of California, Los Angeles
Fernando Garzon	Los Alamos National Laboratory
John Goodenough	University Of Texas At Austin
Avi Gopstein	U.S. Department Of Energy
Harold Gotschall	Technology Insights
Ross Guttromson	Sandia National Laboratories
Bob Higgins	Eaglepitcher
Craig Horne	Enervault

Worksop Report Contributors (Continued)

Robert A. Huggins	Stanford University
David Ingersoll	Sandia National Laboratories
Andrew Jansen	Argonne National Laboratory
Haresh Kamath	Electric Power Research Institute
Dax Kepshire	Sustainx
Dennis Kountz	Dupont
Lan Lam	Csiro Energy Technology
Matthew L. Lazarewicz	Beacon Power Corporation
John Lemmon	Pacific Northwest National Laboratory
Liyu Li	Pacific Northwest National Laboratory
Robert Lin	A123 Systems
Jim Mcdowall	Saft America Inc.
Jeremy Meyers	University Of Texas
John R. Miller	Jme Capacitor
Cortney Mittelsteadt	Giner Inc.
Luis Ortiz	Massachusetts Institute Of Technology
Donald Sadoway	Massachusetts Institute Of Technology
Chet Sandberg	Altair Nanotechnologies, Inc.
Maria Skyllas-Kazacos	University Of New South Wales
Kevin M. Smith	East Penn Manufacturing
Ron Staubley	National Energy Technology Laboratory
Mike Strasik	Boeing
Michael Thackeray	Argonne National Laboratory
Larry Thaller	Retired Consultant
James Voigt	Sandia National Laboratories
Eric Wachsman	University Of Maryland
Chao-Yang Wang	Pennsylvania State University
Jay Whitacre	Cmu/Aquim Energy
Stanley Whittingham	State University Of New York
Rick Winter	Primus Power
Gary Yang	Pacific Northwest National Laboratory
Tom Zawodzinski	Oak Ridge National Laboratory

WORKSHOP FACILITATORS

Ross Brindle	Nexight Group
Jack Eisenhauer	Nexight Group
Mauricio Justiniano	Energetics Incorporated
Lindsay Kishter	Nexight Group
Sarah Lichtner	Nexight Group
Lindsay Pack	Nexight Group

In: Lightning in a Bottle: Electrical Energy Storage
Editor: Fedor Novikov

ISBN: 978-1-61470-481-2
© 2011 Nova Science Publishers, Inc.

Chapter 3

ELECTRIC POWER INDUSTRY NEEDS FOR GRID-SCALE STORAGE APPLICATIONS[*]

United States Department of Energy

ABOUT THIS REPORT

SPONSORED BY
U.S. Department of Energy, Office of Electricity Delivery and Energy Reliability
U.S. Department of Energy, Office of Energy Efficiency and Renewable Energy, Solar Technologies Program
ORGANIZED BY
Sandia National Laboratories

This report was supported by Sandia National Laboratories on behalf of the U.S. Department of Energy's (DOE) Office of Electricity Delivery and Energy Reliability and the DOE's Office of Energy Efficiency and Renewable Energy Solar Technologies Program. This document was prepared by Sarah Lichtner, Ross Brindle, and Lindsay Pack of Nexight Group under the direction of Dr. Warren Hunt, Executive Director, The Minerals, Metals, and Materials Society (TMS).

Notice: This report was prepared as an account of work sponsored by an agency of the United States government. Neither the United States government nor any agency thereof, nor any of their employees, makes any warranty, express or implied, or assumes any legal liability or responsibility for the accuracy, completeness, or usefulness of any information, apparatus, product, or process disclosed, or represents that its use would not infringe privately owned rights. Reference herein to any specific commercial product, process, or service by trade name, trademark, manufacturer, or otherwise does not necessarily constitute or imply its endorsement, recommendation, or favoring by the United States government or

---

[*] This is an edited, reformatted and augmented version of the United States Department of Energy publication, dated December 2010.

any agency thereof. The views and opinions of authors expressed herein do not necessarily state or reflect those of the United States government or any agency thereof

EXECUTIVE SUMMARY

Reliable access to cost-effective electricity is the backbone of the U.S. economy, and electrical energy storage is an integral element in this system. Without significant investments in stationary electrical energy storage, the current electric grid infrastructure will increasingly struggle to provide reliable, affordable electricity, and will jeopardize the transformational changes envisioned for a modernized grid. Investment in energy storage is essential for keeping pace with the increasing demands for electricity arising from continued growth in U.S. productivity, shifts and continued expansion of national cultural imperatives (e.g., emergence of the distributed grid and electric vehicles), and the projected increase in renewable energy sources.

Stationary energy storage technologies will address the growing limitations of the electricity infrastructure and meet the increasing demand for renewable energy use. Widespread integration of energy storage devices offers many benefits, including the following:

- Alleviating momentary electricity interruptions
- Meeting peak demand
- Postponing or avoiding upgrades to grid infrastructure
- Facilitating the integration of high penetrations of renewable energy
- Providing other ancillary services that can improve the stability and resiliency of the electric grid

STRATEGIC PRIORITIES FOR WIDESPREAD DEPLOYMENT OF ENERGY STORAGE TECHNOLOGIES

The widespread deployment of reliable and economically viable energy storage technologies will require support from the U.S. Department of Energy (DOE) and collaboration among energy storage researchers and developers, the electric power industry, and other stakeholders. While some energy storage technologies are now ready for commercial demonstration, the current market structure does not recognize the benefits of energy storage. Other promising technology options require additional research and development to improve performance and reduce costs. A coordinated approach will address challenges of a deficient market structure, limited large-scale demonstrations, insufficient technical progress, lack of standards and models, and weak stakeholder understanding of existing and emerging energy storage technologies.

Figure 1 outlines the high-priority activities and initiatives that are necessary to overcome these challenges and advance the deployment of energy storage devices today through 2030, with particular emphasis on the 1- to 5-year and 5- to 10-year time frames. These activities encompass research and development; studies, modeling, and analyses; demonstrations and data collection; and stakeholder outreach and coordination. While these four activity areas logically propel individual storage technologies from their current status to commercialization, efforts focused on a portfolio of storage options must occur simultaneously to ensure the continued advancement and increased deployment of existing and next-generation energy storage technologies.

- *Research and Development* – Continuous basic and applied research on both new and existing energy storage technologies will provide the electric power industry with more reliable, efficient, and affordable energy storage devices.
- *Studies, Modeling, and Analyses* – Studies, modeling, and analyses can be conducted to assess and simulate the impact and value that energy storage technologies will have at grid scale.
- *Demonstrations and Data Collection* – Large-scale field testing and demonstration of various technologies in multiple applications and regions across the country will validate the performance of energy storage technologies and demonstrate their value to regulators, utilities, and other potential owners.
- *Stakeholder Outreach and Coordination* – Outreach to energy storage stakeholders can promote collaboration on technology development, helping energy storage technologies to reach market penetration more quickly and to operate more efficiently and reliably once successfully integrated into the grid.

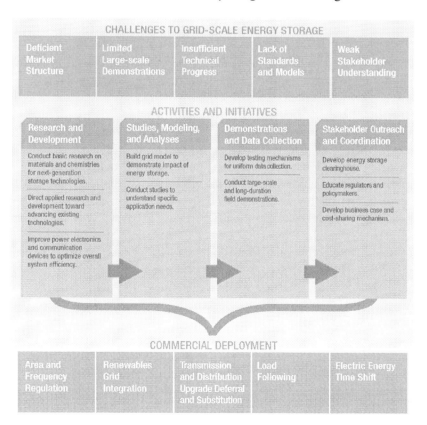

Figure 1. Utility Industry Needs for Grid-Scale Storage Structure and Content.

Committing to these activities will allow DOE, technology developers, and the electric power industry to pursue a coherent market entry strategy for energy storage technologies in grid-scale applications. In the near term, energy storage is most likely to be commercially deployed for the following applications: area and frequency regulation, renewables grid integration, transmission and distribution upgrade deferral and substitution, load following, and electric energy time shift. The use of stationary energy storage devices for these applications has the potential to transform the U.S. electric grid, offering significant benefits to the electric power industry and U.S. citizens who depend on cost-effective, reliable electricity.

INTRODUCTION AND PROCESS

Stationary energy storage at the grid scale promises to transform the electric power industry. Energy storage technologies are a key enabler of grid modernization, addressing the electric grid's most pressing needs by improving its stability and resiliency. Investment in energy storage is essential for keeping pace with the increasing demands for electricity arising from continued growth in U.S. productivity, shifts and continued expansion of national cultural imperatives (e.g., the distributed grid and electric vehicles), and the projected increase in renewable energy sources. The development of cost-effective energy storage technologies will provide the flexibility that the electric grid needs to respond to fluctuating and escalating electricity demands, ensuring that electricity is available when and where it is needed.

Current research and demonstration efforts by the U.S. Department of Energy (DOE), the national laboratories, electric utilities and their trade organizations, storage technology providers, and academic institutions provide the foundation for the extensive effort that is needed to accelerate widespread commercial deployment of energy storage technologies. Many policymakers and other stakeholders are still unaware of the benefits these technologies can provide to a variety of grid applications. In order for grid-scale storage to become a reality, the electric power industry, researchers, policymakers, and other stakeholders need to understand and address the storage needs of the electric power industry, the challenges to the widespread commercial deployment of energy storage devices, and the opportunities these technologies have to modernize the electric grid.

The Minerals, Metals & Materials Society (TMS) organized a workshop to support DOE's contributions to the commercialization of stationary energy storage at grid scale. The DOE Office of Electricity Delivery and Energy Reliability, the DOE Office of Energy Efficiency and Renewable Energy Solar Technology Program, and Sandia National Laboratories sponsored this facilitated workshop that was designed to garner critical information from forward thinkers to develop a path forward for grid-scale energy storage.

Thirty-five stakeholders and experts from across the electric power industry, research, and government communities attended the workshop on June 19–20, 2010 in Albuquerque, New Mexico. The workshop focused its discussions on determining the performance targets that energy storage technologies must meet and the challenges these technologies must overcome to achieve widespread commercialization in grid-scale applications. Participants applied diverse perspectives to identify methods for technology commercialization and implementation, the needs of the electric power industry from a technology-driven perspective, and the needs of the power companies and electric system planners and operators who will use energy storage technologies.

While all energy storage technologies and systems were within the scope of the workshop, the main focus was on technologies for which DOE involvement could accelerate progress toward commercial deployment at grid scale. The time frame under consideration was today through 2030, with particular emphasis on the 1- to 5-year and 5- to 10-year time frames.

Based on the consensus of the workshop participants, this report provides the following guidance to DOE:

- Opportunities and priority applications for grid-scale storage
- Challenges to widespread commercial deployment of storage technologies
- Activities and initiatives for widespread storage adoption

An additional workshop, which immediately followed the workshop on the energy storage needs of the electric power industry, convened experts to identify advanced materials and energy storage devices that can address the needs of the electric power industry. The reports from these workshops will inform future DOE program planning and ultimately help to commercialize energy storage at grid scale.

THE STATE OF THE ELECTRIC GRID: THE CASE FOR STORAGE

Electricity demand in the United States is steadily rising; in 2009, electricity consumption was more than five times what it was 50 years ago.1 The aging electric grid does not have the ability to transmit these large amounts of electricity from the point of generation to the end user as reliably as the U.S. economy requires.2 As wind, solar, and other variable renewable energy sources are deployed in greater quantities, transmission and distribution lines are also unable to accommodate the variable power production that often comes from remote locations. Increasing electricity demands and the shift to renewable energy sources in the United States will require immediate and cost-effective grid updates to provide consumers with electricity at the cost and with the reliability they have come to expect.

Electricity Consumption in the United States

U.S. electricity consumption is projected to continue growing in the years ahead (see Figure 2). U.S. electricity demand is expected to increase at a rate of 1% each year through 2035, at which point the country is expected to consume 5,021 billion kilowatt-hours of electricity.3 This increasing demand has and will continue to put stress on electricity generation, transmission, and distribution infrastructure. To meet the increased electricity demands expected by 2035, an additional 250 gigawatts of generating capacity will have to be added to the electricity generation infrastructure. Much of this additional generating capacity will be derived from renewable energy sources like wind and solar power.4

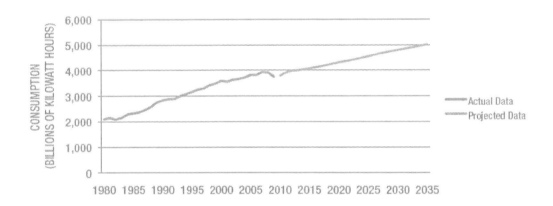

Figure 2. U.S. Net Electricity Consumption[5].

Limitations of the Current Electric Grid

The U.S. electric power grid is a complex system that transfers electricity generated at power plants to substations via 160,000 miles of transmission lines, and then to a variety of consumers throughout the nation via distribution lines (see Figure 3). This system was developed by connecting local grids to form more robust, larger networks that would ensure that nearly all Americans had dependable access to electricity.6 While this methodology worked in the past, widespread development has overburdened the grid in high-demand regions.

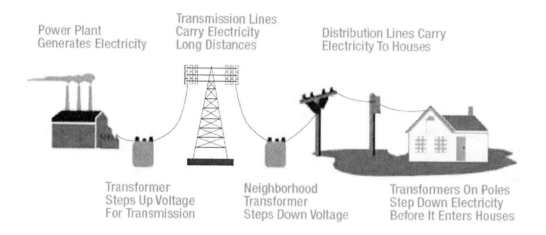

Figure 3. Electricity Transmission from Generation to Point Of Use [7].

As a result of this increase in demand, the grid often experiences interruptions in electric service. The cost of power interruptions to U.S. electricity consumers is approximately $80 billion each year—about one-third of annual electricity costs.8 Many of these interruptions occur due to problems at the distribution level and may be mitigated by distributed energy storage approaches.

Service interruptions exhibit the inefficiencies of current grid networks and emphasize the dire need to modernize the electric grid so it can respond to increasing electricity demands and shifts in generation sources. While building new generation plants and transmission and distribution lines is a costly and time-consuming endeavor, energy storage can optimize the capacity factor of current grid operation. Advanced storage could provide a reliable and cost-effective alternative to infrastructure expansion.

The Shift to Renewable Energy Sources

Perhaps the most significant trend driving the need for grid-scale energy storage is the shift to renewable energy sources, such as wind and solar. While coal has traditionally been the largest fuel source for U.S. electricity generation, emphasis on cleaner energy and decreased reliance on fossil fuels and other nonrenewable sources has placed greater attention on renewable sources for electricity generation. Figure 4 displays the breakdown of electrical power generation by source for 2008.

Currently, renewable sources (including hydroelectric) provide only 9% of the U.S. electricity supply, but federal and state incentive programs and regulations are driving the adoption of renewable energy systems and the purchase of electricity generated from renewable sources (see Figure 5 for projected electricity generation from renewables through 2035). Forty-two states and the District of Columbia have already set specific mandates or goals for a certain percentage of electric power generation and sales to come from renewable energy sources. And utilities across 47 states now offer their consumers the opportunity to purchase electricity generated by renewable energy resources.[10]

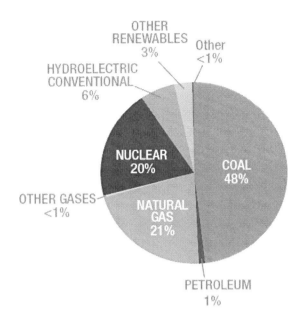

Figure 4. U.S. Electric Power Industry Net Generation by Fuel, 2008.[9]

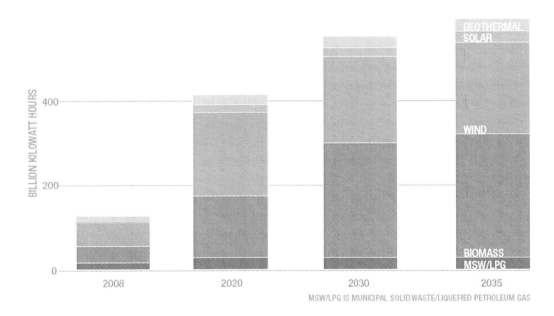

Figure 5. Nonhydroelectric Renewable Electricity Generation, 2008–2035.[11]

While beneficial from an environmental standpoint, the increased demand for renewables-generated power will place increased stress on the electric grid. The variable nature of renewable sources, particularly wind and solar, poses reliability concerns that must be addressed as renewable sources serve a larger role in electricity generation.12 The often remote locations of these sources also causes issues with renewables grid integration and the capacity of transmission and distribution infrastructure.13 Energy storage technologies are well positioned to help offset the intermittent electricity generation from renewable sources and could serve an integral role in the increased adoption of these alternative energy sources.

ENERGY STORAGE OFFERS SOLUTIONS

As previously emphasized, one of the most promising approaches to addressing the growing limitations of the electric grid and the increasing demand for renewable energy is to incorporate stationary energy storage technologies into the U.S. electric grid. With a variety of short-, mid-, and long-term storage options serving multiple applications, energy storage devices can provide multiple simultaneous benefits, including balancing services such as regulation and load following; supply power during brief disturbances to reduce outages and the financial losses that accompany them; defer and substitute transmission and distribution upgrades; and greatly enhance the future resilience of the electric grid while preserving its reliability.

Applications

Stationary energy storage technologies can be applied to a variety of applications that help the electric power industry provide customers with reliable and affordable electricity. By responding to the grid faster than traditional generation sources, operating efficiently at partial load, and being able to vary discharge times depending on application need, energy storage devices can help alleviate momentary interruptions and meet peak demand without requiring major upgrades to grid infrastructure. Some storage applications (e.g., electric supply capacity, area regulation, and reserve capacity) enable compensation from some of the existing ancillary services markets in certain regional transmission organizations. The purpose and benefits of energy storage applications are briefly discussed in the following table.

Table 1. Grid Storage Applications[14]

Electric (Grid-Supplied) Energy Time Shift	Charges the storage plant with inexpensive electric energy purchased during low price periods and discharges the electricity back to the grid during periods of high price
Electric Supply Capacity	Reduces or diminishes the need to install new generation capacity
Load Following	Alters power output in response to variations between electricity supply and demand in a given area
Area Regulation	Reconciles momentary differences between supply and demand within a given control area
Electric Supply Reserve Capacity	Maintains operation when a portion of normal supply becomes unavailable
Voltage Support	Counteracts reactive effects to grid voltage so that it can be upheld or reinstated

Table 1. (Continued)

Transmission Support	Enhances transmission and distribution system performance by offsetting electrical irregularities and interruptions
Transmission Congestion Relief	Avoids congestion-related costs by discharging during peak demand to reduce transmission capacity requirements
Transmission And Distribution Upgrade Deferral And Substitution	Postpones or avoids the need to upgrade transmission and/or distribution infrastructure
Substation On-Site Power	Provides power to switching components and communication and control equipment
Time-Of-Use Energy Cost Management	Reduces overall electricity costs for end users by allowing customers to charge storage devices during low price periods
Demand Charge Management	Reduces charges for energy drawn during specific peak demand times by discharging stored energy at these times
Electric Service Reliability	Provides energy during extended complete power outages
Electric Service Power Quality	Protects on-site loads against poor quality events by using energy storage to protect against frequency variations, lower power factors, harmonics, and other interruptions
Renewables Energy Time-Shift	Stores renewable energy (which is frequently produced during periods of low demand) to be released during periods of peak demand
Renewables Capacity Firming	Addresses issues with ramping from renewable sources by using stored energy in conjunction with renewable sources to provide a constant energy supply
Wind/Solar Generation Grid Integration	Assists in wind- and solar-generation integration by reducing output volatility and variability, improving power quality, reducing congestion problems, providing backup for unexpected generation shortfalls, and reducing minimum load violations

Technologies

Storage technologies currently being researched, developed, and deployed for grid applications include high-speed flywheels, electrochemical capacitors, traditional and advanced lead-acid batteries, high-temperature sodium batteries (e.g., sodium-sulfur and sodium-nickel-chloride), lithium-ion batteries, flow batteries (e.g., vanadium redox and zinc bromine), compressed air energy storage, pumped hydro, and other advanced battery chemistries such as metal-air, nitrogen-air, sodium-bromine, and sodium-ion. While pumped hydro has achieved widespread deployment, all of the suitable pumped hydro locations are currently being used and only meet a small portion of the baseload electricity needs. The wide range of chemistries and structures of these devices enables them to be tailored to meet the discharge duration, capacity, and frequency demands of specific applications.

Storage applications and their associated storage technologies can be loosely divided into power applications and energy management applications, which are differentiated based on storage discharge duration. Technologies used for power applications are usually used for short discharge durations, ranging from fractions of a second to approximately one hour, to address faults and operational issues that cause disturbances such as voltage sags and swells, impulses, and flickers. Technologies used for energy management applications store excess electricity during periods of low demand for use during periods of high demand. These devices are typically used for longer discharge durations exceeding one hour to serve functions that include reducing peak load and integrating renewables. Figure 6 indicates which storage technologies are better suited for power and which are better suited for energy applications. With further research, development, and demonstration, these technologies are more likely to achieve widespread commercial deployment on the U.S. electric grid.

The storage technologies in Figure 6 are currently being demonstrated and deployed across the United States, with many other installations planned for the next five years. The energy and power capacities of some of the current and planned worldwide installations are provided in Figure 7.

STORAGE TECHNOLOGY	MAIN ADVANTAGE (RELATIVE)	DISADVANTAGE (RELATIVE)	POWER APPLICATION	ENERGY APPLICATION
HIGH-SPEED FLYWHEELS	High Power	Low Energy Density	Fully capable and reasonable	Not feasible or economical
ELECTROCHEMICAL CAPACITORS (EC)	Long Cycle Life	Very Low Energy Density	Fully capable and reasonable	Not feasible or economical
TRADITIONAL LEAD ACID (TLA)	Low Capital Cost	Limited Cycle Life	Fully capable and reasonable	Feasible but not quite practical or economical
ADVANCED LA WITH CARBON ENHANCED ELECTRODES (ALA-CEE)	Low Capital Cost	Low Energy Density	Fully capable and reasonable	Fully capable and reasonable
SODIUM SULFUR (NaS)	High Power and Energy Density	Cost and Requirement to Run at High Temperatures	Fully capable and reasonable	Fully capable and reasonable
LITHIUM-ION (LI-ION)	High Power and Energy Density	Cost and Increased Control Circuit Needs	Fully capable and reasonable	Reasonable for this application
ZINC-BROMINE (ZnBr)	Independent Power and Energy	Medium Energy Density	Reasonable for this application	Fully capable and reasonable
VANADIUM REDOX (VRB)	Independent Power and Energy	Medium Energy Density	Reasonable for this application	Fully capable and reasonable
COMPRESSED AIR ENERGY STORAGE (CAES)	High Energy, Low Cost	Special Site Requirements	Not feasible or economical	Fully capable and reasonable
PUMPED HYDRO (PH)	High Energy, Low Cost	Special Site Requirements	Not feasible or economical	Fully capable and reasonable

FULLY CAPABLE AND REASONABLE
REASONABLE FOR THIS APPLICATION
FEASIBLE BUT NOT QUITE PRACTICAL OR ECONOMICAL
NOT FEASIBLE OR ECONOMICAL

Figure 6. Overview of Storage Technologies.[15]

Increasing the amount of deployed storage technologies is critical to the widespread deployment of energy storage systems. Deploying storage technologies in real-world settings provides investors, policymakers, and researchers with the data they need to continue innovating while encouraging the utility buy-in needed to make energy storage work at grid scale. To date, developments in stationary storage technologies have already advanced these technologies to be more cost-effective and secure than current methods for integrating renewables and enhancing grid reliability. With factors such as increasing electricity demand, aging infrastructure, and the shift to renewable energy sources driving the need for a modernized grid now, a strategic approach to the deployment of grid-scale energy storage technologies has the potential to provide a cost-effective, near-term solution.

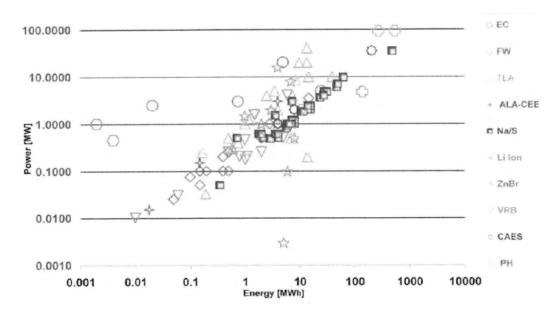

Figure 7. Installed and Planned Energy Storage Systems, April 2010.[16]

OPPORTUNITIES AND PRIORITY APPLICATIONS FOR GRID-SCALE STORAGE

With the increasing penetration of variable renewable energy sources, electricity generation is no longer constant, yet must continue to meet fluctuating electricity demands. This imbalance, along with the current grid limitations and aging infrastructure, has the potential to challenge grid operators as they manage an increasingly dynamic electric grid. Stationary energy storage technologies and devices can be used to serve multiple functions that can increase the reliability and resilience of the electric grid. These technologies can stabilize voltage and frequency, relieve momentary and prolonged stress on the grid, offset the need to build new power plants to meet increasing electricity demand and support increasing variable renewables generation, and store energy for discharge during times of high price or peak demand.

Metrics for Storage Technologies and Applications

To provide the maximum benefit to the electric power industry and electricity end users, storage technologies must meet certain economic, technical performance, and design targets for energy storage applications. While each energy storage application will require different specifications, these interrelated factors must also be considered to ensure the widespread deployment of grid-scale energy storage.

System Economics is the most important metric to the electric power industry. Consumers are accustomed to having electricity when they need it and at affordable prices, which makes the lifecycle cost of storage technologies critical to their widespread adoption. Some stakeholders in the electric power industry believe that an energy storage technology must be competitive with the cost of currently available technologies used for peak electricity generation (e.g., gas turbines) and provide increased efficiency and other benefits that adequately offset capital, operating, and lifetime costs. While this view fails to recognize the full benefits of energy storage, some decision makers in the electric power industry continue to view storage as a peak generation substitute and value it accordingly. In order to achieve widespread implementation at grid scale, the cost of stationary storage devices overall must continue to decline.

The *Technical Performance* of an energy storage device has a significant impact on overall system economics. In addition to being an affordable option, technologies must be able to meet the performance needs of a particular application. These performance needs are application-specific, but include the device's cycle life, response time, rate of charge and discharge, and efficiency. If a system does not meet the needs of its intended application, it is unlikely to be adopted. Advanced technologies must not only lower costs but exceed technical performance requirements when compared to currently available technologies.

System Design, which includes the storage device (e.g., battery, flywheel, regenerative fuel cell, or capacitor), the power conditioning and control systems that allow the system to communicate with the electric grid, and any other ancillary equipment necessary for the device's operation (e.g., auxiliary cooling systems), is also interrelated with the cost and technical performance of an energy storage system. The scalability of a system will depend on many factors, including materials availability, the feasibility of automated manufacturing, the availability of production mechanisms, and system complexity. In addition, the system design must meet the safety standards of the electric power industry, managing any health and safety risks to utility workers and the surrounding community.

Storage technologies that meet the economic, technical performance, and system design requirements of the intended application are well positioned to achieve widespread adoption in the electric power industry.

Metrics and Targets for Priority Storage Applications

There are five storage applications that have the greatest overall potential to benefit power system planning and operations: area and frequency regulation, renewables grid integration, transmission and distribution upgrade deferral and substitution, load following, and electric energy time shift. Area and frequency regulation and certain aspects of renewables grid integration are short-duration power management applications, while transmission and distribution upgrade deferral and substitution, load following, and electric energy time shift (including renewables) are long-duration energy management applications.

The stationary energy storage technologies applied to these applications must meet certain economic, technical performance, and design targets in order to optimize grid functionality. While the metrics and targets will vary depending on the specific energy storage technology or device and the location of the application, they can serve as guidelines for researchers and the electric power industry to assess the value of individual technologies. Many storage technologies currently meet one or several of the proposed metrics. However, in order to achieve widespread commercial deployment, storage systems must meet the set of targets to offer the right combination of performance and cost-effectiveness required for market acceptance. The metrics and targets for storage technologies applied to area and frequency regulation, renewables grid integration, transmission and distribution upgrade deferral and substitution, load following, and electric energy time shift are provided in the following section.

Area and Frequency Regulation (Short Duration)

Energy storage technologies are well suited to resolve momentary differences between supply and demand, as well as fluctuations in grid frequency. The current area and frequency regulation method uses turbines to balance constantly shifting load fluctuations by varying frequency and periodically adjusting generation in response to a signal from the system operator. Spinning reserves are used by electrical plants to quickly ramp up to full output when another generator goes offline; however, these generators must run constantly to assure power quality, which causes significant pollution.

The use of an energy storage technology has the potential to be far faster than regulation by a gas or steam turbine. This faster response time can minimize momentary electricity interruptions, which are more costly than sustained interruptions.[17] These storage technologies can vary output rapidly, changing from no output to full output within seconds. To optimize efficiency and response time, energy storage technologies used for area and frequency regulation must be able to communicate with the grid quickly and efficiently.

Using storage for area and frequency regulation can significantly reduce the costly consequences of power interruptions at high-tech industrial and commercial facilities.

The performance targets of energy storage technologies used for area and frequency regulation emphasize the importance of system cost, system lifetime, discharge duration, response time, and roundtrip efficiency metrics. If a storage technology is able to meet these targets for area and frequency regulation, the technology is well positioned to be adopted by the electric power industry to assist with the recovery from momentary disturbances.

AREA AND FREQUENCY REGULATION (SHORT DURATION)		
Reconciles momentary differences between supply and demand within a given area		
Maintains grid frequency		
METRIC	TARGET	SUPPORTING INFORMATION
SERVICE COST	$20 per MW per hour	The current throughput cost of area and frequency regulation services is $50 per MW per hour.
SYSTEM LIFETIME	10 years	System lifetime is based on 4,500 to 7,000 cycles per year.
DISCHARGE DURATION	15 minutes to 2 hours	Storage technologies should have symmetric charge and discharge rates for this application.
RESPONSE TIME	<1 second	Since this application is intended to reconcile momentary differences, the storage technology must be able to respond to grid signals as fast as technologically possible.
ROUNDTRIP EFFICIENCY	75%–90%	Roundtrip efficiency is the efficiency measured at the transformer of the energy output divided by theenergy input.

Renewables Grid Integration (Short Duration)

The increasing integration of renewable energy into U.S. energy supply can reduce reliance on fossil fuels and emissions from electricity generation. However, the intermittent nature of alternative energy sources introduces generation variability that can cause operational and integration issues when connected to the electric grid on a commercial scale. These issues fall into two general areas: short-duration (e.g., ramp-up/ramp-down) and long-duration (e.g., electricity energy time shift to better match renewable production with demand). This section addresses the short-duration issues; the long-duration issues are addressed later in this document.

Energy storage technologies can support the increased penetration of renewables-generated electricity by smoothing the power from these sources, thereby easing grid operation where large amounts of wind and solar generation have been deployed. To optimize the effectiveness of these operations, storage technologies need the ability to communicate and respond to the grid through the system operator.

Performance targets for storage technologies used for renewables grid integration address roundtrip efficiency, system lifetime, capacity, and response time metrics. If a storage technology is able to meet these targets for the area and frequency regulation application, the technology is more likely to be adopted by the electric power industry.

RENEWABLES GRID INTEGRATION (SHORT DURATION)		
Offsets fluctuations of short-duration variation of renewables generation output		
METRIC	TARGET	SUPPORTING INFORMATION
ROUNDTRIP EFFICIENCY	75%–90%	Roundtrip efficiency is the efficiency of the energy input measured at the transformer divided by the energy output.
SYSTEM LIFETIME	10 years	System lifetime will vary by technology and the number of cycles per year, but 10 years with high cycling would be a sufficient technology lifetime. Low-cost, shorter-lived storage technologies that can be recycled cost-effectively may offer another pathway to achieving system cost-effectiveness.
CAPACITY	1 MW–20 MW	The capacity need of a storage technology will depend on the size and intermittency of the renewables operation (e.g., a large wind farm with periods of strong wind and no wind has high potential to contributeto the grid but will be more effective with storage).
RESPONSE TIME	1–2 seconds	Fast system response times will allow storage to respond to changes in renewable operation to minimize generation fluctuations.

Transmission and Distribution Upgrade Deferral and Substitution (Long Duration)

The increasing demand for electricity requires additional transmission and distribution infrastructure to transport electricity from power plants to the customer. Building new transmission lines from power plants to substations and new distribution lines from substations to customers is a costly and time-consuming process. Further, transmission lines currently experience very low capacity utilization because they are designed for high reliability at peak conditions.

Relatively small amounts of storage can help to postpone or eliminate the need to build new transmission and distribution lines. The ability to store power generated during times of low demand can reduce the stress on power plants trying to generate power to meet demand, as well as the stress on transmission and distribution infrastructure. Storing electricity closer to the point of use lowers the congestion of the grid during peak demand.

The performance targets for energy storage technologies applied to transmission and distribution upgrade deferral and substitution emphasize the importance of system cost, discharge duration, capacity, reliability, and system lifetime. Safety is also an important metric for storage systems used for transmission and distribution deferral and substitution; these units must be safe to locate throughout the grid and must have cyber security protection consistent with current utility practice. These targets must be met to justify investing in energy storage technologies instead of investing in upgrades to the electricity and distribution infrastructure. If a storage technology is able to meet these targets for transmission and distribution upgrade deferral and substitution, the technology is well positioned to be adopted by the electric power industry as a more cost-effective and efficient way to modernize the electric grid and accommodate increasing electricity demands.

TRANSMISSION AND DISTRIBUTION UPGRADE DEFERRAL AND SUBSTITUTION (LONG DURATION)		
Delays or avoids the need to upgrade transmission and/or distribution infrastructure using relatively small amounts of storage		
Reduces loading on existing equipment to extend equipment life		
METRIC	TARGET	SUPPORTING INFORMATION
COST	$500 per kWh	The cost of transmission and distribution upgrade deferral and substitution should be comparable to or less than transmission costs. However, because storage can be deployed incrementally whereas transmission upgrades are generally large, storage has an advantage in present costs.
DISCHARGE DURATION	2–4 hours	Storing power for several hours will help to offset fluctuations in electricity demand.
CAPACITY	1 MW–100 MW	Transmission lines need storage capacity of greater than 100 MW.
RELIABILITY	99.9%	Storage systems used for transmission and distributionupgrade deferral and substitution need to be as reliable as transmission lines.
SYSTEM LIFETIME	10 years	While the system lifetime should be 10 years, the system must be easy to transport every 4 to 5 years.

Load Following (Long Duration)

Fluctuations in power demand require load following, which is the changing of power output in response to the changing balance between electricity supply and demand in a given area. Generation-based resources, such as gas turbines, are typically used for this service, but the partial-load operation required by load following uses more fuel and results in more emissions than full output. Varying output also usually results in increased maintenance needs.

Energy storage technologies are ideal for load following because they are able to insulate the rest of the grid from rapid and substantial changes in net supply in comparison to demand. Many types of storage technologies can operate at partial output or input without compromising system performance. Additionally, storage technologies can quickly respond to either increasing or decreasing load by discharging or charging. Storage systems used for load following also have the potential to be used for other applications, including electric energy time shift.

The performance targets for energy storage technologies used for load following emphasize the importance of system capital cost, operations and maintenance cost, and discharge duration. The need to quickly respond to fluctuating demands also requires the storage device to have the ability to communicate and respond to the grid through the system operator. If a storage technology is able to meet these targets for the load following application, the technology is well positioned to be adopted by the electric power industry to assist with abrupt fluctuations in electricity demand.

LOAD FOLLOWING (LONG DURATION)		
Changes power output in response to the changing balance between energy supply and demand		
Operates at partial output or input without compromising performance or increasing emissions		
Responds quickly to load increases and decreases		
METRIC	TARGET	SUPPORTING INFORMATION
CAPITAL COST	$1,500 per kW or $500 per kWh for 3-hour duration	This cost is the upfront cost of the unit.
OPERATIONS AND MAINTENANCE COST	$500 per MWh	A slightly higher operating cost should be acceptable if utilities provide justification.
DISCHARGE DURATION	2–6 hours	The discharge duration should be about 2 hours for capacity firming and 4–6 hours for load following.

Electric Energy Time Shift (Long Duration)

The cost of electricity varies along with daily cycles of changing electricity demand. Electricity prices are higher when electricity is in high demand or when supply is low than they are when the demand for electricity is lower or when electricity supply is high. Energy storage can take advantage of lower electricity prices by charging a storage device during times of low price and then discharging this electricity when electricity prices are high. Often the extreme prices (high or low) occur not at peak times or at night, but at times of high rates of change in load or renewable generation.

Electric energy time shift is often referred to as arbitrage. But unlike arbitrage in the financial sense, energy arbitrage does not occur simultaneously; instead it involves the purchase and sale of electricity at different times to benefit from a price discrepancy. For example, electricity generated from wind at night or solar power in the morning can be purchased during these off-peak times and sold later during on-peak hours. Storage devices used for electric energy time shift, including pumped hydro plants, compressed air energy storage facilities, and large battery installations, can typically store large amounts of electricity to optimize the gain from electricity price differentials and offset the disadvantages of intermittent renewable energy sources by shifting this energy to times when it is needed most.

Performance targets for electric energy time shift focus on system capital cost, operations and maintenance cost, discharge duration, efficiency, and response time. The environmental impact of storage devices used for electric energy time shift is also an important factor to consider. If a storage technology is able to meet these targets, the technology is well positioned to be adopted by the electric power industry as a way to take advantage of the fluctuating price of and demand for electricity.

ELECTRIC ENERGY TIME SHIFT (LONG DURATION)		
Stores inexpensive energy during low demand periods and discharges the energy during times of high demand (often referred to as arbitrage)		
Accommodates renewables generation at times of high grid congestion by storing energy and transmitting it when there is no congestion		
METRIC	TARGET	SUPPORTING INFORMATION
CAPITAL COST	$1,500 per kW or $500 per kWh	$250 per kWh is a utility-set metric that may not reflect the full value of storage technologies. $500 per kWh is a sufficient metric to make storage technologies competitive with a gas turbine integrated plant.
OPERATIONS AND MAINTENANCE COST	$250–$500 per MWh	Lower operations and maintenance costs will allowstorage technologies to offer the greatest economic advantages for electric energy time shift, creating greater market pull for these technologies.
DISCHARGE DURATION	2–6 hours	The price and demand for electricity may fluctuate over several hours.
EFFICIENCY	70%–80%	70%–80% is an acceptable baseline efficiency for electric energy time shift. If the efficiency of the system is only 70%, the storage system will have to incorporate other benefits to have sufficient value. The efficiency of pumped hydro and compressed air energy storage is likely to be on the lower end of this range.
RESPONSE TIME	5–30 minutes	The price of electricity will remain low or high for several hours, which decreases the need for an instantaneous response from a storage device. However, technologies with fast response can provide some frequency response and load following simultaneously with time shifting.

CHALLENGES TO WIDESPREAD COMMERCIAL DEPLOYMENT OF STORAGE TECHNOLOGIES

While the need and opportunities for energy storage are evident, a number of gaps and limitations currently constrain the adoption of grid-scale storage technologies by the electric power industry. These challenges must be overcome to achieve the widespread commercial deployment of stationary energy storage technologies and to realize the opportunities for storage in area and frequency regulation, renewables grid integration, transmission and distribution upgrade deferral and substitution, load following, and electric energy time shift applications.

The current challenges preventing the widespread commercial deployment of energy storage technologies include the following:

- Deficient market structure
- Limited large-scale demonstrations

- Insufficient technical progress
- Lack of standards and models
- Weak stakeholder understanding

These challenges are interrelated and will need to be addressed from both an individual and high-level systems approach in order to achieve commercial deployment.

Deficient Market Structure

The current electric power industry structure divides wholesale markets into generation, transmission, and distribution categories. Since storage can support all of these functions, it is challenging to classify storage from a regulatory standpoint and assess its value in comparison to traditional infrastructure. The pricing mechanism of storage also depends on its classification. Without an accurate pricing mechanism and long-term contracts, it is difficult to ensure stakeholders that they will be compensated for the various benefits energy storage provides to the grid. This compensation is necessary for stakeholders to receive a return on their investment.

Limited Large-Scale Demonstrations

One main reason that storage technologies are not yet ready to be commercialized is that they lack performance data from real-world, large-scale demonstrations. Many of the current testing sites for storage cannot accommodate grid-scale systems, which limits the size and scale of simulations that can evaluate device cost, efficiency, durability, reliability, and safety. This data is needed to validate storage devices and prove the benefits of grid-scale storage. Without additional testing and demonstrations, it will be difficult to define storage applications for specific devices and convince stakeholders and regulators of storage benefits. Additionally, large-scale demonstrations are needed to help build manufacturing infrastructure for both batteries and power electronics, develop the experience needed for large-scale project financing, and demonstrate the full-scale integration of storage with grid operations.

Insufficient Technical Progress

The energy crisis in the 1970s spurred the research and development of various energy storage technologies. But as oil prices stabilized, interest and funding of these projects dropped off, resulting in many technologies not reaching a maturity level conducive to commercial deployment.

The high cost of many of today's storage technologies, such as electrochemical capacitors and sodium-sulfur batteries, and the current complexity of storage technologies, such as high-speed flywheels and battery units, are major obstacles to production scale-up and integration of storage devices at grid scale. Additionally, the limited storage duration and energy capacity of technologies such as high-speed flywheels, electrochemical capacitors, and lithium-ion batteries is too short to meet the current needs of the electric power industry. Device efficiency and lifetime are also significant issues—technologies such as compressed air energy storage and traditional lead-acid batteries are not efficient enough to convince the electric power industry and regulators of the value of energy storage technologies.

To be integrated into the electric grid, storage devices also require control systems and power electronics that increase the reliability and safety of storage devices by increasing the speed and efficiency at which storage devices respond to grid signals. These system components are not currently advanced enough to ensure seamless interoperability between storage devices and the electric grid. Without sufficient power electronics, control systems, and other communication systems, even the most advanced storage technology will be unable to be reliable and secure when integrated into the grid.

Lack of Standards and Models

Limited demonstration data and the immaturity of storage technologies has also led to a lack of standards and models that can help storage system developers and the electric power industry design and integrate reliable and high-performing energy storage technologies. Despite recent progress in developing dispatch algorithms for regulation service, operators lack a comprehensive dispatch strategy and operational models for different storage applications; this makes it difficult to assess the impact (e.g., forced outages from renewables integration) and benefits of grid-scale storage across wholesale and resale markets and at local and national scales.

System planners and engineers need to be able to understand the value of grid-scale storage and consistently and accurately evaluate storage integration against other supply, delivery, and demand-side options. The lack of testing protocols and simulation models prevents cross comparison and more accurate performance specifications regarding energy efficiency, cost, and other valuable attributes of energy storage systems of varying sizes and applications. Additionally, the lack of standards for power electronics, communications, and control systems and protocols poses challenges to ensuring successful and safe interoperability between storage devices and the electric grid.

Weak Stakeholder Understanding

Energy storage technologies and devices have a substantial opportunity to transform and modernize the U.S. electric grid, but the benefits of grid-scale storage are not well understood by stakeholders. Without this understanding, energy storage technologies will be unable to achieve the support and level of deployment necessary to actualize substantial changes to the electric grid.

Much of the electric power industry, including utilities, grid operators, and energy storage developers, is unaware of the value of energy storage technologies and the applications to which they can be applied. This lack of understanding prevents the industry from demonstrating storage technologies in grid-scale real-world settings, which is critical to producing the understanding needed to validate the benefits of storage technologies (e.g., storage costs and return on investment, energy efficiency, and discharge duration) and to eventually attain widespread industry operating experience and acceptance.

The lack of industry understanding also inhibits regulators from considering energy storage. To allocate resources toward classifying storage within the generation, transmission, and distribution market structure of the electricity pricing system, regulators will require evidence of viable business models for energy storage. This model must demonstrate the potential of energy storage to capture diffuse revenue streams, modernize the electric grid, and reduce greenhouse gas emissions. If this evidence is not available, regulators will not be able to provide the resources needed to commercially deploy grid-scale energy storage.

With the industry and regulators lacking a clear understanding of the benefits of energy storage, it follows that the public may be difficult to convince. Public acceptance issues can prevent the siting of energy storage installations, such as large compressed air energy storage facilities, pumped hydro, or high-temperature battery installations. Required permitting could also make siting difficult, as installations may require complex interagency approvals and long permitting processes.

LEAD-CARBON BATTERY STACKS AT EAST PENN MANUFACTURING PLANT
DOE FUNDING FROM THE AMERICAN RECOVERY AND REINVESTMENT ACT

20-MEGAWATT BEACON POWER FLYWHEEL FACILITY
DOE FUNDING FROM THE AMERICAN RECOVERY AND REINVESTMENT ACT

ACTIVITIES AND INITIATIVES FOR WIDESPREAD STORAGE ADOPTION

Overcoming the challenges to widespread deployment and meeting the performance targets of applications such as area and frequency regulation, renewables grid integration, transmission and distribution upgrade deferral and substitution, load following, and electric energy time shift is critical to the effective integration of advanced energy storage technologies into the grid. To help achieve this goal, DOE should pursue a range of activities and initiatives in the areas of research and development; studies, modeling, and analyses; demonstrations and data collection; and stakeholder outreach and coordination. By tackling the highest priority initiatives in each of these areas, DOE, technology developers, and the electric utility industry can pursue a coherent market entry strategy for energy storage technologies in grid-scale applications.

Research and Development

Continuous basic and applied research on both new and existing energy storage technologies will provide the electric power industry with more reliable, efficient, and affordable energy storage devices. To pursue a complete systems approach, researchers also need to focus on device power electronics to ensure optimum interoperability between energy storage devices and the electric grid.

Conduct Basic Research on New Materials and Chemistries

The complex nature of electrochemistry offers a vast number of potential materials combinations for use in energy storage devices. Current energy storage devices focus only on a small portion of these extensive possibilities, so it is very likely that more effective, less expensive, and more robust combinations may exist. To capitalize on these opportunities, researchers need to explore new electrochemical combinations and catalysts that may be more ideal for energy storage applications. New battery couples, redox chemistries, bifunctional redox catalysts, and phase-change materials could help design next-generation technologies with the potential to revolutionize the electric power industry.

Direct Applied Research and Development Toward Existing Technologies

While new chemistries have the potential to expand the number of energy storage devices available, research can also help fine-tune existing energy storage technologies to make energy storage available to the electric grid in the near term. Applied research and development of existing technologies could help energy storage materials and device experts to advance storage devices such as metal-air batteries, adiabatic compressed air energy storage, and electrochemical capacitors, among others. Using more advanced materials for energy storage technologies, including lithium ion and advanced lead-acid batteries, could help increase the capacity, efficiency, and energy density of these technologies, making them more suitable for grid-scale operation.

Improve Power Electronics and Communication Systems

To achieve widespread commercialization of energy storage technologies, researchers need to take a systems approach by working to advance power electronics, controls, and communications components that help to integrate storage devices into the electric grid. The capabilities of these devices need to be extended so they can reduce system response time for power application to less than 100 milliseconds, decrease system cost, and heighten system reliability, which will ultimately increase the value of energy storage technologies.

Distributed and digital controls could also support the integration of intermittent renewables and the flexibility of the grid. Regardless of the methods used, collaboration between system integrators and storage device researchers and developers on storage systems is pertinent to ensure that researchers deliver devices that can meet the interoperability needs of grid operators.

RESEARCHERS NEED TO ADVANCE POWER ELECTRONICS, CONTROLS, AND COMMUNICATIONS COMPONENTS THAT HELP TO INTEGRATE STORAGE DEVICES INTO THE ELECTRIC GRID.

Studies, Modeling, and Analyses

To better understand the intricacies of energy storage technologies operating at grid scale, studies, modeling, and analyses can be conducted on a smaller scale to simulate and assess the impact and value of these technologies.

Build a Model to Demonstrate the Impact of Energy Storage on the Grid

Advanced energy storage has the ability to impact the grid in ways that have not yet been quantified due to the sheer size and intricate nature of the grid. It is critical to develop models that can be used to conduct a strategic, wide-area analysis of storage systems to investigate the effects and value of energy storage for use in grid applications. These models should assess the impact of energy storage on generation, transmission, distribution, and end-use applications and compare the cost (e.g., installation and maintenance), emissions, materials availability, efficiency, system lifetime, and cycle life of energy storage to other conventional and advanced solutions. More importantly, the models need to demonstrate the value of energy storage under varying load conditions in comparison to other types of transmission.

Conduct Studies to Understand Specific Application Needs

Studies, modeling, and analyses can help the electric power industry understand the needs of specific grid applications in order to convey application requirements, particularly cost, energy capacity, response speed, and reliability, to researchers and equipment developers. In the case of area and frequency regulation, studies should be conducted to determine the necessary response speed and minimum energy capacity required of the energy storage technologies used. Additionally, studies for renewables grid integration should quantify costs (in dollars per megawatt-hour) at various renewable penetration rates (e.g., 20%, 33%, and 40%). The results of these studies can be used to characterize the benefits of storage in specific grid applications and to formulate application standards that must be considered during technology development.

Demonstrations and Data Collection

An action plan that focuses heavily on real-world technology demonstrations at grid scale is essential to propel research and development efforts toward technology commercialization. Promising concepts and devices will not be able to achieve commercialization without available data that can validate their performance and demonstrate their value to regulators, utilities, and other potential owners.

Develop Testing Mechanisms for Uniform Data Collection

The results of studies, modeling, and analyses efforts must be measured and reported in a consistent manner to enable comparison across technologies and methods. Standards (i.e., monitoring, validation, and communications) are needed for storage and other distributed energy resources to verify system costs and performance and enable uniform data collection. Developing a DOE test center and an energy storage systems screening house could help to establish the methodology and diagnostic tools necessary to assess cycle life, efficiency, robustness, and other performance characteristics of energy storage devices. Researchers can use modular testing to obtain real-world results before a test bed is available that is able to accommodate grid-scale testing. Testing obtained using uniform methods can better enable the electric power industry to develop potential dispatch strategies for various technologies in both wholesale and retail applications.

Conduct Large-Scale and Long-Duration Field Demonstrations

Large-scale field testing and demonstration of various technologies in multiple applications and regions across the country are the foundation for data and analysis that will support technology commercialization. Longer-duration demonstrations can better quantify the real-world cycle life, shelf life, lifecycle cost, efficiency, impact on the grid, and performance of energy storage technologies. Large-scale demonstrations of energy storage technologies used for the priority grid applications identified earlier in this document can confirm whether a technology has the energy capacity and response speed necessary for specific applications. For example, DOE needs to support demonstrations of commercial-scale distributed battery systems (greater than 100-megawatt capacity, with a discharge duration of 4 hours) for use in transmission and distribution upgrade deferral and substitution as well as renewables integration. These demonstrations need to demonstrate the performance of the system over the battery lifetime. A large-scale demonstration of a 10-megawatt, 4-hour battery should be used to supply uninterruptible power (i.e., load management) for a data center and should also show the ease at which systems of this size can be relocated. These demonstrations should also include power conditioning systems to accurately evaluate the impact of the entire system on circuit operations. The results from demonstrations, particularly from high-risk, high-potential technologies and pilot deployments, will generate publishable data that can help communicate the capabilities of energy storage technologies and better quantify the value propositions of these technologies for grid-scale applications.

Stakeholder Outreach and Coordination

To ensure continued support of the commercialization of energy storage technologies, it is critical for DOE to engage stakeholders and communicate the value of storage by sharing advances in energy storage research, development, and demonstrations. Outreach to renewable energy stakeholders, system integrators, regulators, researchers and equipment developers, government agencies, and other members of the electric power industry can promote collaboration on technology development, which will help energy storage technologies reach market penetration more quickly and operate more efficiently and reliably once successfully integrated into the grid. Increased support from informed stakeholders can encourage additional stakeholder experience and acceptance through the exchange of lessons learned, best practices, and demonstration results.

Develop Energy Storage Clearinghouse

A DOE-sponsored clearinghouse of energy storage could serve as an authoritative source for the current status of energy storage research and commercialization. A detailed map should be developed to indicate the locations of deployed energy storage, along with technology specifications, performance data, and photographs. A database of application needs, technology specifications, and lessons learned from previous energy storage efforts should be used to assess research needs, identify technologies that meet the load needs and conditions of a particular area, and direct future energy storage research efforts.

Educate Regulators and Policymakers

Because regulators and policymakers develop the tariffs and rules that shape the energy storage market, it is critical to provide them with the most up-to-date and comprehensive information about energy storage technologies and applications through training, briefings, and other educational materials. To help regulators and policymakers make informed decisions about the role of energy storage, DOE should work with system operators, the Federal Energy Regulatory Commission, and the North American Electric Reliability Corporation to define and assess alternative product definitions, market rules, and dispatch signals to facilitate maximum participation and benefits from non-generation resources like energy storage. In particular, market rules should be in agreement across the United States to enable the development of a robust market for electrical energy storage.

Develop Business Case and Cost-Sharing Mechanism

Stakeholders must understand the value of energy storage technologies and anticipate a return on their investment for storage to receive the necessary funding to reach commercial deployment. To help encourage the financing of energy storage, DOE should develop a cost-sharing mechanism that also lowers individual stakeholder risk. The partial funding of real-world deployments of these technologies can demonstrate how storage can be used to solve operational problems experienced by utilities, load-serving entities, and independent power producers. This cost sharing will drive up production and automation while also reducing manufacturing costs, ultimately allowing more storage devices to be deployed across the grid in the near term.

REVOLUTIONIZING THE ELECTRIC GRID

This workshop report will guide the U.S. Department of Energy's investments in the research, development, and commercialization of energy storage technologies for grid-scale applications. Fulfilling the research and development; studies, modeling, and analyses; demonstrations and data collection, and stakeholder outreach and coordination activities and initiatives outlined in this workshop report will help the electric power industry to address the challenges currently preventing the widespread adoption of advanced energy storage. These activities and initiatives will also help the electric power industry to capitalize on opportunities for grid-scale storage, particularly for the applications of area and frequency regulation, renewables grid integration, transmission and distribution upgrade deferral and substitution, load following, and electric energy time shift.

Energy storage technologies are the solution to meeting growing electricity demands, accommodating proposed renewable energy increases, and deferring infrastructure upgrades. Storage will reduce U.S. energy dependence on foreign imports and provide electricity with fewer emissions than ever before. The nation's ability to implement the advanced and efficient grid of the future hinges on the development and deployment of cost-effective, widespread energy storage technologies. The improved grid reliability and stability resulting from the incorporation of stationary energy storage technologies in these applications will secure dependable, affordable access to electricity for nearly all U.S. citizens in the decades to come.

REFERENCES

[98] U.S. Department of Energy, Energy Information Administration, "Electricity End Use, Selected Year, 1949–2009," http://www.eia.gov/emeu/aer/pdf/pages/sec8_37.pdf (accessed August 30, 2010).

[99] U.S. Department of Energy, *National Transmission Grid Study* (May 2002), http://www.ferc.gov/industries/electric/indus-act/transmission-grid.pdf.

[100] U.S. Department of Energy, Energy Information Administration, *Annual Energy Outlook 2010 with Projections to 2035* (May 2010), http://www.eia.doe.gov/oiaf/aeo/electricity.html.

[101] U.S. Department of Energy, Energy Information Administration, "Electricity Demand," *Annual Energy Outlook 2010 with Projections to 2035* (May 2010), http://www.eia.doe.gov/oiaf/aeo/electricity.html.

[102] U.S. Department of Energy, Energy Information Administration, "Electricity End Use, Selected Year, 1949–2009," http://www.eia.gov/emeu/aer/pdf/pages/sec8_37.pdf (accessed August 30, 2010); Ibid., "Table 8. Electricity Supply, Disposition, Prices, and Emissions," Forecasts and Analysis: U.S. Data Projections, http://www.eia.doe.gov/oiaf/forecasting.html (accessed September 16, 2010).

[103] U.S. Department of Energy, Energy Information Administration, "Electricity Grid Basics," *Guide to Tribal Energy Development* (December 18, 2008), http://www1.eere.energy.gov/tribalenergy/guide/electricity_grid_basics.html.

[104] U.S. Department of Energy, Energy Information Administration, "Residential Electricity Prices: A Consumer's Guide" (January 2008), http://www.eia.doe.gov/bookshelf/brochures/rep/index.html.

[105] Kristina Hamachi LaCommare and Joseph H. Eto, Ernest Orlando Lawrence Berkeley National Laboratory, *Cost of Power Interruptions to Electricity Consumers in the United States* (February 2006), http://eetd.lbl.gov/ea/ems/reports/58164.pdf.

[106] U.S. Department of Energy, Energy Information Administration, Form EIA-923, "Power Plant Operations Report" and predecessor form(s) including Energy Information Administration, Form EIA-906, "Power Plant Report;" and Form EIA-920, "Combined Heat and Power Plant Report,"
[107] http://www.eia.doe.gov/energyexplained/index.cfm?page=electricity_in_the_united_states (accessed September 16, 2010).
[108] U.S. Department of Energy, Energy Information Administration, "Incentives," *Renewable Sources* (January 11, 2010),
[109] http://www.eia.doe.gov/energyexplained/index.cfm?page=renewable_home#tab3 (accessed August 30, 2010).
[110] U.S. Department of Energy, Energy Information Administration, *Annual Energy Outlook 2010 with Projections to 2035* (May 11, 2010),
[111] http://www.eia.doe.gov/oiaf/aeo/electricity.html.
[112] U.S. Department of Energy, Energy Information Administration, "Basics," *Renewable Sources* (January 11, 2010),
[113] http://www.eia.doe.gov/energyexplained/index.cfm?page=renewable_home (accessed August 30, 2010).
[114] Ibid.
[115] Energy Storage Association, "Applications," (April 2010).
[116] Energy Storage Association, "Technologies," (April 2010).
[117] Ibid.
[118] Kristina Hamachi LaCommare and Joseph H. Eto, Ernest Orlando Lawrence Berkeley National Laboratory, *Understanding the Cost of Power Interruptions to U.S. Electricity Consumers* (September 2004), http://certs.lbl.gov/pdf/55718.pdf.

WORKSHOP REPORT CONTRIBUTORS

Jacquelyn Bean	National Energy Technology Laboratory
John Boyes	Sandia National Laboratories
Ralph Braccio	Booz Allen Hamilton
Ed Cazalet	Megawatt Storage Farms
Lynn Coles	National Renewable Energy Laboratory
Andrew Cotter	National Rural Electric Cooperative Association Cooperative Research Network
Eric Cutter	Energy And Environmental Economics
Fouad Daghen	National Grid
Jim Eyer	Distributed Utility Associates
Eva Gardow	First Energy
Harold Gotschall	Technology Insights
Bob Higgins	Eaglepicher Technologies
David Ingersoll	Sandia National Laboratories
Mark Johnson	Advanced Research Projects Agency-Energy
Shaun Johnson	New York Independent System Operator

Workshop Report Contributors (Continued).

Haresh Kamath	Electric Power Research Institute
Robert King	Good Company Associates
Michael Kintner-Meye	Pacific Northwest National Laboratory
Hal Laflash	Pacific Gas & Electric
Matthew Lazarewicz	Beacon Power
Carl Lenox	Sunpower Corp.
Dave Marchese	Haddington Ventures
Mariko Mcdonagh	Southern California Edison
Jim Mcdowall	Saft America Inc
Drew Mcguire	Southern Company
Shaneshia Mcnair	Cps Energy
Ali Nourai	Kema
Frank Novachek	Xcel Energy
Kimberly Nuhfer	National Energy Technology Laboratory
Arnie Quinn	Federal Energy Regulatory Commission
Brad Roberts	S&C Electric Company
Mark Rowson	Sacramento Municipal Utility District
Tom Sloan	Kansas State Legislator
Kevin Smith	East Penn Manufacturing
David Walls	Navigant Consulting
Byron Washom	University Of California, San Diego
Steve Willard	Pnm Resources
Rick Winter	Primus Power
Gary Yang	Pacific Northwest National Laboratory

Facilitators

Ross Brindle	Nexight Group
Jack Eisenhauer	Nexight Group
Lindsay Kishter	Nexight Group
Sarah Lichtner	Nexight Group
Lindsay Pack	Nexight Group
Rich Scheer	Scheer Ventures

In: Lightning in a Bottle: Electrical Energy Storage
Editor: Fedor Novikov

ISBN: 978-1-61470-481-2
© 2011 Nova Science Publishers, Inc.

Chapter 4

ENERGY STORAGE: PROGRAM PLANNING DOCOUMENT[*]

United States Department of Energy

EXECUTIVE SUMMARY

Energy storage systems have the potential to extend and optimize the operating capabilities of the grid, since power can be stored and used at a later time. This allows for flexibility in generation and distribution, improving the economic efficiency and utilization of the entire system while making the grid more reliable and robust. Additionally, alternatives to traditional power generation, including variable wind and solar energy technologies, may require back-up power storage. Thus, modernizing the power grid may require a substantial volume of electrical energy storage (EES).

The Office of Electricity Delivery & Energy Reliability (OE) is taking leadership on energy storage to ensure that the technologies live up to their potential, and will assist in bringing these solutions into the commercial market. OE's vision of the Program is based on four principles:

- ➢ Understanding where to put energy storage on the system for the best value
- ➢ Developing and driving performance targets—e.g., cost, safety, cycle life
- ➢ Partnering to get the most value, specifically using research, development and demonstrations to understand the value chain, and then using feedback from demonstrations to guide the science in ways for improvements, and
- ➢ Focusing on engineering that is necessary for manufacturing.

OE is providing assistance in three main areas: research, demonstrations/deployments, and systems analysis.

---

[*] This is an edited, reformatted and augmented version of the United States Department of Energy, Office of Electricity Delivery & Energy Reliability publication, dated February 2011.

> Research is focused on technologies that store electricity in chemicals or batteries, including sodium (Na) based batteries, lithium-ion (Li-ion) battery applications for grid scale systems, advanced lead-acid batteries, and flow batteries. Research is also supported for other categories of technologies, such as superconducting magnetic energy storage (SMES), ultra-capacitors, advanced materials for flywheels, and geological aspects of compressed air energy storage systems (CAES).
> Demonstrations, including deployments within commercial-scale systems, are focused on advancements in the above mentioned batteries, flywheels, CAES, etc. There is some work, also related to advancements in power electronics and control systems as applied to energy storage.
> Systems analysis is mainly on the effective integration of storage options within energy balancing areas/authorities.

The energy storage program at OE is designed to advance all these areas and technologies. The Program is positioning to reach the Department's 2015 target of reducing the cost of energy storage by 30%. Assuming a funding level of approximately $200 million over the next five years (2011 to 2015), the Program has set for itself a number of objectives:

Near-term Objectives (within 5 years)	Longer-term Objectives
Related to Research and Development • Ensure that new, promising technologies are added to the pipeline for research and verification • Focus at the device level, and the elements needed to fully integrate devices into systems • Test technologies so that device reliability will be raised sufficiently resulting in demonstration use by utilities • Work with SBIRs, EFRCs, and ARPA-E, and through university solicitations, to mine sources of ideas. Initiate efforts in discovering new materials and chemistries to lead new EES technologies • Develop and optimize EES redox flow batteries, Na-based batteries, lead-carbon batteries, Li-ion batteries to meet the following performance and cost targets: System capital cost: under $250/kWh Levelized cost: under 20 ¢/kWh/cycle System efficiency: over 75% Cycle life: more than 4,000 cycles • Develop and optimize power technologies to meet capital cost targets under $1,750/kW **Related to Demonstrations** • Capitalize on the significant momentum gained by Recovery Act demonstration projects	• Simultaneously filter out less-promising technologies while keeping the research pipeline filled so that the level of innovation remains robust • Maintain momentum for promising technologies identified in the research phase • Develop new technologies based on previously discovered materials and chemistries to meet the following targets: System capital cost: under $150/kWh Levelized cost: under 10 ¢/kWh/cycle System efficiency: over 80% Cycle life: more than 5,000 cycles • Develop and optimize power technologies to meet capital cost targets under $1,250/kW • Maintain a pipeline of viable demonstration projects that advance the knowledge of storage systems as well as verify the reliability of ESS

Near-term Objectives (within 5 years)	Longer-term Objectives
• Working with utilities, storage providers, and renewable energy integrators to target specific demonstrations that will have high leverage (and cost sharing) • Analyze current Recovery Act demonstrations early to understand what is working and what can be improved and feed that information back into current demonstration decisions; don't wait for the Recovery Act demonstrations to be "complete" in order to use results and feedback	• Re-deploy electric vehicle batteries after their useful life in vehicles as part of a grid storage system • Demonstrate configurations for test-bed distributed generations systems and storage-connected houses ("community energy storage") • Demonstrate configurations for test-bed storage linked with demand response

To implement the above objectives, OE has set a number of specific milestones for each the next five years, as depicted on the following two pages.

Energy Storage Program Milestones

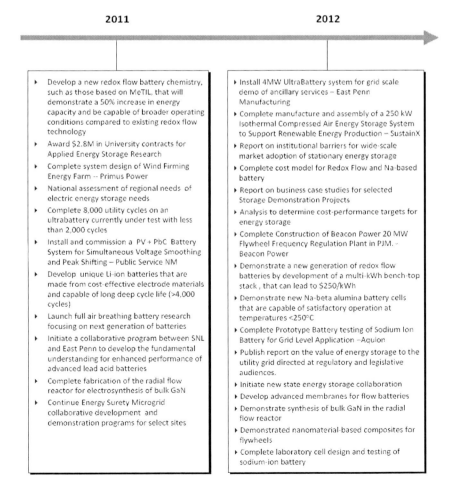

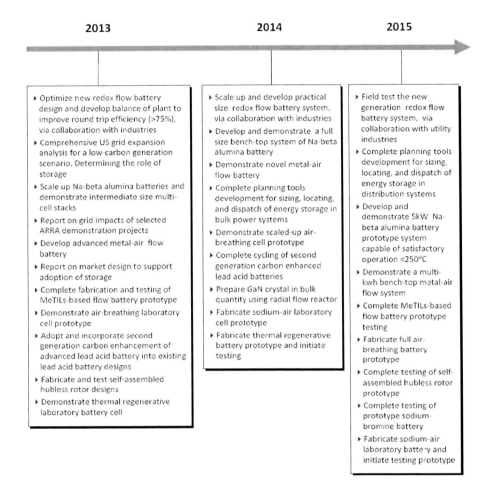

1.0. INTRODUCTION TO THE OE STORAGE PROGRAM

Modernizing the power grid will help the nation meet the challenge of a more technologically advanced, reliable system, one able to handle the projected volumes of energy required. There are estimates that by 2050 North America will need somewhere between 15 and 20 terawatt-hours of electric power annually.[1] This growth will increase pressure on finding the sources of power generation required, as well as ensuring that systems have lower carbon emissions, be more economical and commercially viable, and be environmentally sustainable.

Energy storage systems have the potential to extend and optimize the operating capabilities of the grid, since power can be stored and used at a later time. This allows for flexibility in generation and distribution, improving the economic efficiency and utilization of the entire system while making the grid more reliable and robust. Additionally, alternatives to traditional power generation, including variable wind and solar energy technologies, may require back-up power storage. Thus, modernizing the power grid may require a substantial volume of electrical energy storage (EES); the total requirement for U.S. bulk storage over the next 5-10 years has been estimated at between 10 and 100 gigawatts.[2]

In addition to large-scale grid applications, energy storage can be effective in designing smart microgrids on the residential or commercial building level, with storage as part of an integrating controller system contributing to grid stability. Convergence of a number of issues, such as community energy storage, second use of vehicular batteries, smart charging for plug-in electric vehicles, and the increasing use of rooftop solar will pose challenges to grid reliability while providing opportunities for the development of optimized storage systems.

In the three years since the inception of the Office of Electricity Delivery & Energy Reliability (OE) the Office has been working to stimulate investment in electric and energy infrastructure, advance the state of scientific development in supply and demand side electric technologies, identify barriers to continued reliable electric service, deepen consideration of security and resiliency measures in infrastructure planning, and expand partnerships with State and private-sector stakeholders. Supporting energy storage efforts is central to OE's mission as it facilitates the creation, advancement, and deployment of the new technologies that will ensure a truly modern and robust grid capable of meeting the demands of the 21st century.

1.1. The Grid Energy Storage Value Proposition

For the past decade, industry, utilities, and the Department have known that energy storage can be an important element of future grids, and energy storage is becoming a more pressing issue. A number of industry changes are happening:

- States are requiring utilities to include energy storage in their portfolio
- Energy storage is being looked to for increasing the penetration of renewable energy on the grid to meet renewable portfolio standards
- Energy storage is being assessed for enhancing existing—and aging—electric system capital assets, thereby increasing the reliability of electricity transmission and distribution
- Smart Grid deployment may require wide-scale energy storage in order to achieve its full potential and deliver on its value proposition, and
- Energy storage will be needed to support the electrification of the transportation sector.

Energy storage is poised to grow from $1.5 billion in 2010 to a $35 billion industry by 2020.[3] To help shape the impact of energy storage, the Department has a mission to help facilitate the development of a robust manufacturing base of advanced energy storage devices in the U.S. Cost effective energy storage will be a key element in ensuring the successful deployment of the grid of the future, and the Department has set a target of reducing the cost of energy storage by 30% by 2015.

Application	Description	CAES	Pumped Hydro	Flywheels	Lead-Acid	NaS	Li-ion	Flow Batteries
Off-to-on peak intermittent shifting and firming	Charge at the site of off peak renewable and/or intermittent energy sources; discharge energy into the grid during on peak periods	◐	◐	○	●	●	●	●
On-peak intermittent energy smoothing and shaping	Charge/discharge seconds to minutes to smooth intermittent generation and/or charge/discharge minutes to hours to shape energy profile	○	○	◐	●	●	●	●
Ancillary service provision	Provide ancillary service capacity in day ahead markets and respond to ISO signaling in real time	◐	◐	◐	◐	◐	◐	◐
Black start provision	Unit sits fully charged, discharging when black start capability is required	◐	◐	○	●	●	●	●
Transmission infrastructure	Use an energy storage device to defer upgrades in transmission	○	○	○	●	●	●	●
Distribution infrastructure	Use an energy storage device to defer upgrades in distribution	○	○	○	●	●	●	●
Transportable distribution-level outage mitigation	Use a transportable storage unit to provide supplemental power to end users during outages due to short term distribution overload situations	◐	○	○	◐	●	●	●
Peak load shifting downstream of distribution system	Charge device during off peak downstream of the distribution system (below secondary transformer); discharge during 2-4 hour daily peek	○	○	○	●	●	●	●
Intermittent distributed generation integration	Charge/Discharge device to balance local energy use with generation. Sited between the distributed and generation and distribution grid to defer otherwise necessary distribution infrastructure upgrades	○	○	◐	●	●	●	●
End-user time-of-use rate optimization	Charge device when retail TOU prices are low and discharge when prices are high	◐	◐	○	◐	●	●	◐
Uninterruptible power supply	End user deploys energy storage to improve power quality and/or provide back up power during outages	○	○	○	●	●	●	●
Micro grid formation	Energy storage is deployed in conjunction with local generation to separate from the grid, creating an islanded micro-grid	○	○	○	●	●	●	●

Definite suitability for application ●; Possible use for application ◐; Unsuitable for application ○

Figure 1-1. Principal Energy Storage Applications and Technologies.

Figure 1-1 summarizes 12 of the most common types of value propositions—or applications—associated with specific energy storage technologies; there are also 7 types of technologies presented.4 It is worth noting that not all energy storage technologies are at the same level of commercial readiness. Equally important, specific technologies can provide more than one type of value. As a result, determining the exact value proposition for energy storage requires an analysis of each system. As more technologies are demonstrated and deployed, the determination of value will become easier.

1.2. Grid Energy Storage at DOE

OE is taking leadership on energy storage to ensure that the technologies live up to their potential, and will assist in bringing these solutions into the commercial market. OE's vision of the Program is based on four principles:

- Understanding where to put energy storage on the system for the best value
- Developing and driving performance targets—e.g., cost, safety, cycle life
- Partnering to get the most value, specifically using research, development and demonstrations to understand the value chain, and then using feedback from demonstrations to guide the science in ways for improvements, and
- Focusing on engineering that is necessary for manufacturing.

OE is providing assistance in three main areas: research, demonstrations/deployments, and systems analysis.

- Research is focused on technologies that store electricity in chemicals or batteries, including sodium (Na) based batteries, lithium-ion (Li-ion) battery applications for grid scale systems, advanced lead-acid batteries, and flow batteries. Research is also supported for other categories of technologies, such as superconducting magnetic energy storage (SMES), ultra-capacitors, advanced materials for flywheels, and geological aspects of compressed air energy storage systems (CAES).
- Demonstrations, including deployments within commercial-scale systems, are focused on advancements in the above mentioned batteries, flywheels, CAES, etc. There is some work, also related to advancements in power electronics and control systems as applied to energy storage.
- Systems analysis is mainly on the effective integration of storage options within energy balancing areas/authorities.

Additionally, this program, the Energy Storage Systems Program (ESSP), provides neutral third-party testing for developers of energy storage components and systems. These bench, prototype, and field testing capacities provide industry with critical feedback on technology and design throughout the development process, and enable more effective integration of new storage technologies into the grid.

The energy storage program at OE is designed to advance all these areas and technologies. The Program is positioning to reach the Department's 2015 target of reducing the cost of energy storage by 30%. Assuming a funding level of approximately $200 million over the next five years (2011 to 2015), the Program has set for itself a number of objectives:

Near-term Objectives (within 5 years)	Longer-term Objectives
Related to Research and Development	
• Ensure that new, promising technologies are added to the pipeline for research and verification • Focus at the device level, and the elements needed to fully integrate devices into systems • Test technologies so that device reliability will be raised sufficiently resulting in demonstration use by utilities • Work with SBIRs, EFRCs, and ARPA-E, and through university solicitations, to mine sources of ideas. Initiate efforts in discovering new materials and chemistries to lead new EES technologies. • Develop and optimize EES redox flow batteries, Na-based batteries, lead-carbon batteries, Li-ion batteries to meet the following performance and cost targets: System capital cost: under $250/kWh Levelized cost: under 20 ¢/kWh/cycle System efficiency: over 75% Cycle life: more than 4,000 cycles • Develop and optimize power technologies to meet capital cost targets under $1,750/kW	• Simultaneously filter out less-promising technologies while keeping the research pipeline filled so that the level of innovation remains robust • Maintain momentum for promising technologies identified in the research phase • Develop new technologies based on previously discovered materials and chemistries to meet the following targets: System capital cost: under $150/kWh Levelized cost: under 10 ¢/kWh/cycle System efficiency: over 80% Cycle life: more than 5,000 cycles • Develop and optimize power technologies to meet capital cost targets under $1,250/kW
Related to Demonstrations	
• Capitalize on the significant momentum gained by Recovery Act demonstration projects • Working with utilities, storage providers, and renewable energy integrators to target specific demonstrations that will have high leverage (and cost sharing) • Analyze current Recovery Act demonstrations early to understand what is working and what can be improved and feed that information back into current demonstration decisions; don't wait for the Recovery Act demonstrations to be "complete" in order to use results and feedback	• Maintain a pipeline of viable demonstration projects that advance the knowledge of storage systems as well as verify the reliability of ESS • Re-deploy electric vehicle batteries after their useful life in vehicles as part of a grid storage system • Demonstrate configurations for test-bed distributed generations systems and storage-connected houses ("community energy storage") • Demonstrate configurations for test-bed storage linked with demand response

To implement the above objectives, OE has set a number of specific milestones for each the next five years, as depicted in Figure 1-2, on the following two pages.

Energy Storage Program Milestones

2011

- Develop a new redox flow battery chemistry, such as those based on MeTIL, that will demonstrate a 50% increase in energy capacity and be capable of broader operating conditions compared to existing redox flow technology
- Award $2.8M in University contracts for Applied Energy Storage Research
- Complete system design of Wind Firming Energy Farm -- Primus Power
- National assessment of regional needs of electric energy storage needs
- Complete 8,000 utility cycles on an ultrabattery currently under test with less than 2,000 cycles
- Install and commission a PV + PbC Battery System for Simultaneous Voltage Smoothing and Peak Shifting – Public Service NM
- Develop unique Li-ion batteries that are made from cost-effective electrode materials and capable of long deep cycle life (>4,000 cycles)
- Launch full air breathing battery research focusing on next generation of batteries
- Initiate a collaborative program between SNL and East Penn to develop the fundamental understanding for enhanced performance of advanced lead acid batteries
- Complete fabrication of the radial flow reactor for electrosynthesis of bulk GaN
- Continue Energy Surety Microgrid collaborative development and demonstration programs for select sites

2012

- Install 4MW UltraBattery system for grid scale demo of ancillary services – East Penn Manufacturing
- Complete manufacture and assembly of a 250 kW Isothermal Compressed Air Energy Storage System to Support Renewable Energy Production – SustainX
- Report on institutional barriers for wide-scale market adoption of stationary energy storage
- Complete cost model for Redox Flow and Na-based battery
- Report on business case studies for selected Storage Demonstration Projects
- Analysis to determine cost-performance targets for energy storage
- Complete Construction of Beacon Power 20 MW Flywheel Frequency Regulation Plant in PJM. - Beacon Power
- Demonstrate a new generation of redox flow batteries by development of a multi-kWh bench-top stack, that can lead to $250/kWh
- Demonstrate new Na-beta alumina battery cells that are capable of satisfactory operation at temperatures <250°C
- Complete Prototype Battery testing of Sodium Ion Battery for Grid Level Application –Aquion
- Publish report on the value of energy storage to the utility grid directed at regulatory and legislative audiences.
- Initiate new state energy storage collaboration
- Develop advanced membranes for flow batteries
- Demonstrate synthesis of bulk GaN in the radial flow reactor
- Demonstrated nanomaterial-based composites for flywheels
- Complete laboratory cell design and testing of sodium-ion battery

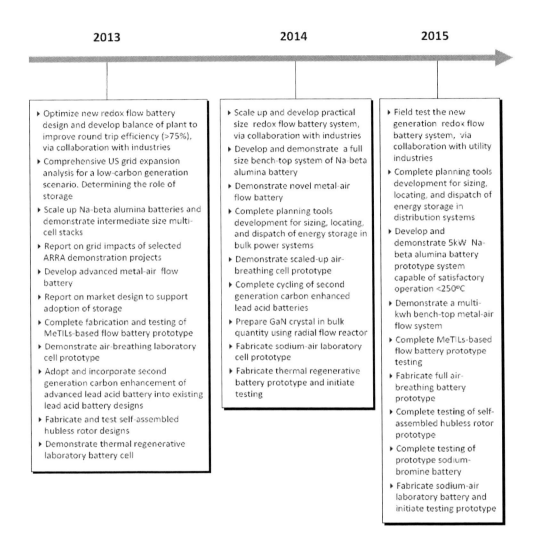

1.3. DOE's Partnership Strategies

Building and maintaining effective public–private partnerships is one of the key strategies for achieving the objectives of the ESSP. The strategy is to engage world-class professionals from key public and private organizations and help support and leverage research and development so that it meets the goals of the DOE and nation. Partners include electric utilities and manufacturers of energy storage devices, electricity consumers, project developers, and State and regional agencies.

Examples of electric utility stakeholders include investor-owned and public utilities; electric cooperatives; and Federal utilities such as the Tennessee Valley Authority, Bonneville

Power Administration, and Western Area Power Administration. Partners also include the California Energy Commission and New York State Energy Research and Development Authority, who are partnering with major pioneering storage installations. ESSP works closely with industry partners, and many of its projects (and all of its Recovery Act demonstrations projects) are cost-shared at a significant level.

The program leverages Federal resources and partners with co-sponsors on technical research initiatives led by the nation's most technically capable organizations and individuals, thus achieving the best returns on taxpayer investments. Research partners include the following:

- Universities
- Industry research organizations (e.g., Electric Storage Association (ESA))
- National laboratories (e.g., PNNL, Sandia, ORNL, NETL)
- The Office of Science
- Programs within DOE (e.g., Energy Frontier Research Centers, Solar Energy Technology Program, Vehicle Technology Program, ARPA-E)
- Potential manufacturers
- Utilities
- State energy research and development agencies
- Federal agencies such as:
  - Department of Homeland Security
  - Department of Defense
  - Federal Energy Regulatory Commission
  - Other agencies that will benefit from energy storage development.

The engagement of public–private partnerships takes several forms:

- Technical exchanges achieved through periodic conferences, workshops, annual peer reviews, informal meetings, and joint RD&D planning sessions in addition to the work being executed by the various National Laboratories.
- Communications and outreach through websites, webcasts, and publications in technical journals to foster information sharing and technology transfer.
- Cost-shared RD&D projects that leverage resources and focus on accomplishing tasks of mutual interest. This is an important element of the effort as it ensures that the parties outside the government space are also committed to ensuring the success of energy storage. It also signals the willingness of the private parties in taking over and sponsoring the energy storage efforts beyond the limitations of government.
- Competitive solicitations to engage the nation's top RD&D performers in projects to design fabricate, laboratory test, field test, and demonstrate new technologies, tools, and techniques.
- Small Business Innovative Research (SBIR) grants, which can be used by Federal agencies to nurture innovative concepts from small businesses.

Universities, industry, and National Laboratories will play a key role in the EES Program activities. Targeted capabilities at each will be applied to program management and implementation, as well as to research needs that require specific scientific and engineering talent.

2.0. CHALLENGES AND NEEDS

Increased recognition of the role of energy storage has stimulated increased Research, Development, and Demonstration (RD&D) efforts addressing both new battery materials and chemistries, less expensive and more robust designs, and enhanced grid storage systems analysis. In addition there have been full-scale demonstration projects. Many challenges remain and those that are at the center of this program are cost, reliability, value proposition, competitive environment, and regulatory environment.

Cost

- ➢ Challenges—Most of current EES technologies are not competitive in capital cost and/or life cycle cost for broad market penetration. The capital cost is at least $500/kWh in terms of energy or $2,500/kW in terms of power. Pumped hydro storage with a low life cycle cost (¢/kWh/cycle) may be an exception, but is limited by site selection, large initial investment, long construction time, and environmental concerns. The same may apply for underground compressed air storage.
- ➢ Needs—The performance and costs of a storage technology largely depend on the materials and chemistries that make up the system. A desirable system has to be made from cost-effective materials and components that can demonstrate targeted properties. It is critical to optimize existing materials and chemistries and modify current technologies to improve their performance and reduce costs. Additionally, there is a need to discover new materials—without constraints as to their availability—and components that can lead to new technologies meeting the performance and cost requirements of grid storage applications.

Reliability

- ➢ Challenges—As a utility asset, storage technologies are required for a life at least 10 years, often longer, and a deep cycle life (e.g., more than 4,000 cycles). Minimum or no maintenance is preferable. Long term reliability and durability are not typically proven for technologies entering the market. There is a lack of extensive testing data for the current storage technologies.
- ➢ Needs—There are few highly-reliable storage technologies that have been widely developed for non-grid applications and potentially suitable for the grid applications. Of particular interest are technologies such as Li-ion batteries and modified versions of lead-acid batteries (e.g., leadcarbon) both of which were initially developed for vehicle applications, but may also be suitable for grid applications. There remain questions, however, whether these technologies are suitable for various stationary

applications and what markets, if any, they might serve economically. The evaluation may offer insights or knowledge for further modification to improve their suitability as Li-ion and lead-acid lifetimes and reliabilities would need to improve significantly to meet stationary needs, and that is one of the real challenges.

Value Proposition

- ➢ Challenges—Energy storage can provide multiple values to the grid that have not been fully revealed because of market designs or lack of understanding of the benefits of fast responding bidirectional electricity flows. Significant grid/storage analysis is required to ensure that technology development is targeted at high-value applications, to understand and highlight the value of energy storage, and to build confidence and familiarity in this new resource.
- ➢ Needs—There is a need for multiple long-term energy storage demonstration projects to establish proven success. Utilities are somewhat risk averse and they need confidence that technologies perform as required before adding them to their systems. Such field demonstrations help define grid requirements for EES technologies, and provide valuable knowledge for further development and optimization of EES technologies. Current simulation and analysis tools are inadequate to assess the value of storage on the "evolving" grid, particularly for management of transient and highly dynamic conditions. We require new methodologies for optimal sizing, placements, and control strategies to maximize the value of storage.

Competitive Environment

- ➢ Challenges—The non-vehicle energy storage sector is still, generally, commercially immature. There is not enough manufacturing capability, particularly in the U.S., to deliver the projected quantity of EES systems to meet future demands. Like Li-ion batteries, many of existing EES technologies were invented in the U.S., but other countries are leading in their development and commercialization.
- ➢ Needs—It is important to demonstrate the reliability of systems so that a significant demand will build, requiring manufacturing capability and industries in the U.S., enhancing our international competiveness and creating jobs.

Regulatory Environment

- ➢ Challenges—Although there have been numerous changes to the utility market over the past decades, including deregulation, understanding the differences between markets is an ongoing challenge. Furthermore, the highly integrated nature of the evolving grid will likely encourage a use of assets in ways that defy traditional classifications. Storage can potentially play multiple roles in the grid of the future,

but only if institutional processes appropriately represent the contribution storage can provide.
➢ Needs—The contribution of energy storage is not universally understood within the policy and regulatory community. Outreach based on demonstrations, is necessary to inform policy makers of the attributes of energy storage that might be beneficial to the utility system and to customers in general. There is also a need to address the factors that inhibit beneficial deployment of energy storage and which, if any, policy or regulatory steps might be undertaken to facilitate such deployment.

3.0. ANALYSIS, RESEARCH & DEVELOPMENT AND DEMONSTRATION PROGRAM

OE's Program adopts an integrated strategy (schematically shown in Figure 3-1) to carry out technology development. The strategy involves collaboration among National Laboratories, universities, and industries forming two main teams. The research team led by National Laboratories with participation from universities and industries (e.g., through SBIR) focus on developing key materials and components, cells and prototypes, as well as carrying out grid analytics. A second, vertical, team formed of battery manufacturers and utility industries lead system design and demonstration, as well as commercialization. The vertical team feeds back issues or concerns to the research team that solve the problems and help optimize technologies. The research team has the responsibility to monitor R&D conducted by other DOE offices to ensure scientific and technology achievements are effectively translated into EES efforts.

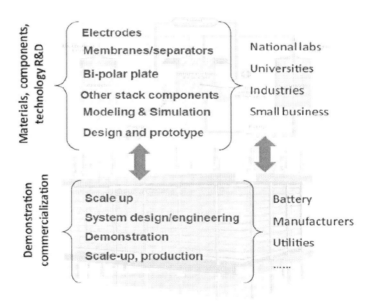

Figure 3-1. Vertically integrated efforts to develop and commercialize ESS—In this case for redox flow batteries.

3.1. Current State of Electric Storage

To understand OE's program, and its objectives, it is important to understand the current state of electric storage. Although generic in nature, the overlay of potential storage applications and functional characteristics shown in Figure 3-2 is a useful guide. From fast reacting high power applications for frequency regulation and power conditioning, which can prevent transient phenomena such as power spikes, to slow response high energy applications of peak shaving and diurnal shifting, which can prevent brownouts or even match renewable generation to load, the diversity of storage needs are significant.

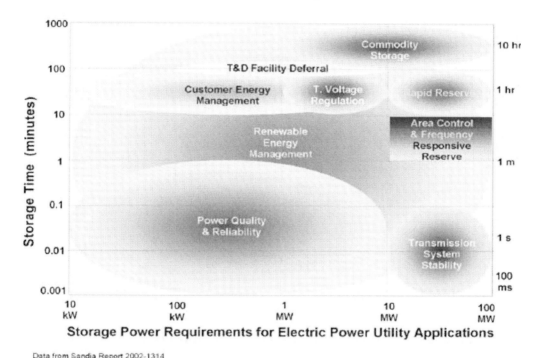

Figure 3-2. Regime and scale of storage applicationss.

Currently, energy storage is utilized in the grid primarily for diurnal energy storage, to enable use of cheap baseload energy, available at night, for serving high daytime loads that would otherwise require expensive peaking power plants. These plants (primarily pumped hydroelectric plants) are effective, and have some capacity to provide grid support for shorter-term imbalances (10s of minutes). However, these plants tend to be large, capital intensive, and with specific siting requirements.

A number of existing EES technologies can be potential candidates for other grid applications. These EES technologies store electricity either directly in charges or via energy conversion into a different form of energy. Super-capacitors are an example of a direct charge type of storage that features high power but low energy. Most of EES technologies involve energy conversion from electricity to kinetic energy, potential energy, or chemical potential. Flywheels are an EES technology that store electricity in kinetic energy. Like direct storage, flywheels are characterized by high power but low energy, and thus are most suitable for the category of power management. Compressed-air and pumped-hydro storage are capable of a very large amount of electricity energy, allowing hours or even longer duration, but are not well suited to power quality management. The largest group of EES technologies is the one that stores electricity in chemical (or electrochemical) potential driven by an external voltage. These EES technologies include redox flow batteries, Na-based batteries, and Li-ion batteries. With some limitations, batteries are capable of uptake and release of electrical energy rapidly or over a longer time period, and thus are potentially applicable to both power and energy applications depending on performance characteristics of a specific technology. The chart in Figure 3-3, again generic in nature, includes an overlay of some current energy storage technologies on the operational and application characteristic space already presented in Figure 3-2. While not inclusive of all energy technologies, this overlay shows the diversity of applications even for technologies within a single class.

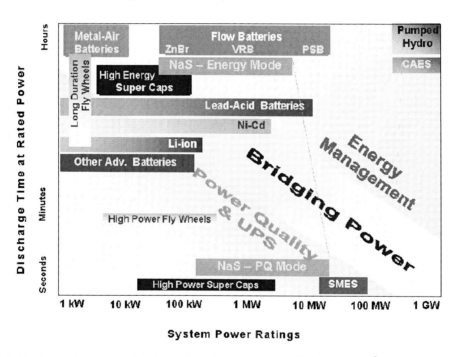

Figure 3-3. Regimes of storage technologies based on power and discharge time. [6]

3.2. The Research & Development Component of the Program

The Research & Development (R&D) component of the Program currently aims to understand the performance promise of multiple technologies. The Program takes this view because the likely needs of different technologies for varied stationary markets that have different requirements in performance, costs, siting, etc. Also at this time it is too early in the industry's maturity to focus on one or two "selected" technologies.

The discussion that follows highlights the technologies—and the challenges— that are the focus of the R&D component of the Program:

- Redox flow batteries
- Sodium based batteries
- Lithium-ion batteries
- Advanced lead-acid batteries
- Compressed air energy storage
- Flywheel storage

Redox flow batteries (RFBs) store electrical energy in two redox couples that are typically soluble in liquid electrolytes contained in external tanks. The energy conversion from electrical energy to chemical (or electrochemical) potentials, or vice versa, during charging and discharging, respectively, is realized when the liquid electrolytes flow through a cell stack where electrode reactions occur. RFBs are essentially regenerative fuel cells that feature many potential advantages. They offer the capability of storing energy or power (up to multi-MWhs or MWs, respectively) in a simple module design. The design allows for adjustable power/energy for varied durations up to over 10 hours or even longer, and a response of sub-seconds, due to the intimate interfaces between the flowing electrolytes and electrodes. [Ref., "Advanced Energy Storage for Green Grid," Yang, ZG, et al., Chemical Reviews, 2011.] RFB also have no structural and mechanical stresses during charge and discharge, potentially allowing a long cycle life. Flow battery systems are finally beginning to be demonstrated in select grid storage applications. Nevertheless, shortcomings in the technology still exist, where several areas of investigation have been identified. To assist in long-term technology viability and to facilitate widespread technology adoption, the research can be broadly broken down into two strongly connected areas: 1) new materials/chemistries and component development; and 2) cell, stack, and system design and development. The new materials/chemistries development activities include efforts focused on development of new electrolytes, new electrochemical redox couples and advanced membranes. When developed, these activities will provide industry with the tools, materials solutions, and validated demonstrations necessary to further advance, lower cost, high-reliability flow battery systems.

Sodium based batteries (SBB) are electrochemical devices that store electricity in Na or Na-compoundsbased electrodes. Sodium-based battery chemistries have attractive attributes for large-scale energy storage. Sodium is readily available in the U.S. and is an inexpensive raw material. The sodium-sulfur battery chemistry is one of the more mature, commercially viable technologies for large-scale electrical energy storage, and the sodium-based Zebra battery system is nearing commercialization with the recent, significant investments made by industry in production facilities. Some general advantages of SBBs for the grid applications[7] lie in the fact that Na is abundant and low cost, compared with other elements (e.g., lithium). Na-beta alumina batteries in particular demonstrate an energy efficiency of over 90%, with energy density comparable to that of Li-ion batteries. Like other electrochemical devices, the Na-based batteries offer a quick response in sub-seconds and they can be capable of storing multi-MW/MWhs electricity in a system based on module design. Power/energy ratios can be adjusted by cell and stack designs (e.g., electrode thickness, planar designs) for both power and/or energy applications. The Na-S battery was initially developed Ford in the late 1960s and has been commercialized by NGK Insulator, Inc. in Japan. A number of MWh systems have been demonstrated on the electrical grid. The concept of ZEBRA was proposed in 1978 and has been planned for commercialization by MES-DEA, FZ Sonick SA, GE, and others.8 However, there remain issues for the Na-beta alumina batteries to fully meet the performance and cost requirement metrics for broad market penetration. At the recent DOE sponsored workshop on "Advanced Materials and Devices for Stationary Electrical Energy Storage Applications," applied research needs for sodium-based batteries were identified that will assist in long-term technology viability and facilitate widespread technology adoption. Among these research needs are developing the materials and architectures to reduce operating temperatures from those needed to operate the NaS or Zebra chemistries. Na-NaSICON and Na-ion batteries to reduce manufacturing costs and operating temperatures of Na batteries are at the R&D stage.

Lithium-ion batteries store electrical energy in electrodes made of Li-intercalation (or insertion) compounds. During charge and discharge, Li+ ions transfer across a liquid organic electrolyte between one host structure and the other, with concomitant oxidation and reduction processes occurring at the two electrodes. The Li-ion technologies offer high energy and power density, along with a nearly 100% columbic efficiency, and thus have found great success for applications in mobile electronics. These technologies are considered the most promising options for hybrid and electrical-vehicle applications. Given the favorable electrochemical performance in energy/power densities and high energy efficiency, along with advancement in system design and manufacturing, the Li-ion batteries are also being examined for wide-scale stationary applications.9 The emphases for stationary applications are on costeffectiveness, long calendar and cycle life, etc. Overall, Li-ion technologies have not yet been fully demonstrated to meet the performance and economic matrix for the utility sector. Significant advancements are needed in materials, processing, design, and system integration for the technologies to achieve broad market penetration.

Advanced lead-acid batteries, based on a mature technology, are attractive for large-scale energy storage because of their low cost. However, their limited cycle life presents a significant barrier to widespread implementation. The addition of select carbon materials to the negative plate of valve regulated lead acid (VRLA) batteries has been demonstrated to increase cycle life by an order of magnitude or more. In addition, on cycling the battery capacity increases, and in fact exceeds the performance of the batteries when new. The mechanism by which carbon extends battery life is generally accepted to be through reduction/elimination of sulfation of the electrode—although it's not always clear why different carbon compounds behave differently; whereas the underlying mechanism responsible for improvement in capacity on cycling is not known. By developing an understanding of the fundamental physical, chemical, and electrochemical mechanisms behind both aspects of enhanced performance, the possibility to significantly improve lead acid batteries by allowing intentional design and fabrication of electrode structures having superior performance exists. Furthermore, the possibility to extend this fundamental understanding and approach to enhanced performance to other battery chemistries exist. Engineering enhanced performance of lead carbon batteries will ultimately lead to reduced life cycle cost.

Compressed Air Energy Storage is a technically mature energy storage option. Infusion of CAES into the U.S. depends on host rock availability at an appropriate depth. A review of the geology of the U.S. to assess and determine the geologic potential for CAES in the U.S is needed. Porosity, permeability, and degree of saturation of storage formations will be considered in determinations of number of boreholes, diameter and spacing systems for CAES. This will have direct input to determinations of feasibility and upfront costs of CAES, as well as set practical limits on accessible geology. For this technology to achieve wide market penetration, geo-mechanical modeling tools are needed to further the development of more efficient, safe, and cost effective underground storage systems.

Flywheel kinetic energy stored increases with the square of the angular velocity of the wheel. Materials limitations currently constrain the speed with which flywheels can operate. In current prototypes, significant obstacles in transverse strain behavior, long-term mechanical creep and micro-fracture propagation, as well as composite processing/manufacturing and system requirements have been encountered for the materials now employed. This has resulted in limits to the operational efficiency of the flywheels, both in terms of leading edge technology and overall system costs. Since the kinetic energy stored in these devices is proportional to the speed squared, even incremental improvements will result in substantial extra energy storage and more attractive operational dynamics. With improved the material selection criteria and manufacturing improvements, U.S. based companies will be able to position themselves as global leaders in this evolving technology, and to develop their own supply chain of higher quality and better understood materials by removing the current dependency on foreign assignments.

Pumped-hydro storage (PHS) is a mature technology that stores a bulk of electricity (up to few GWs) by pumping water from a reservoir up to a another reservoir at a higher elevation. When electricity is needed, water is released from the higher reservoir through a hydroelectric turbine into a low reservoir to generate electricity. PHS operates at about 76-85% efficiency, depending on design, and has very long lives on the order of 50 years. With its low life cycle cost, the technology has been the major technology deployed in the US for the grid applications. The first PHS plant was built in Europe in the 1890s, and currently, the United States has 38 plants installed, supplying 19 GW of electricity. But PHS is limited by site selection and requires a large initial investment and long construction periods up to 7 or 8 years as well as a reaction time up to 10 minutes. Research is not needed for the technology, which is mature, but for testing the limits and multiple values of system applications.

Advanced concepts for electrochemical energy storage are being investigated as part of the OE R&D program to revolutionize the state-of-the-art in performance and/or cost of batteries and battery systems. Perhaps the most far-reaching concept under development is the "N2-O2" battery, which uses a novel nitrogen redox reaction at the anode of an air-breathing system. If successful, such a chemistry would have extremely high-energy density and low cost, as both anode and cathode materials can be harvested from the air. The development of this concept requires conducting cutting-edge applied science that is of benefit to a wide range of existing technological problems facing the energy and energy efficiency sectors. Additionally, various metal-air systems are being studied to push the boundaries of existing technology and maintain the pipeline of innovation.

For any of the energy storage technologies to gain traction and be adopted by the commercial sector they need to first demonstrate that they are at a stage of development where they can deliver on their performance promises in a viable, reliable and cost effective manner. Hence, projects that demonstrate the actual capabilities of these technologies are critical to further understanding the field. The short term feedback from these demonstration plants would be to shed light on the actual construction costs related to these facilities along with their operational costs. These demonstration facilities will also provide a wealth of data on the operational challenges that will be faced by energy storage technologies. In essence they will serve as filter showing which of the storage technologies are better suited for commercial widescale deployment and would also help determine new applications for energy storage.

3.3. The Demonstration Component of the Program

In the past the ESS Program focused primarily on demonstrating components and systems to bring them to commercial viability. Drivers for this narrowly defined program focus were twofold: the desire to create a pull for the technology from potential users, and low funding levels. At the outset, the utility industry and potential customers of energy storage systems did not understand the technology, its applications, costs, reliability or benefits. Likewise, system developers were uninformed regarding the needs of the utilities. To address these limitations, collaborative partnerships were formed that included technology developers and potential users, as well as other partners who had the potential to benefit, such as system integrators, state agencies like California Energy Commission (CEC) and New York Energy Research and Development Authority (NYSERDA), who are interested in supporting demonstrations of near commercial energy storage systems, and other international programs like the Commonwealth Scientific and Industrial Research Organisation (CSRIO) in Australia and Spanish energy storage research programs. In these programs DOE supplied both limited funding and technical expertise and oversight responsibilities, principally through Sandia National Laboratories, and DOE continues to support these programs as they have been successful in that:

- technologies are reaching the marketplace
- utility industrial partners recognize the need for energy storage, and
- industrial and commercial users of energy storage are beginning to deploy technologies, better understand their use and value, and are sharing lessons learned.

To further the collaboration between DOE and the energy storage community, DOE focused on the demonstration aspect through the "Recovery Act." The collection, analysis, and dissemination of data relevant to energy storage systems demonstration projects will help the community to increase the collective knowledge of energy storage systems, their operations, uses, benefits and potential issues. These demonstration projects shed light on the actual costs of production and operation of energy storage devices along with the challenges involved with their integrating with the electric grid. The information obtained will help further define the research areas that the community should be focusing on to further improve these technologies. Hence, in addition to the current partnerships, DOE provided partial funds to 16 energy storage demonstration projects in an effort to further bridge the gap between these technologies and the market place.

Figure 3-4 shows the geographical location of all OE sponsored "Recovery Act" storage demonstration projects, followed by a listing of the ongoing energy storage demonstration projects being executed in various locations.

Energy Storage: Program Planning Document 281

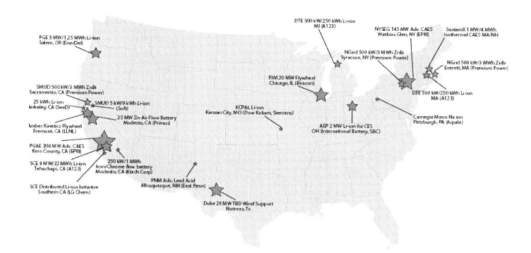

Figure 3-4. Current DOE sponsored storage demonstrations.

ARRA Energy Storage Demonstrations

AWARDEE (TECHNOLOGY)	SIZE-POWER (ENERGY)	APPLICATION	TITLE-DESCRIPTION	LOCATION	UTILITY	FUNDING: ARRA (TOTAL)
1. BATTERY STORAGE FOR UTILITY LOAD SHIFTING OR FOR WIND FARM DIURNAL OPERATIONS AND RAMPING CONTROL							
DUKE ENERGY BUSINESS SVCS. (TBD)	24MW (15MW slow)	Renewables and Demand	Notrees Wind Storage – Deploy a wind energy storage demonstration project at the Notrees Wind power Project in western Texas. The project will demonstrate how energy storage and power storage technologies can help wind power systems address intermittency issues by building a 24 megawatt (MW) hybrid-energy storage system capable of optimizing the flow of energy.	Goldsmith, TX	Duke	$21,806,219 ($43,612,445)
PRIMUS POWER (ZINC-CHLORIDE FLOW BATT.)	25MW (75MWhr)	Renewables	Wind Firming EnergyFarm™ – Deploy a 25 MW – 75 MWh EnergyFarm for the Modesto Irrigation District in California's Central Valley, replacing a planned $76M / 50 MW fossil fuel plant to compensate for the variable nature of wind energy providing the District with the ability to shift on-peak energy use to off-peak periods.	Alameda, San Ramon, Modesto, CA	Modesto Irrig. Dist.	$14,000,000 ($46,700,000)
SOUTHERN CALIF. EDISON CO. (LITHIUM-ION BATT.)	8MW (4 hrs)	Renewables	Tehachapi Wind Energy Storage Project – Deploy and evaluate an 8 MW utility-scale lithium-ion battery technology to improve grid performance and aid in the integration of wind generation into the electric supply. The project will evaluate wider range of applications for li-ion batteries that will spur broader demand for the technology, bringing production to a scale that will make this form of large energy storage more affordable.	Tehachapi, CA	So. Calif. Edison	$24,978,264 ($54,856,495)
					ARRA Sub-Total:	$60,784,483
					(*Project Value Sub-Total*):	$145,168,940

AWARDEE (TECHNOLOGY)	SIZE-POWER (ENERGY)	APPLICATION	TITLE-DESCRIPTION	LOCATION	UTILITY	FUNDING: ARRA (TOTAL)
2. FREQUENCY REGULATION ANCILLARY SERVICES							
BEACON POWER (FLYWHEELS)	20MW (5MWhr)	Frequency	Beacon Power 20MW Flywheel Frequency Regulation Plant – Design, build, test, commission, and operate a utility-scale 20 MW flywheel energy storage frequency regulation plant in either Hazle Township, PA or Chicago Heights, Illinois, and provide frequency regulation services to the grid operator, the PJM Interconnection. The project will also demonstrate the technical, cost and environmental advantages of fast response flywheel-based frequency regulation management.	Tyngsboro, MA; Hazle Township, PA or Chicago Heights, IL	PPL Corp. (PA site); Midwest Energy (IL site)	24,063,978 ($48,127,957)
					ARRA Sub-Total:	$24,063,978
					(*Project Value Sub-Total*):	$48,127,957

3. DISTRIBUTED ENERGY STORAGE FOR GRID SUPPORT

Recipient	Size	Application	Description	Location	Utility	ARRA Funding (Project Value)
CITY OF PAINESVILLE (VANADIUM-REDOX BATT.)	1MW (6-8MWhr)	Coal Efficiency	**Painesville Municipal Power Vanadium Redox Battery Demonstration Program** - Demonstrate 1 MW vanadium redox battery (VRB) storage system at the 32 MW municipal coal fired power plant in Painesville. The project will provide operating data and experience to help the plant maintain its daily power output requirement more efficiently while reducing its carbon footprint.	Painesville, Parma, OH; Johnstown, PA; Alexandria, VA; Evansville, IN; Devens, MA	Painesville Municipal Power	$4,242,570 ($9,666,324)
DETROIT EDISON CO. (LITHIUM-ION BATT.)	25kW (20 units of 50 kWhr each)	Frequency, Demand and Renewables,	**Detroit Edison's Advanced Implementation of A123s Community Energy Storage Systems for Grid Support** – Demonstrated proof of concept for aggregated Community Energy Storage Devices in a utility territory. The project is comprised of the following major research objectives: 1) The 20 Community Energy Storage (CES) devices across a utility territory; 2) The installation and use of a centralized communication across the service territory; 3) The integration of a renewable resource with energy storage; 4) The creation of algorithms for dispatching CES devices for peak shaving and demand response; 5) The integration and testing of secondary-use electric vehicle batteries; and 6) The use of Energy storage devices to provide ancillary services to the power grid.	West Lebanon, Hanover, NH; Saxonville, MA	Detroit Edison	$4,995,271 ($10,887,258)
EAST PENN MFG. CO. (ULTRACAPACITOR / LEAD-ACID BATT.)	3MW (1-4MWhr)	Frequency / Demand	**Grid-Scale Energy Storage Demonstration for Ancillary Services Using the UltraBattery Technology** - Demonstrate the economic and technical viability of a 3MW grid-scale, advanced energy storage system using the lead-carbon UltraBattery technology to regulate frequency and manage energy demand.	Lyons Station, PA	Met-Ed	$2,543,746 ($5,087,269)
PREMIUM POWER CORP. (ZINC-BROMINE BATT.)	5-500 kW (6 hrs)	Renewables & Micro-grid	**Premium Power Distributed Energy Storage System Demonstration for National Grid and Sacramento Municipal Utility District** - Demonstrate competitively-priced, multi-megawatt, long-duration advanced flow batteries for utility grid applications. This three-year project incorporates engineering of fleet control, manufacturing and installation of five 500-kW/6-hour TransFlow 2000 energy storage systems in California and New York to lower peak energy demand and reduce the costs of power interruptions.	North Reading, MA; Syracuse, NY; Sacramento, Rancho Cordova, CA	National Grid & Sacramento Municipal Utility Dist.	$6,062,552 ($12,514,660)
PUBLIC SVC. CO. OF NM (PNM) (ADVANCED LEAD ACID BATT.)	500kW (2.5MWhr)	Renewables and Modeling	**PV Plus Storage for Simultaneous Voltage Smoothing and Peak Shifting** - Demonstrate how a 2.5MWh Advanced Lead Acid flow battery along with a sophisticated control system turns a 500kW solar PV installation, into a reliable, dispatchable distributed generation resource. This hybrid resource will mitigate fluctuations in voltage normally caused by intermittent sources such as PV and wind and simultaneously store more energy for later use when customer demand peaks.	Albuquerque, NM	PNM	$2,505,931 ($6,313,433)

4. COMPRESSED AIR ENERGY STORAGE (CAES)

Recipient	Size	Application	Description	Location	Utility	ARRA Funding (Project Value)
IBERDROLA USA (NY STATE ELEC. & GAS CORP.) (CAES)	150MW (2-8 hrs)	Peaking	**Advanced CAES Demonstration Plant (150MW) Using an Existing Salt Storage Cavern** - Demonstrate an advanced, less costly 150 MW Compressed Air Energy Storage (CAES) technology plant using an existing salt cavern. The project will be designed with an innovative smart grid control system to improve grid reliability and enable the intergration of wind and other intermittent renewable energy sources.	Watkins Glen, NY	Iberdrola USA	$29,561,142 ($125,006,103)
PACIFIC GAS & ELECTRIC CO. (CAES)	300MW (10 hrs)	Renewables, Spinning Reserve, VARS	**Advanced Underground CAES Demonstration Project Using a Saline Porous Rock Formation as the Storage Reservoir** - Build and validate the design, performance, and reliability of an advanced, underground 300 MW Compressed Air Energy Storage (CAES) plant using a saline porous rock formation located near Bakersfield, CA as the storage reservoir.	Kern County, CA	Pacific Gas & Electric	$25,000,000 ($355,956,300)

ARRA Sub-Total: $54,561,142
(Project Value Sub-Total): **$480,962,403**

5. DEMONSTRATIONS OF PROMISING ENERGY STORAGE TECHNOLOGIES

Company	Size	Category	Description	Location	Partner	ARRA Funding (Project Value)
AQUION ENERGY, INC. (SODIUM-ION BATT.)	10-100 kWhr	Renewables	Demonstration of Sodium Ion Battery for Grid Level Applications - Partner with Carnegie Mellon University to demonstrate a new, low cost, long-life, highly efficient, environmentally friendly, stationary energy storage battery that uses a proven and fully novel cell chemistry. Specifically, an aqueous sodium-ion based electrolyte is used in conjunction with simple highly scalable electrode materials housed in low cost packaging.	Pittsburgh, PA	AES Duke Energy	$5,179,000 ($10,359,827)
AMBER KINETICS, INC. (FLYWHEELS)	50 kW (50kWhr)	Frequency	Amber Kinetics Flywheel Energy Storage Demonstration - Develop and demonstrate an innovative flywheel technology for use in grid-connected, low-cost bulk energy storage applications. This demonstration effort, which partners with AFS Trinity, will improve on traditional flywheel systems, resulting in higher efficiency and cost reductions that will be competitive with pumped hydro technologies.	Fremont, CA	(in-house)	$3,694,660 ($10,003,015)
KTECH CORP. (IRON-CHROMIUM REDOX FLOW BATT.)	250kW (1MWhr)	Renewables	Flow Battery Solution for Smart Grid Renewable Energy Applications - Demonstrate a prototype flow battery system with an intermittent renewable energy source - a helios dual-axis tracker photovoltaic system. The project will combine a proven redox flow battery chemistry with a unique, patented design to yield an energy storage system that meets the combined safety, reliability, and cost requirements for distributed energy storage.	Albuquerque, NM; Sunnyvale, Snelling, CA	(none)	$4,764,284 ($9,528,567)
SEEO, INC. (LITHIUM-ION BATT.)	(25kWhr)	CES	Solid State Batteries for Grid-Scale Energy Storage - Develop and deploy a 25kWh prototype battery system based on Seeo's proprietary nanostructured polymer electrolytes. This new class of advanced lithium-ion rechargeable battery will demonstrate the substantial improvements offered by solid state lithium-ion technologies for energy density, battery life, safety, and cost. These batteries would be targeted for utility-scale operations, particularly Community Energy Storage projects.	Berkeley, Van Nuys, CA	PG&E	$6,196,060 ($12,392,120)
SUSTAINX (CAES)	1MW (4MWhr)	Renewables – both	Demonstration of Isothermal Compressed Air Energy Storage to Support Renewable Energy Production - Design, build, and deploy a utility-scale, low-cost compressed air energy storage system to support the integration of renewable energy sources onto the grid. The 1 MW/4hr system will store potential energy in the form of compressed air in above-ground industrial pressure facilities. The technology utilizes isothermal gas cycling coupled with staged hydraulic compression and expansion to deliver an efficient and cost-effective energy storage solution.	W. Lebanon, Hanover, NH; Saxonville, MA	AES Energy Storage	$5,396,023 ($10,792,045)

ARRA Sub-Total: $25,230,027
(Project Value Sub-Total): $53,075,574

ARRA TOTAL FUNDING: $184,989,700

(*PROJECT VALUE TOTAL: $771,803,818*)

In addition to the ARRA demonstration projects there are several other storage related efforts being executed in the U.S. to further the successful deployment of these technologies, such as:

- *State Energy Agency Support (California Energy Commission (CEC), New York State Energy Research and Development Authority (NYSERDA)):* DOE supports CEC and NYSERDA in areas including project formation, assisting in drafting procurement documents, proposal evaluation, technical oversight of projects and third-party independent testing and analysis. Sandia assists CEC and NYSERDA in

developing and implementing projects as well as in testing those projects after commissioning.
- *Integrate Energy Storage into Energy Surety Microgrids:* To address current shortcomings of back-up power reliability for military facilities, DOE is investigating approaches to locate more secure and robust power sources near critical loads and better manage existing power generation and loads to improve the reliability and security of electric power at military bases. The approach, called the Energy Surety Microgrid (ESM), is an alternate energy delivery methodology developed to ensure that the reliability of the electric infrastructure at a military facility will fully satisfy critical mission needs.
- *Smart Grid, Microgrid and Storage Technical Assistance for the Hawaii Clean Energy Initiative:* This project provides technical assistance in the smart grid, microgrid, and storage areas to support the Hawaii Clean Energy Initiative. The support will be through providing expertise in smart grid and microgrid technology implementation, storage design and integration, analysis of integration of high percentage of renewables and their effect on the grid, and regulatory analysis. The project provides technical assistance in storage design and integration for Hawaii-based utilities, developers, State agencies, and others that are actively working on Hawaii's current storage projects.

In addition to participating with DOE's Loan Guarantee Program Office, and monitoring current demonstration projects, and sharing lessons learned, the Program will assess the benefits of and resources for future demonstration projects.

3.4. The Analysis Component of the Program

Quantitative analytics is a critical component of integrated RD&D that will apply science and advanced engineering and manufacturing approaches toward deployment of stationary energy storage technologies.

Quantitative analytics enables OE and its partners to:

➢ sharpen cost and performance targets for various stationary storage applications
➢ identify and characterize markets for stationary storage
➢ articulate the value propositions, and develop business cases
➢ assess the impact of grid operational differences across the regions, as well as different market designs on energy storage requirements, and address the role an impact of storage on the regional grid characteristics and operational requirements
➢ provide insight on operational and economic factors that help shape energy storage performance targets, development of RD&D agendas and schedules, and
➢ inform policy decision makers regarding the role of storage in grid operations and economics particularly for emerging grid environments.

Additionally, the analysis component addresses the needs of transmission and distribution (T&D) planners to have adequate planning and analysis tools for the T&D upgrades and expansions. Furthermore, the analysis component performs economics and system impact studies of high-value storage demonstrations to quantify the performance and system impacts and benefits. Specific analyses over the next few years would include the following:

Impact Assessments

1) *A national assessment of the role of energy storage:* This will quantify the role of energy storage technologies as the grid transitions to a greener, reliable, and highly efficient energy infrastructure. The assessment will estimate the potential market size of energy storage for various grid services as storage competes against other generation technologies (such as gas turbines), demand response, and transmission assets. The assessment will be performed for a 2030 time horizon and various scenario definitions of variable renewable energy penetrations and load growth assumptions. The assessment will be based on a capacity expansion planning process that treats generation, load resources, storage and transmission assets by their cost performance characteristics and chooses the least-cost technology portfolio mix.
2) *Reliability assessment of storage technology:* A study that quantifies power system reliability as a function of the types of resources contained within the system or subsystem. For example, a utility may consider four options to mitigate the effects of distribution line capacity limits; 1) upgrade existing distribution infrastructure, 2) add generation near the end of the feeder, 3) install demand response technologies, or 4) add storage near the end of the feeder. This study should quantify the differences in grid system reliability metrics for multiple scenarios.

Case Studies

3) 3Business case development of ARRA demonstration projects: Business cases will be developed based on technical performance of a set of demonstration projects that are funded by ARRA and the prevailing market designs of the respective region in which the demonstration takes place. Expected demonstration sites are in Hawaii and Texas.
4) 4Development of business case analysis templates: Templates will be created that guide the technical and economic evaluation process of a specific storage technology. The templates can be used to help utilities and storage providers determine economic viability of a particular storage technology.
5) Lessons-learned from ARRA demonstration projections: A consolidated report should be released detailing the project span from conception to operation for all recent DOE funded ARRA demonstration projects. The study should be comprehensive in nature and should include issues such as policy barriers, costs, cost recovery, environmental impacts, unexpected costs and benefits, opportunities for improvement, warnings and potential pitfalls, differences in perspectives from owner/operator to utilities, etc. The report should be comprehensive to benefit capital financiers, owner/operators, utilities, and public utility commissioners.

Market Design Studies

6) *Market design study for ancillary services:* A study to characterize the technical balancing needs of the system(s) and to offer market formulation/modification suggestions which will extract maximum value from many different types of resources, including fast ramping storage, combustion turbines, various types of demand response, self regulation of renewable resources. This study will explore how current market rules need to be modified or new rules established to reward grid resources for the value they provide to the system.

7) *Value gap analysis for storage:* A study to assess the transparency of costs and benefits by assessing storage for multiple value streams, whether they are financially recoverable or not. Here, value is defined as filling a system need which displaces the need for another resource to fulfill the need. In cases where value is provided, but financial recovery is not possible, an exploration of possible remedies should be presented, consistent with preservation of regulatory and free market principles. Policy considerations (existing and those needed) must be addressed as well.

Cost Modeling

8) *Detailed cost and "state of health modeling" of energy storage:* To provide transparency of the cost composition of an entire battery system, component-based cost models will be developed for stationary energy storage systems. The goal of the cost models is to assess in detail if and where advancements in new materials and chemistry, manufacturing processes of materials, novel cells and systems designs and or other technological breakthroughs must occur in order for the storage device to meet future cost/performance targets. The cost models will consider the potential supply constraints of materials at low market penetration and at full scale deployment as well as technological advances and expected during ramp-up of production as well as cost reductions by applying economies of scale in material procurement, manufacturing, and deployment. The first activity will develop the cost model for a redox flow battery. In following years, cost models for other storage chemistries will be developed. To increase its value the cost model will need to be maintained and constantly updated as new advancements from within the program as well as outside the program and DOE funding are achieved. In addition, "state of health modeling" will be initiated to estimate the performance characteristics of the battery throughout its life. The modeling will be based on measurements of internal resistance and diffusion coefficients.

Codes and Standards

9) *Development of industry standards for energy storage technologies:* Industry requires specifications of standards for characterizing the performance of energy storage under grid conditions and for modeling behavior. Discussions with industry professionals indicate a significant need for standards in the following area:

- Infrastructure planning: Planning engineers perform stability analyses to assure safe operating conditions of the grid. Dynamic models of grid components are used to simulate transient behavior to disturbances. There is a need in the planning community for dynamic models for archetypical storage technologies (e.g., flywheels, batteries) that are sufficiently generic and yet realistic and verified by a standards body
- Protocols for grid-relevant performance testing: While generic battery cell characterizations exist, no performance testing procedures exist that would test performance, durability, and state of health under conditions that represent realistic grid services. The automotive industry has developed a set of standardized drive cycles for testing of the entire drive trains and including the batteries. The analog of a drive cycle for various grid services (e.g., regulation services, load following, arbitrage, and combinations thereof) are required for assuring the user community with realistic and reliable performance expectations. Statistical analyses will be performed to determine the diversity of operating conditions that energy storage equipment would need to provide for selected grid services. DOE will need to engage with IEEE standards committees to explore the appropriate committee for initiating standards developments.

4.0. PORTFOLIO DEVELOPMENT AND MANAGEMENT

Principal areas of portfolio development and program management that are integral to the Program include:

➢ Communication of the program
➢ Analysis of the program
➢ Evaluation, and
➢ Technology transfer

These management areas combine to assure that industry, the public, and government are effectively served by the Program. This Program follows a multi-step planning and management process designed to ensure that all funded technical R&D projects are chosen based on their qualifications in meeting clearly defined criteria. This process entails the following:

➢ Competitive solicitations for financial assistance awards and National Laboratory RD&D.
➢ Peer reviews of proposals in meeting the Funding Opportunity Announcement goals, objectives, and performance requirements.
➢ Peer reviews of in-progress projects on the scientific merit, the likelihood of technical and market success, the actual or anticipated results, and the cost effectiveness of research management. The Program and its in-progress RD&D projects will be reviewed through this external review process once every two years with evaluation results feeding back to planning and portfolio management.

> Stage-gate reviews to determine readiness of a technology or activity to advance to its next phase of development, pursue alternative paths, or be terminated; these readiness reviews will be conducted on an as-needed schedule based on project progression in meeting the established stage-gate criteria.
> OE internal review of the Program annually to ensure continuous improvements and proper alignment with priorities and industry needs.

The value of RD&D projects, individually and collectively, to achieving the program goal and targets will be made transparent by applying this management process consistently throughout the Program. Moreover, this value that is supported by rigorous analysis and evaluation will be transparent in Program communications to the industry, the public, and other stakeholders.

ACRONYMS AND ABBREVIATIONS

ARPA-E	Advanced Research Projects Agency-Energy
ARRA	American Recovery and Re-investment Act
CAES	Compressed Air Energy Storage Systems
CEC	California Energy Commission
CSRIO	The Commonwealth Scientific and Industrial Research Organisation
DOE	Department of Energy
EES	Electrical Energy Storage
EFRCs	Energy Frontier Research Centers
EPRI	Electric Power Research Institute
ESM	Energy Surety Microgrid
ESSP	Energy Storage Systems Program
IEEE	Institute of Electronics and Electrical Engineers
Li-ion	Lithium Ion
Na	Sodium
NYSERDA	New York State Energy Research and Development Authority
OE	Office of Electricity Delivery and Energy Reliability
PHS	Pumped Hydro Storage
R&D	Research and Development
RD&D	Research Development and Demonstration
RFB	Redox Flow Batteries
SBB	Sodium Based Batteries
SBIRs	Small Business Innovation Research
SMES	Superconducting Magnetic Energy Storage
T&D	Transmission and Distribution
VRLA	Valve Regulated Lead Acid

End Notes

[1] U.S. DOE, "National Electric Delivery Technologies Vision and Roadmap," November 2003.

[2] Ibid.

[3] Pike Research, "Energy Storage on the Grid," August 2010.

[4] Figure is an OE created diagram derived from industry presentations, including material from Southern California Edison.

[5] Graphical data representation obtained from *http://electricitystorage.org/ Sandia Report, 2002-1314*

[6] Graphical representation of the data contained in Sandia Report 2002-1314. *http://prod.sandia.gov/techlib/accesscontrol. cgi/2002/021314.pdf*

[7] Lu, X., Xia, G.-G., Lemmon, J.P., Yang, Z.G., J. Power Sources, 2009, 195, 2431.

[8] Dustmann, C.H., *Advances in ZEBRA batteries.* Journal of Power Sources, 2004, 127, 85; Sudworth, J.L., *Sodium/nickel chloride (ZEBRA) battery.* Journal of Power Sources, 2001, 100, 149.

[9] Yang, Z.G., Choi, D., Krisit, S., Rosso, K.M., Wang, D., Zhang, J.-G., Graff, G., Liu, J., J. Power Sources, 2009, 192, 588.

In: Lightning in a Bottle: Electrical Energy Storage
Editor: Fedor Novikov

ISBN: 978-1-61470-481-2
© 2011 Nova Science Publishers, Inc.

Chapter 5

SOLAR ENERGY GRID INTEGRATION SYSTEMS: ENERGY STORAGE (SEGIS-ES)[*]

*Dan T. Ton, Charles J. Hanley,
Georgianne H. Peek, and John D. Boyes*

NOTICE

Prepared by
Sandia National Laboratories
Albuquerque, New Mexico 87185 and Livermore, California 94550

Sandia is a multiprogram laboratory operated by Sandia Corporation, a Lockheed Martin Company, for the United States Department of Energy's National Nuclear Security Administration under Contract DE-AC04-94AL85000.

Approved for public release; further dissemination unlimited.

Issued by Sandia National Laboratories, operated for the United States Department of Energy by Sandia Corporation.

This report was prepared as an account of work sponsored by an agency of the United States Government. Neither the United States Government, nor any agency thereof, nor any of their employees, nor any of their contractors, subcontractors, or their employees, make any warranty, express or implied, or assume any legal liability or responsibility for the accuracy, completeness, or usefulness of any information, apparatus, product, or process disclosed, or represent that its use would not infringe privately owned rights. Reference herein to any specific commercial product, process, or service by trade name, trademark, manufacturer, or otherwise, does not necessarily constitute or imply its endorsement, recommendation, or favoring by the United States Government, any agency thereof, or any of their contractors or

---

[*] This is an edited, reformatted and augmented version of the United States Department of Energy publication, Sandia Report SAND2008-4247, dated July 2008.

subcontractors. The views and opinions expressed herein do not necessarily state or reflect those of the United States Government, any agency thereof, or any of their contractors.

Printed in the United States of America. This report has been reproduced directly from the best available copy.

ABSTRACT

This paper describes the concept for augmenting the SEGIS Program (an industry-led effort to greatly enhance the utility of distributed PV systems) with energy storage in residential and small commercial applications (SEGIS-ES). The goal of SEGIS-ES is to develop electrical energy storage components and systems specifically designed and optimized for grid-tied PV applications. This report describes the scope of the proposed SEGIS-ES Program and why it will be necessary to integrate energy storage with PV systems as PV-generated energy becomes more prevalent on the nation's utility grid. It also discusses the applications for which energy storage is most suited and for which it will provide the greatest economic and operational benefits to customers and utilities. Included is a detailed summary of the various storage technologies available, comparisons of their relative costs and development status, and a summary of key R&D needs for PV-storage systems. The report concludes with highlights of areas where further PV-specific R&D is needed and offers recommendations about how to proceed with their development.

1. EXECUTIVE SUMMARY

In late 2007, the U.S. Department of Energy (DOE) initiated a series of studies to address issues related to potential high penetration of distributed photovoltaic (PV) generation systems on our nation's electric grid. This Renewable Systems Interconnection (RSI) initiative resulted in the publication of 14 reports and an Executive Summary that defined needs in areas related to utility planning tools and business models, new grid architectures and PV systems configurations, and models to assess market penetration and the effects of high-penetration PV systems. As a result of this effort, the Solar Energy Grid Integration Systems Program (SEGIS) was initiated in early 2008. SEGIS is an industry-led effort to develop new PV inverters, controllers, and energy management systems that will greatly enhance the utility of distributed PV systems.

This paper describes the concept for augmenting the SEGIS Program with energy storage (SEGIS-ES) in residential and small commercial (≤100 kW) applications. Integrating storage with SEGIS in these applications can facilitate increased penetration of distributed PV systems by providing increased value to both customers and utilities. Depending on the application, the systems can reduce customer utility bills, provide outage protection, and protect equipment on the load side from the negative effects of voltage fluctuations within the grid. With sufficient penetration, PV-Storage systems are expected to reduce emissions related to generation and will be critical to maintaining overall power quality and grid reliability as grid-tied distributed PV generation becomes more common.

Although electrical energy storage is a well-established market, its use in PV systems is generally for stand-alone systems. The goal of SEGIS-ES is to develop electrical energy storage components and systems specifically designed and optimized for grid-tied PV applications. The Program will accomplish this by conducting targeted research and development (R&D) on the applications most likely to benefit from a PV-Storage system (*i.e.,* peak shaving, load shifting, demand response, outage protection, and microgrids) and developing PV-Storage technologies specifically designed to meet those needs. Designing optimized systems based on existing storage technologies will require comprehensive knowledge of the applications and the available storage technologies, as well as modeling tools that can accurately simulate the economic and operational effect of a PV-Storage system used in that application.

This paper describes the scope of the proposed SEGIS-ES Program; why it will be necessary to integrate energy storage with PV systems as PV-generated energy becomes more prevalent on the nation's utility grid; and a discussion of the applications for which energy storage is best suited and for which it will provide the greatest economic and operational benefits to customers and utilities.

Because selecting and optimizing a storage technology for an application will be critical to the success of any PV-Storage system, this paper also provides a detailed summary of the various storage technologies available and compares their relative costs and development status (*e.g.,* mature, emerging, *etc.*).

Finally, the paper highlights areas where further, PV-specific R&D is needed and offers recommendations about how to proceed with the proposed work.

2. VISION

The U.S. infrastructure for electricity generation and delivery is undergoing a revolution that will lead to increased efficiency, improved reliability and power quality for customers, 'smart' communications to match generation and loads, and the development of distributed generation from local and renewable resources. The high penetration of PV and other renewable energy technologies into the infrastructure will be enabled by developing managed, efficient, reliable, and economical energy storage technologies that will eliminate the need for back-up utility baseload capacity to offset the intermittent and fluctuating nature of PV generation.

These dispatchable storage technologies will bring added benefits to utilities, home-owners, and commercial customers through greater reliability, improved power quality, and overall reduced energy costs.

3. PROGRAM OBJECTIVE

The SEGIS Program will develop advanced energy storage components and systems that will enhance the performance and value of PV systems, thereby enabling high penetration of PV-generated electricity into the nation's utility grid. Through its RSI initiative, the DOE Solar Energy Technology Program is identifying needs and developing technologies to facilitate the high penetration of distributed electricity generation. The need for improved energy storage has been highlighted as a key factor for achieving the desired level of PV generation.

The electrical energy storage industry is well established and offers a variety of products for vehicle, uninterruptable power supply (UPS), utility-scale, and other storage applications. The design and development of storage products specifically for PV applications, however, is almost nonexistent. Traditional PV-Storage systems have been used for off-grid applications that required some amount of autonomy at night and/or during cloudy weather.

However, the objective of this Program is to develop energy storage systems that can be effectively integrated with new, grid-tied PV and other renewable systems, which will provide added value to utilities and customers through improved reliability, enhanced power quality, and economical delivery of electricity.

4. PROGRAM SCOPE

In late 2007, DOE began a series of studies to address issues related to the potentially high penetration of distributed PV generation systems on our nation's electricity grid. The RSI initiative resulted in the publication of 14 reports and an Executive Summary that defined needs in areas related to utility planning tools and business models, new grid architectures and PV system configurations and models to assess market penetration and the effects of high-penetration PV systems.[1] As a result of this effort, the SEGIS program was initiated in early 2008. SEGIS is an industry-led effort to develop new PV inverters, controllers, and energy management systems that will greatly enhance the utility of distributed PV systems.

SEGIS-ES is closely related to the SEGIS Program, a three-year program with a goal to develop new commercial PV inverters, controllers, and energy management systems with new communications, control, and advanced autonomous features.[2] The heart of the SEGIS hardware, the inverter/controller, will manage generation and dispatch of solar energy to maximize value, reliability, and safety, as the nation moves from 'one-way' energy flow in today's distribution infrastructure to 'two-way' energy and information flow in tomorrow's grid or microgrid infrastructure.

The applicable markets for the SEGIS Program[3] are defined in Table 1, which shows the size of the PV system in watts, or power output. Storage systems are typically rated in terms of energy capacity (*i.e.,* watt-hours), which is highly dependent on the application for which the storage is being used. The applications are discussed later in this document.

Table 1. Target Market Sectors for SEGIS PV Systems

Residential	Less than 10 kW, single-phase
Small Commercial	From 10 to 50 kW, typically three-phase
Commercial	From 50 to 100 kW, three-phase

SEGIS-ES is focused on developing commercial storage systems for distribution-scale PV in the market sectors shown in Table 1; specifically, PV systems designed for applications up to 100 kW that can be aggregated into multi-megawatt systems.

Integrating electrical energy storage into homes or commercial buildings is also a key focus of SEGIS-ES. New storage systems developed under the Program will play an important role in the development of independent microgrids – either individual buildings or communities of buildings – so microgrid-scale storage, on the order of one megawatt of distributed generation, is within the scope of this effort.

Storage systems developed through SEGIS-ES will interface with SEGIS products to further enhance PV system value and economy to customers. Products to be developed through SEGIS-ES include, but are not limited to:

- Battery-based systems using existing technologies that are enhanced or specifically designed for PV applications, including the development of PV-Storage hybrid systems;
- New energy storage system controllers that interface with SEGIS hardware to optimize battery use in order to obtain the highest possible system efficiency and battery life;
- Non-battery storage systems (*e.g.*, electrochemical capacitors [ECs], flywheels) designed specifically for PV applications; and
- New devices that integrate with building infrastructure.

SEGIS-ES does not address:

- Development of PV modules;
- Development of new battery technologies (although collaboration with the DOE Office of Basic Energy Sciences Energy Frontier Research Centers' Funding Opportunity is encouraged);
- Utility-scale storage systems or storage at the level of large distribution feeders (Although these efforts are key to achieving high penetration of distributed generation, they will be addressed through other Program activities);
- PV inverters or related power conditioning devices; and
- Non-solar-related storage system development, smart appliances, and utility portals.

5. THE NEED FOR ENERGY STORAGE IN HIGH-PENETRATION PV SYSTEMS

PV systems are a small part of today's electricity infrastructure and have little effect on the overall quality or reliability of grid power. Nevertheless, state and federal efforts are currently underway to greatly increase the penetration of PV systems on local and regional utility grids to achieve goals related to emissions reduction, energy independence, and improved infrastructure reliability. However, when PV penetration reaches sufficiently high levels (*e.g.*, 5 to 20% of total generation), the intermittent nature of PV generation can begin to have noticeable, negative effects on the entire grid.

Figure 1 illustrates the transient nature of PV generation as clouds pass over a typical residential system during the course of a day. Both the magnitude and the rate of the change in output are important: in mere seconds, the PV system can go from full output to zero (essentially), and back again. At high levels of PV penetration, this intermittency can wreak havoc on utility operations and on load-side equipment, due to fluctuations in grid voltage and power factor. Fluctuations at this scale simply cannot be allowed.

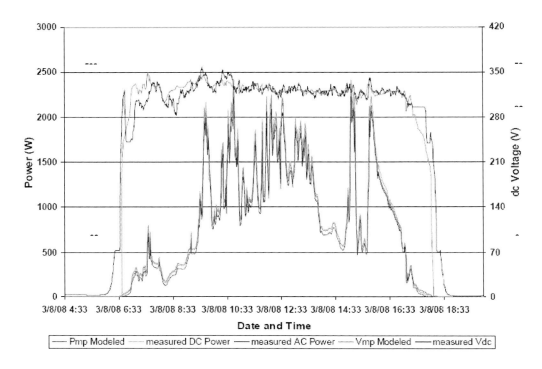

Figure 1. Measured and modeled PV system output on a day with frequent passing clouds.

To some degree, the distributed nature of PV can help mitigate negative consequences of high PV penetration; over large regions, the effects of intermittent generation on the grid will be less noticeable. Nevertheless, utilities must continue to address worst-case possibilities.

When transients are high, area regulation will be necessary to ensure that adequate voltage and power quality are maintained. When PV generation is low, some type of back-up generation will be needed to ensure that customer demand is met. Additionally, because most utilities require an amount of 'spinning reserve' power that typically is equal to the power output of the largest generating unit in operation, the amount of spinning reserve necessary will increase with the amount of distributed PV generation that is brought online. Without such measures, the benefits of high PV penetration are partially lost—carbon and other emissions are offset through PV-produced electricity; but, utility infrastructure is not reduced and power quality is not necessarily improved.

As the graph in Figure 2 illustrates, high PV penetration might reduce intermediate fossil fuel generation; but, without storage, PV will do little, or nothing, to reduce a utility's overall conventional generation due to the higher requirements for spinning reserve.

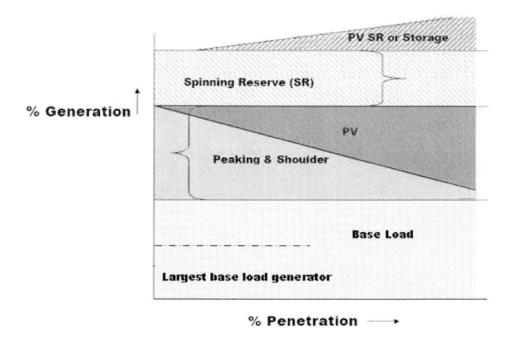

Figure 2. The need for additional spinning reserve or storage to back up intermittent PV generation at increasing levels of penetration.

As a whole, the utility grid must evolve in several ways to accommodate any increased penetration of PV and other distributed and intermittent electricity generation sources; including improved flexibility, better load management, integration of storage technologies, and even limited curtailment for extreme events. Several efforts are underway to define the next-generation grid infrastructure, which will include those characteristics.[4]

A recent study that specifically focused on the current grid and high-penetration PV called energy storage the 'ultimate solution' for allowing intermittent sources to address utility baseload needs. The report stated that "a storage system capable of storing substantially less than one day's worth of average demand could enable PV to provide on the order of 50% of a system's energy."[5] This paper focuses on incorporating storage as part of the overall 'systems' solution.

Successfully integrating energy storage with distributed PV generation in grid-connected applications involves much more than selecting an adequately sized system based on one of the many, commercially available technologies. Optimal integration of storage with grid-tied PV systems requires a thorough understanding of the following:

- The application for which the storage is being used and the benefits integrated storage provides for that application;
- The available storage technologies and their suitability to the application;
- The requirements and constraints of integrating distributed generation and electrical energy storage with both the load (residential, commercial, or microgrid) and the utility grid;

- The power electronics and control strategies necessary for ensuring that all parts of the grid-connected distributed generation and storage system work; and
- The requirements to provide service to the load and to maintain or improve grid reliability and power quality.

The complexity of an integrated PV-Storage system is illustrated in Figure 3, which shows SEGIS-based generation integrated with electrical energy storage for a residential or small commercial system.

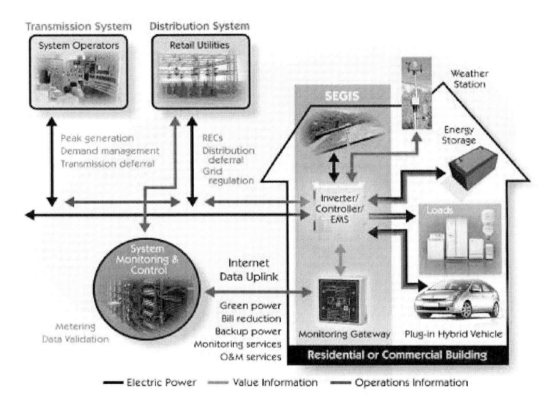

Figure 3. The relationship between SEGIS, electric energy storage, the customer, and the utility in an optimal configuration.[6]

6. APPLICATIONS OF ENERGY STORAGE IN HIGH-PENETRATION PV SYSTEMS

Integrated PV-Storage systems provide a combination of operational, financial and environmental benefits to the system's owner and the utility through peak shaving and reliability applications.[7]

Peak Shaving, *Load Shifting*, and *Demand Response* are variations on a theme—supplying energy generated at some point in time to a load at some later time. The rate structure and interactions between the utility and the customer determine which application is being addressed.

Peak Shaving: The purpose of this application is to minimize demand charges for a commercial customer or to reduce peak loads experienced by the utility. Peak shaving using PV-Storage systems requires that the PV provide all required power above a specified threshold and, if PV is not available, then provide adequate energy storage to fill the gap. Failure to peak shave on one day can have severe economic consequences in cases where customers' rates are based on monthly peak demand. Thus, reliability of the PV-storage system is a key element. If PV is unavailable to meet the load, the system controller must be able to dispatch power from an energy storage system in order to implement peak shaving.

Load Shifting: Technically, load shifting is similar to peak shaving, but its application is useful to customers purchasing utility power on a time-of-use (TOU) basis. Many peak loads occur late in the day, after the peak for PV generation has passed. Storage can be combined with PV to reduce the demand for utility power during late-day, higher-rate times by charging a storage system with PV-generated energy early in the day to support a load later in the day.

Pacific Gas and Electric (PG&E) offers the experimental rate structure shown in Table 2 for residential customers. In this schedule, peak rates apply between 2 p.m. and 7 p.m. on weekdays and super-peak rates apply between 2 p.m. and 7 p.m. for no more than 15 days in a calendar year during critical events (as designated by the independent system operator, or ISO) and emergencies. Thus, customers with a PV-Storage system could use PV to charge the storage device earlier in the day (*i.e.,* during peak insolation) and then use the storage system to supply all or part of the load when peak or super-peak rates are in effect. With rate structures such as these in effect, a PV-Storage system could potentially provide significant economic benefits to residential and small commercial customers.

Table 2. PG&E Rate Schedule E-3 – Experimental Residential Critical Peak Pricing Service[8]

Total Energy Rates ($/kWh)	Super Peak	Peak	Off Peak
Summer Baseline Usage	0.67439	0.23096	0.08039
Winter Baseline Usage	0.50997	0.31197	0.10497

Demand Response: Demand response is rapidly becoming a viable load management tool for electric utilities. During high-demand periods, demand response allows the utility to control selected high-load devices, such as heating, ventilation, and air conditioning (HVAC) and water heating, in a rolling type of operation. Utility rate structures are currently changing to accommodate this new operational strategy by reducing rates for customers who choose to be included in the demand response program. For both residential and small commercial customers, using an appropriately sized PV-Storage system should allow the implementation of demand response strategies with little or no effect on local operations.

While both residential and commercial PV-Storage systems have the inherent capability to manage demand response requirements, control systems capable of reacting to demand response must be developed. Specifically, control systems must dispatch the PV-Storage system, as necessary, to manage the loads curtailed by the demand response program. Consequently, at least one-way communication with the utility might be required.

Outage Protection, *Grid Power Quality Control,* and *Microgrids* increase the reliability of the electricity grid and are not as subject to regulatory and rate-based actions.

Outage Protection: An important benefit of a PV-Storage system is the ability to provide power to the residential or small commercial customer when utility power is unavailable (*i.e.,* during outages). To provide this type of protection, it is necessary to intentionally island the residence or commercial establishment in order to comply with utility safety regulations designed to prevent the back-feeding of power onto transmission and distribution (T&D) lines during a blackout. Islanding requires highly reliable switching equipment for isolating the local loads from the utility prior to starting up local generation.

Islanding capability, whether utility or customer-controlled, is mutually beneficial to both the utility and the customer, because it allows the utility to shed loads during high demand periods while protecting the customer's loads if the utility fails. To realize the full benefit of these capabilities, however, new controllers are needed to respond to both utility and customer needs. Additionally, new regulations will be needed to define how these controllers will be managed safely to benefit both the utility and the customer.

Grid Power Quality Control: In addition to outage protection, power quality ensures constant voltage, phase angle adjustment, and the removal of extraneous harmonic content from the electricity grid. On the customer side, this function is currently supplied by UPS devices. A UPS must sense, within milliseconds, deviations in the AC power being supplied and then take action to correct those deviations. A common deviation is a voltage sag in which the UPS supplies the energy needed to return the voltage to the desired level.

UPS functions can be added to PV-Storage systems in the power conditioning system by designing it to handle high power applications and including the necessary control functions. UPS functionality can be combined with peak shaving capability in the same system.

Microgrids: Microgrids have the potential to significantly increase energy surety,[9] and their incorporation into the larger grid infrastructure is expected to become an increasingly important feature of future distribution systems. Renewable generation and energy storage are essential to achieving highly sustainable, highly reliable microgrids. When operating separately from the local utility (*i.e.,* when 'islanded'), microgrids with PV-Storage systems will use PV-generated electricity to supply power to the load.

Energy storage is essential to ensure stable operation of the Microgrid, through management of load and supply variations, and for keeping voltage and frequency constant. Successful integration into the larger utility grid infrastructure of microgrids that include PV-Storage systems will provide many operational benefits to utilities and customers. However, microgrids will require a high level of system control, a detailed knowledge of the load(s) being served, and thoughtful design of the PV-Storage system.

Table 3 summarizes current and future applications that can be addressed by integrating energy storage with distributed PV generation.

Table 3. Applications for Storage-integrated PV

Residential			
Homeowner-owned Systems		Utility-owned Systems	
Current: • Save solar energy for evening use in TOU operations • Back-up power (UPS)	Future: • With time of day residential rates, load shifting • Lower cost than utility • Smart grid interface	Current: • Solar community – ride-through during cloud cover • Distributed generation • Congestion reduction	Future: • Smart grid applications (*e.g.*, distributed energy management, microgrid islanding, peak shaving/shifting.) • High penetration ramp control (short-term spinning reserve) • Emission reduction, carbon credits (with high penetration)
Commercial			
Business-owned Systems		Utility-owned Systems	
Current: • Peak shaving to reduce TOU and/or demand charges • Power quality and UPS	Future: • Carbon credits • Microgrid generation and islanding • Smart grid/building management interfaces	Current: • Distributed generation • Congestion reduction • Improved power quality	Future: • Microgrid generation and islanding • Emission reduction, carbon credits (with high penetration)

The economic benefits that can be realized from PV-Storage systems are a function of the application, the size of the system, the sophistication of the system's electronic control equipment, the customer's rate structure, and the utility's generation mix and operating costs. Systems that include UPS features are expected to mitigate the costs of power quality events and outages.

Results of a recent study also suggest that adding PV generation to a planned UPS installation is attractive because of the synergy between PV and storage in the UPS market. In other words, sites where customers have already decided to purchase load protection via energy storage might be an attractive near-term target for PV developers.[10]

In general, however, most financial benefits will result from reduced peak-demand and TOU charges for consumers and the avoided costs of maintaining sufficient peak and intermediate, power generating capability plus spinning reserve for utilities. By facilitating an optimal mix of generation options, it is expected that the cost to the utility of adding additional generating capacity and the associated T&D equipment can be reduced, as can the costs associated with upgrading existing T&D equipment to meet new demand.

In the future, additional financial benefits could accrue to the end user by selling power back to the utility and to the utility by selling carbon credits realized by aggregating PV generation as a market commodity. Ideally, rate structures for PV-Storage systems could be designed to benefit both system owners and utilities. To fully realize all of the potential, economic benefits will, however, require an advanced control system that includes communications between the utility, the PV-Storage system, and (possibly) the customer.

Finally, at high levels of penetration, PV systems offer significant environmental benefits. One such benefit is that they create no emissions while generating electricity. Another is that they can be installed on rooftops and on undesirable real estate, such as brown fields, which can reduce a utility's need to acquire land for construction of new, large-scale generating facilities, not to mention the associated local opposition to such acquisitions, and the environmental consequences of large-scale industrial construction. Adding electrical energy storage to distributed PV generation also produces no emissions and, by allowing PV-generated electricity to be used at times when PV would normally not be available (*e.g.*, at night or when it is cloudy), allows greater benefits to be realized than with PV systems alone.

7. CURRENT ELECTRICAL ENERGY STORAGE TECHNOLOGIES AND R&D

Energy storage devices cover a variety of operating conditions, loosely classified as 'energy applications' and 'power applications'. Energy applications discharge the stored energy relatively slowly and over a long duration (*i.e.*, tens of minutes to hours). Power applications discharge the stored energy quickly (*i.e.*, seconds to minutes) at high rates. Devices designed for energy applications are typically batteries of various chemistries. Power devices include certain types of batteries, flywheels, and ECs. A new type of hybrid device, the lead-carbon asymmetric capacitor, is currently being developed and is showing promise as a device that might be able to serve both energy applications and power applications in one package.

Figure 4 illustrates several battery and capacitor technologies in relation to their respective power/energy capabilities.[11] The traditional lead-acid battery stands as the traditional benchmark. The plot shows that significantly greater energy and power densities can be achieved with several rechargeable battery technologies.

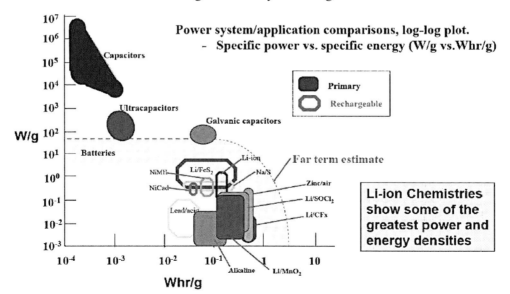

Figure 4. Specific power vs. specific energy of several energy storage technologies.[12]

To date, the advantages of lead-acid technology, such as low cost and availability, have made it the default choice for energy storage in most PV applications. Indeed, new developments in valve-regulated lead-acid (VRLA) technology might revolutionize this well-established technology.

A number of lead-acid battery manufacturers, such as East Penn in the U.S. and Furukawa in Japan, are manufacturing prototype batteries for hybrid electric vehicles (HEVs) that promise to overcome the main disadvantages of VRLA batteries by using special carbon formulations in the negative electrode. The added carbon inhibits hard sulfation, which minimizes or eliminates many common failure mechanisms (*e.g.*, premature capacity loss and water loss). In cycling applications, the new VRLA technology could dramatically lower the traditional battery energy costs by increasing cycle life, efficiency, and reliability.

Traditionally, nickel-cadmium (NiCd) batteries have been the replacement for lead-acid; but, due to various operational and environmental issues, industry is moving away from this technology as newer and better technologies are developed. Indeed, even in the portable electronics market, lithium-ion (Li-ion) batteries are rapidly replacing NiCd.

Additionally, a new Li-ion technology, the Li-iron phosphate (Li-FePO) cell, is rapidly becoming a prime contender for the next generation of HEV batteries, replacing existing nickel-metal hydride (Ni-MH) technology. This Li-ion technology is proving to be much safer than the previous generation and is capable of higher power levels, which makes it a better candidate for HEV applications.

A lesser known technology, sodium/nickel-chloride (Na/NiCl), has been developed by Zebra Technologies in Europe for motive applications, and is currently being considered for some stationary applications, such as peak shaving, in the U.S. Other advanced battery technologies (*e.g.*, sodium/sulfur, or Na/S) are currently targeting utility-scale (> 1MW) stationary applications.

Although these technologies are not currently being considered for use in the smaller applications discussed here, future advances in their developments might increase the technical and economic viability for such applications. Table 4 summarizes the battery technologies that have been identified as potential candidates for integration with grid-tied distributed PV generation in residential and small commercial systems.

Table 5 provides a summary of non-battery technologies that can be integrated with grid-tied distributed PV generation. Although they are still in the early commercial stage of development, hybrid lead-carbon asymmetric capacitors are also targeting the peak-shaving market and low-speed flywheels are currently being used in many UPS applications.

ECs are ideal for high-power, short-duration applications because they are capable of deep discharge and have a virtually unlimited cycle life. Due to these advantages, a great deal of research is being focused on developing ECs that can be used for small-scale stationary energy storage. Any of these battery or non-battery technologies can be appropriate for residential and small-commercial integrated PV and storage systems in the near future.

In addition to the efforts of the technology manufacturers, DOE (through several program offices) is conducting research and executing pilot programs to improve the utilization of electrical energy storage for stationary applications. In particular, since the late 1970s, the DOE Energy Storage Systems Program (DOE/ESS) has worked with the utility industry to develop stationary energy storage systems for utility applications.

In the 1990s, DOE/ESS shifted the focus of its development of advanced storage technologies to include an emphasis on integrating storage devices with power electronics and communications equipment for use in specific applications. Over the past decade, the Program has gained valuable practical experience by partnering with storage technology manufacturers, power electronics and monitoring equipment manufacturers, systems integrators, electric utilities, and their customers to demonstrate integrated electric energy storage systems of all types and sizes. Lessons learned from the Program's demonstrations and research provide uniquely applicable experience for successfully incorporating electrical energy storage with distributed PV generation.

The DOE Vehicle Technologies Program, in partnership with the automotive industry, manages and conducts research on battery technologies for EVs and HEVs (*e.g.*, lithium-aluminum-iron-sulfide, Ni-MH, Li-ion, and lithium-polymer). Li-ion systems come closest to meeting all of the technical requirements for vehicle applications; but, they face four barriers: calendar life, low-temperature performance, abuse tolerance, and cost. Technology advances that address these barriers will have direct applicability to PV-Storage systems for stationary applications.

Finally, the DOE Office of Basic Energy Sciences conducted a comprehensive workshop on April 2-4, 2007, that set R&D priorities for improving the energy density of several storage technologies. Principal barriers identified at the workshop were related to reducing cost, increasing power and energy density, lengthening lifetime, increasing discharge times, improving safety, and providing reliable operation through one-to-ten thousand rapid charge/discharge cycles.

Several associated R&D efforts are underway. One such effort is the announcement of a funding opportunity to establish Energy Frontier Research Centers (EFRCs) specifically focused on "addressing fundamental knowledge gaps in energy storage."[13] The SEGIS-ES program will be closely coordinated with any developments to come from these programs.

8. THE COSTS OF ELECTRICAL ENERGY STORAGE

Both current and projected costs for battery and other storage systems are related to capital costs (first costs) and are based on the overall energy capacity of those systems. Table 6 shows the current and projected first capital costs of energy storage systems based on technologies identified as suitable for residential and small-commercial PV-Storage systems.

Table 6 was compiled from the results of a literature review and discussions with technology leaders at national laboratories and in industry. Recent increases in the prices of materials, such as lead, for existing battery technologies have led to increased system costs. These trends are likely to continue, possibly driving the prices for established technologies even higher.

Unless noted, the system costs include the storage device and the power conditioning system necessary for turning DC output from the storage device into 60-Hz AC power suitable for delivery to the load. For these systems, capital costs will be lowered by combining the power electronics for both the PV and storage components.

Table 4. Battery Technologies for Electric Energy Storage in Residential and Small-commercial Applications

Technology	Advantages	Disadvantages	Commercial Status	Current R&D	Applications
Flooded Lead-acid	• Cost effective • Mature technology • Relatively efficient	• Low energy density • Cycle life depends on battery design and operational strategies when deeply discharged • High maintenance • Environmentally hazardous materials	• Globally commercial • Over $40B in all applications • Estimated $1B in utility applications worldwide	Focused on reducing maintenance requirements and extending operating life.	• Motive power (forklifts, carts, etc.) and deep-cycling stationary applications • Back-up power • Short-duration power quality • Short-duration peak reduction
VRLA	• Cost effective • Mature technology	• Traditionally have not cycled well • Have not met rated life expectancies	• Globally commercial • Over $40B in all applications • Estimated $1B in utility applications worldwide	Improving cycle-life and extending operating life, such as using carbon-enhanced negative electrodes.	• Limited motive power applications (e.g., electric wheelchairs) • Back-up power • Short-duration power quality • Short-duration peak reduction
NiCd	• Good energy density • Excellent power delivery Long shelf life • Abuse tolerant • Low maintenance	• Moderately expensive • "Memory Effect" • Environmentally hazardous materials	• Globally commercial • Over $1B in all applications • Over $50M in utility applications worldwide	None identified.	• Aircraft cranking, aerospace, military and commercial aircraft applications • Utility grid support • Stationary rail • Telecommunications back-up power • Low-end consumer goods
NiMH	• Good energy density • Low environmental impact • Good cycle life	• Expensive	• Globally commercial for small electronics • Emerging market for larger applications	Bipolar design.	• EVs, HEVs • Small, low-current consumer goods

Table 4. (Continued).

Technology	Advantages	Disadvantages	Commercial Status	Current R&D	Applications
Li-ion	• High energy density • High efficiency	• High production cost • Scale-up proving difficult due to safety concerns	• 50% of global small portable market	Batteries for use in EVs and HEVs are currently being developed.	• Small consumer goods
Li-FePO4	• Safer than traditional Li-ion • High power density • Lower cost than traditional Li-ion	• Lower energy density than other Li-ion technologies	• High-volume production began in 2008	Focused on improving performance and safety systems.	• Small consumer goods and tools • EVs, HEVs
Na/S	• High energy density • No emissions • Long calendar life • Long cycle life when deeply discharged • Low maintenance • Integrated thermal and environmental management	• Relatively high cost • Requires powered thermal management (heaters) • Environmentally hazardous materials • Rated output available only in 500-kW/600-kWh increments	• Recently commercial (2002) in Japan • Estimated $0.4B in utility/industrial applications worldwide	Focused on increasing manufacturing yield and reducing cost.	• Utility grid-integrated renewable generation support • Utility T&D system optimization • Commercial/industrial peak shaving • Commercial/industrial backup power
Zebra Na/NiCl	• High energy density • Good cycle life • Tolerant of short circuits • Low-cost materials	• Only one manufacturer • High internal resistance • Molten sodium electrode • High operating temperature	• Globally commercial for traction applications.	Focused on cost reduction and systems for stationary applications.	• EVs, HEVs, and locomotives • Peak shaving

Technology	Advantages	Disadvantages	Commercial Status	Current R&D	Applications
Vanadium Redox	• Good cycle life • Good AC/AC Efficiency • Low temperature/low pressure operation • Low maintenance • Power and energy are independently scaleable	• Low energy density	• Commercial production since 2007	Focused on cost reduction.	• Firming capacity of renewable resources • Remote area power systems • Load management • Peak shifting
Zinc/bromine (Zn/Br)	• Low temperature/low pressure operation • Low maintenance • Power and energy are independently scaleable	• Low energy density • Requires stripping cycle • Medium power density	• Emerging commercial products	Focused on system integration.	• Back-up power • Peak shaving • Firming capacity of renewables • Remote area power • Load management

Table 5. Non-battery Technologies for Electric Energy Storage in Residential and Small-commercial Applications

Storage Type	Advantages	Disadvantages	Commercial Status	Current R&D	Applications
Lead-carbon asymmetric capacitors (hybrid)	• Rapid recharge • Deep discharge • High power delivery rates • Long cycle life • Low maintenance	• Lower energy density than batteries • Lower power density than other ECs	• Non-commercial prototypes	Laboratory prototypes Field demonstration planned FY08 in NY.	• Peak shaving • Grid buffering
Electrochemical Capacitors	• Extremely long cycle life • High power density	• Low energy density • Expensive	• Commercialized in US, Japan, Russia, and EU, emerging elsewhere • Over $30 million in all applications • $5 million in utility applications by 2006	Devices with energy densities over 20 kWh/m3 are under development.	• HEVs • Portable electronics • Utility power quality • T&D stability
Flywheels	• Low maintenance • Long life • Environmentally inert	• Low energy density • High cost	• Commercialized in US, Japan, Europe, emerging elsewhere • Projected to sell over 1,000 systems per year, estimated rated capacity of 250 MW • Retail value exceeding $50 million by 2006	Focused on low cost commercial flywheel designs for long duration operation.	• Aerospace • Utility power quality • T&D stability • Renewable support • UPS • Telecommunications

Table 6. Energy Storage System Capacity Capital Costs[14 15 16 17 18 19 20 21 22]

Technology	Current Cost ($/kWh)	10-yr Projected Cost ($/kWh)
Flooded Lead-acid Batteries	$150	$150
VRLA Batteries	$200	$200
NiCd Batteries	$600	$600
Ni-MH Batteries	$800	$350
Li-ion Batteries	$1,333	$780
Na/S Batteries	$450	$350
Zebra Na/NiCl Batteries	$800[1]	$150
Vanadium Redox Batteries	20 kWh=$1,800/kWh; 100 kWh =$600/kWh	25 kWh=$1,200/kWh 100 kWh=$500/kWh
Zn/Br Batteries	$500	$250/kWh plus $300/kW[2]
Lead-carbon Asymmetric Capacitors (hybrid)	$500	<$250
Low-speed Flywheels (steel)	$380	$300
High-speed Flywheels (composite)	$1,000	$800
Electrochemical Capacitors[3]	$356/kW	$250/kW

[1] €600/kWh

[2] The battery system includes an integrated PCS; the PCS price will vary with the rated system output.

[3] Electrochemical capacitors are power devices used only for short-duration applications. Consequently, their associated costs are shown in $/kW rather than $/kWh.

Determining life-cycle costs depends on a number of factors related to system design, component integration, and overall use. Accurate prediction of life-cycle costs also depends on developing reasonably predictive models for PV-integrated storage. More modeling and analytical work are needed to determine the incremental, levelized cost of energy (LCOE) and the incremental value of increased benefits that storage will bring to PV systems.

9. SUMMARY OF KEY R&D NEEDS FOR PV-STORAGE SYSTEMS

Achieving high-penetration of PV-Storage systems on the nation's utility grid will require overcoming certain technological and economic obstacles. In addition to the specific gaps described below, the successful implementation of optimal, small-scale PV-Storage systems will require further development, testing, and demonstration of complete systems of varying complexities and costs.

9.1. Storage Technologies

To meet the needs of SEGIS-based systems, it will be necessary to develop battery and other storage systems that, although state-of-the-art, are enhanced or specifically designed for use with grid-tied PV systems. It should be noted that any advances in storage technology will be of value to grid-tied PV-Storage systems because they further the understanding of the technology, which facilitates selection of the most appropriate technology for the application and, ultimately, reduces the costs of the storage components.

As previously stated, the main R&D needs for storage technologies address the following aspects of their use:

- Increasing power and energy densities;
- Extending lifetimes and cycle-life;
- Decreasing charge-discharge cycle times;
- Ensuring safe operation; and
- Reducing costs.

Typically, batteries do not work effectively under partial state of charge (PSOC) conditions. PSOC operation occurs when a battery is less than fully discharged and then less than fully recharged before again being discharged. Current research into carbon-enhanced, lead-acid batteries shows high potential for significantly improving PSOC operation. Nevertheless, PSOC operation is not fully understood for all battery chemistries. Charge and discharge profiles for grid-connected PV-Storage applications should be tested on the most promising technologies. To improve PSOC operation, further development and optimization of batteries of various chemistries is also needed.

9.2. Control Electronics

To achieve long lifetimes, maximum output, and optimal efficiency, batteries must be charged and discharged according to the recommendations of the manufacturer. For example, traditional lead-acid batteries require a long (multiple-hour), low-current finish charge to remove sulfation from the lead plates. If finish charging is not completed properly, battery lifetime is shortened and capacity is reduced. This finish charge is very difficult to accomplish with only a PV-based generation source.

Advanced battery management systems can be developed to address some of the charge/discharge issues. The U.S. Coast Guard is sponsoring an effort to develop the Symons Advanced Battery Management System (ABMAS) for off-grid, PV-StorageGenerator hybrid systems.[23] Initial results using the ABMAS system show a 25% reduction in fuel use and improved battery charging and discharging profiles, thus promising increased battery lifetime. Similar management systems are needed for grid-connected PV-Storage systems and applications.

By themselves, energy storage devices (batteries, flywheels, etc.) do not discharge power with a 60-Hz AC waveform, nor can they be charged with 60-Hz AC power. Instead, a power conditioning system is necessary to convert the output. Under the SEGIS initiative, the DOE Solar Energy Program is currently developing integrated power conditioning systems for PV systems. These systems include inverters, energy management systems, control systems, and provisions for including energy storage. It is anticipated that charging and discharging control algorithms for different battery technologies will be included in the SEGIS control package. In the case of lead-acid and NiCd batteries, this will be relatively straightforward.

Other technologies (*e.g.*, Li-ion, vanadium redox, and Zn/Br batteries or flywheels) require more complex safety and control systems. These systems are typically sold by the battery manufacturer as part of an integrated, 'plug-and-play' energy storage system that includes the storage device, an inverter, and proprietary control and safety systems. To achieve the most economical total system using these technologies, SEGIS system manufacturers and manufacturers of these energy storage products could cooperate to design a fully integrated product with minimal duplicated functionality.

9.3. Comprehensive Systems Analysis

Successful development of SEGIS-based PV-Storage systems will require comprehensive systems analysis, including economic and operational benefits and system reliability modeling. Systems must be analyzed based on the requirements of the application. The analysis should include an investigation of all of the possible storage technologies suitable for use in the application and the operational/cost/benefits tradeoffs of each. The analysis must include a methodology for determining the life-cycle costs of PV-Storage systems using conventional industry metrics. The methodology will be used to determine benefit/cost tradeoffs for specific applications and system configurations.

Software-based modeling and simulation tools represent a key component of successful systems analysis. PV system designers use various models to evaluate the needs for, and effects of, various technologies. The system-level modeling software packages that are currently available to designers include Solar Advisor Model (SAM), Hybrid Optimization Model for Electric Renewables (HOMER), PV Design Pro, and HybSim. For the most part, these models do not accommodate storage well. HybSim, funded by DOE's Energy Storage Systems Program, focuses on integrating storage, diesel generation, and wind or PV-generation.

Ideally, models and simulation tools for grid-tied PV-Storage systems will be able to:

- Fully evaluate the benefits of a given PV-Storage system by modeling solar energy production, building loads, and energy storage capabilities relative to capital cost, maintenance, and the real-time cost of alternate energy sources (utility power);
- Accurately simulate residential, commercial, and utility systems and provide recommendations for how to operate, dispatch, and control the PV-Storage system to optimize its economic performance under various loads and rate structures; and
- Provide detailed models of the interrelationships between the various system components and operating parameters, including the physical relationships, operating rules, regulations, and business decision-making criteria to aid in comprehensive systems analysis and identify relationships that might create unexpected vulnerabilities or provide additional robustness.

10. SUMMARY – THE PATH FORWARD

To address the technology gaps described above and to ensure that grid-tied PV-Storage systems meet the needs of customers, utilities, and all other stakeholders, a three-pronged approach is recommended:

- Comprehensive systems analysis and modeling,
- An industry-led R&D effort focused on commercialization of new integrated systems, and
- Development of appropriate codes and standards that facilitate broader market penetration of PV-Storage systems and address all related safety concerns.

These three aspects of SEGIS-ES are discussed in greater detail below.

10.1. Systems Analysis and Modeling

The RSI studies resulted in a series of reports that addressed the myriad issues related to high penetration of PV on utility infrastructure and business models, technical system design, and economic effects. A similar set of studies is proposed to fully investigate the role of energy storage in this environment. These analytical studies will include the development of new modeling tools and will address the following:

- Development of models that explore several aspects of PV-Storage system integration, including system technical performance optimization; grid operational performance, stability, and reliability; cost/benefits; life-cycle costs; *etc.* Models will also address advantages and disadvantages of distributed versus aggregated storage systems (*e.g.*, community-scale *vs.* residential), and the integration of PV-Storage systems with building loads, operating rules and regulations and business decision-making criteria to identify relationships that might create unexpected vulnerabilities or provide additional robustness.
- Investigation of integrating Energy Management Systems (EMS) with PV-Storage systems to optimally manage power for commercial facilities, including development of predictive algorithms for loads and PV output to effectively manage storage.
- Exploration of the role and potential for plug-in hybrid electric vehicles (PHEVs) to provide grid and PV generation support. Because they are mobile devices, using PHEVs for grid support or as energy storage devices to support residential/small-commercial distributed PV generation presents unique challenges for system integrators. Consequently, we recommend investigating how PHEV-based storage can best be aggregated to support distributed PV generation and determine the operational requirements and system specifications necessary for doing so.

10.2. Partnered Industry Research and Development

An industry-led effort will be initiated to strengthen ties among manufacturers and installers in the storage industry with appropriate partners and stakeholders in the PV industry (including utilities), to achieve the following goals:

- Development of new components and integrated PV-Storage systems for grid-connected applications by identifying the requirements and constraints of integrating distributed generation and electrical energy storage with both the load (residential, commercial, or microgrid) and the utility grid. This effort will include development of the power electronics and control strategies necessary to ensure that all parts of the grid-connected distributed generation and storage system work as expected to provide service to the load and maintain or improve grid reliability and power quality.
- Test and verification of promising battery technologies using charge/discharge profiles specifically designed for grid connected PV-Storage applications, in order to develop and optimize the PSOC operation of the battery chemistries.
- Provide the training (cross training, in some instances,) necessary for successfully installing, operating, maintaining, and troubleshooting these highly integrated systems.

10.3. Codes and Standards Development

Ultimately, high levels of penetration of grid-tied distributed generation and storage will affect the utility grid and those who use it in many significant ways. Consequently, codes standards, and regulations for integrating these systems with the grid will be needed to facilitate this integration. Additionally, safety guidelines and regulations that specifically address the complexities of these systems will need to be developed and implemented.

The development of this regulatory environment will be a concerted effort that will build on the current codes and standards infrastructures that exist for the PV, energy storage, construction, and utility industries; and will lead to a comprehensive set of guidelines that will facilitate the greater market penetration of PV-Storage systems.

REFERENCES

[1] Kroposki, B.; R. Margolis, G. Kuswa; J. Torres; W. Bower; Ton, D., Renewable Systems Interconnection: Executive Summary, 2008. Downloadable at http://www1.eere.energy.gov/solar/solar_america/pdfs/42292.pdf
[2] Ton, D.; Cameron, C.; Bower, W. Solar Energy Grid Integration Systems "SEGIS". Concept Paper.
[3] ibid. [1].
[4] Maire, J.; Von Dollen, D. Profiling and Mapping of Intelligent Grid R&D Programs. Report 1014600 to the IEEE Working Group o Distribution Automation. December 2006.

[5] Denholm, P.; Margolis, R. Evaluating the Limits of Solar Photovoltaics (PV) in Electric Power Systems Utilizing Energy Storage and Other Enabling Technologies. April 2007.
[6] ibid. [1].
[7] Manz, D.; Schelenz, O.; Chandra, R.; Bose, S.; de Rooij, M.; Bebic, J. Enhanced Reliability of Photovoltaic Systems with Energy Storage and Controls. RSI Study. NREL/SR-581-42299. February2008.
[8] Pacific Gas and Electric Company. www.pge.com/tariffs/ERS.shtml
[9] Boyes, J.; Menicucci, D. "Energy Storage: The Emerging Nucleus". Distributed Energy. January/February 2007.
[10] Hoff, T.; Perez, R.; Margolis, R. Maximizing the Value of Customer-sited PV Systems Using Storage and Controls.
[11] Roth, E.P.; Doughty, D. Thermal Response and Flammability of Li-ion Cells for HEV Applications. SAND2005-1791P.
[12] ibid. [10].
[13] Basic Research Needs for Electrical Energy Storage. Report of the Basic Energy Sciences Workshop for Electrical Energy Storage. July 2007.
[14] Schoenung, S.; Eyer, J. Benefit/Cost Framework for Evaluating Modular Energy Storage. SAND2008-0978.
[15] Tiax, LLC. Energy Storage: Role in Building PV-based Systems. Final report to DOE EERE. March 2007.
[16] E-mail communications with Tom Hund of Sandia National Laboratories and Jim McDowall of SAFT America, Inc. March 14, 2008.
[17] E-mail communication with Brian Beck of VRB Power Systems, Inc. April 11, 2008.
[18] E-mail communications with Peter Gibson and Doug Alterton of Premium Power Corporation. April 15, 2008.
[19] Dickinson, E.; Clark, N. "Development of High-performance Electrodes Containing Carbon for Advanced Batteries and Asymmetric Capacitors". DOE Energy Storage Systems Program FY08 Quarter 1 Report (October through December 2007). April 2008.
[20] Eckroad, S.; Gyuk, I. EPRI-DOE Handbook of Energy Storage for Transmission & Distribution Applications. December 2003.
[21] E-mail communications with Ib Olsen of Gaia Power Technologies, Inc. April 15, 2008.
[22] E-mail communications with Harold Gotschall, Technology Insights. May 6, 2008.
[23] Corey, G. "Optimizing Off-grid Hybrid Generation Systems." EESAT 2005 Conference Proceedings.

CHAPTER SOURCES

Chapter 1 - This is an edited, reformatted and augmented version of a Sandia National Laboratories publication, Sandia Report Sand2010-0815, dated February 2010.

Chapter 2 - This is an edited, reformatted and augmented version of a United States Department of Energy publication, dated December 2010.

Chapter 3 - This is an edited, reformatted and augmented version of a United States Department of Energy publication, dated December 2010.

Chapter 4 - This is an edited, reformatted and augmented version of a United States Department of Energy, Office of Electricity Delivery & Energy Reliability publication, dated February 2011.

Chapter 5 - This is an edited, reformatted and augmented version of a United States Department of Energy publication, Sandia Report SAND2008-4247, dated July 2008.

INDEX

A

abuse, 304
access, vii, 42, 45, 49, 65, 121, 122, 123, 182, 194, 211, 229, 232, 236, 257
access charges, 49, 65, 121, 122, 123
accommodation, 183
accounting, 20, 86, 123, 157, 182
accounting standards, 157
acid, 25, 34, 51, 97, 160, 197, 199, 207, 208, 217, 240, 251, 253, 262, 267, 272, 277, 278, 303, 305, 308, 309
acquisitions, 302
actuators, 51
adaptation, 158
additives, 208
adjustment, 128, 300
advancement, 195, 208, 211, 215, 218, 227, 233, 265, 278
advancements, 205, 262, 267, 278, 286
advocacy, 147
aerospace, 199, 305
aesthetics, 15
age, 125, 126
agencies, 134, 146, 147, 155, 190, 256, 270, 271, 280, 284
aggregation, 4, 5, 9, 18, 130, 133, 134, 148, 150, 156, 191
air emissions, 16, 41, 44, 89, 124, 126, 144, 145, 148, 153, 159, 177, 178, 185
algorithm, 85, 87
alternative energy, 239, 245
amortization, 18
APC, 104
appropriate technology, 309
aquifers, 26
arbitrage, 39, 248, 249, 287
assessment, 97, 285

assets, 13, 16, 121, 141, 154, 265, 273, 285
authorities, 262, 267
automate, 13
automation, 257
autonomy, 294
aversion, 147, 156, 191
awareness, 190

B

banks, 171
barriers, 9, 228, 265, 285, 304
base, 93, 97, 147, 190, 265
batteries, 25, 28, 29, 30, 34, 51, 90, 97, 136, 149, 152, 153, 176, 194, 195, 197, 199, 200, 206, 207, 208, 210, 211, 213, 214, 215, 216, 217, 218, 219, 224, 225, 226, 227, 240, 250, 253, 262, 263, 265, 267, 268, 272, 273, 274, 276, 277, 278, 279, 287, 289, 302, 303, 307, 309, 310
beneficiaries, 11, 79, 144
biomass, 57
boreholes, 279
breakdown, 237
break-even, 182
breathing, 279
bromine, 10, 25, 199, 240, 307
building blocks, 32, 156
building code, 190
business environment, 128
business model, viii, 252, 292, 294, 311

C

cables, 50, 154
cadmium, 10, 25, 303
candidates, 61, 78, 275, 303

capital intensive, 275
carbon, 126, 194, 197, 206, 207, 208, 210, 224, 226, 262, 264, 268, 278, 296, 301, 302, 303, 305, 307, 308, 309
carbon dioxide, 224, 226
carbon emissions, 264
carbon materials, 207, 278
carbon monoxide, 126
cash, 11, 23
cash flow, 11, 23
catalyst, 218
cathode materials, 279
CEC, 9, 14, 74, 75, 76, 161, 163, 176, 181, 184, 185, 280, 283, 288
CEE, 200
ceramic, 215
challenges, 3, 4, 5, 12, 15, 38, 58, 66, 67, 95, 115, 124, 128, 130, 133, 134, 137, 138, 144, 145, 147, 148, 150, 152, 153, 154, 155, 156, 157, 169, 184, 190, 194, 210, 217, 222, 232, 234, 249, 250, 251, 253, 257, 265, 272, 273, 277, 280, 311
chemical, 25, 31, 218, 275, 277, 278
chemical reactions, 25, 31
chemicals, 34, 262, 267
chromium, 217
citizens, 229, 234, 257
City, 147, 190
classes, 119
classification, 250
CO2, 126, 128, 178
coal, 31, 58, 67, 126, 191, 237
Coast Guard, 309
collaboration, 232, 233, 254, 256, 274, 280, 295
combustion, 25, 31, 40, 62, 83, 88, 89, 108, 110, 113, 125, 126, 153, 286
commercial, 1, 9, 37, 45, 69, 75, 79, 100, 123, 124, 137, 139, 140, 147, 151, 165, 178, 179, 181, 182, 184, 191, 193, 197, 200, 203, 206, 207, 214, 217, 231, 232, 234, 235, 241, 244, 245, 249, 250, 256, 257, 261, 262, 265, 267, 279, 280, 291, 292, 293, 294, 295, 297, 298, 299, 300, 303, 304, 305, 306, 307, 310, 311, 312
commodity, 301
communication, 11, 51, 97, 149, 161, 178, 240, 251, 300, 313
communication systems, 251
communities, 134, 146, 155, 197, 234, 295
community, 146, 147, 158, 190, 202, 228, 244, 263, 265, 268, 274, 280, 287, 301, 311
compatibility, 130, 133, 218

compensation, 48, 73, 149, 164, 168, 169, 171, 239, 250
competitiveness, 16, 78, 145, 208
complement, 3, 4, 152
complexity, 19, 111, 202, 203, 204, 244, 250, 298
compliance, 43, 177
composition, 210, 219, 286
compounds, 212, 278
computer, 13, 170
computer technology, 13
conception, 285
conditioning, 9, 10, 26, 31, 32, 47, 59, 65, 105, 125, 138, 139, 154, 171, 177, 202, 243, 256, 275, 295, 299, 300, 304, 309
conductivity, 206, 215
conductor, 12, 212
conductors, 12
configuration, 88, 298
conflict, 48
consensus, 5, 157, 235
conservation, 151, 152
construction, 124, 145, 179, 215, 272, 279, 302, 312
consumer goods, 305, 306
consumers, 13, 149, 198, 236, 237, 270, 301
consumption, 198, 235, 236
containers, 25
contaminant, 206
contamination, 206, 217
contingency, 11, 50, 170
convention, 13, 20, 164
convergence, 150
cooling, 26, 202, 243
cooperation, 155, 193
coordination, 149, 155, 163, 233, 253, 257
copper, 144, 206
correlation, 38, 111
cost effectiveness, 287
costs of production, 280
covering, 103
creep, 279
crystalline, 206
crystalline solids, 206
customer service, 134, 146
customers, 13, 16, 18, 52, 53, 68, 75, 79, 95, 100, 103, 105, 119, 122, 123, 129, 144, 149, 155, 156, 158, 164, 190, 202, 239, 240, 246, 274, 280, 292, 293, 294, 295, 299, 300, 301, 304, 311
cycles, 29, 30, 48, 69, 207, 210, 212, 214, 245, 246, 248, 262, 268, 272, 287, 304
cycling, 211, 212, 221, 222, 246, 278, 303, 305

D

damping, 43
data center, 256
data collection, 233, 253, 255, 257
database, 219, 256
decision makers, 128, 202, 243, 284
degradation, 29, 171, 206, 212, 215, 227
demonstrations, 14, 157, 159, 184, 194, 201, 232, 249, 250, 253, 255, 256, 257, 261, 263, 267, 268, 271, 273, 274, 277, 280, 281, 285, 304
Department of Defense, 271
Department of Energy, v, viii, 1, 2, 9, 13, 14, 149, 159, 160, 162, 163, 172, 176, 193, 194, 195, 197, 228, 229, 231, 232, 234, 257, 258, 259, 261, 288, 291, 292
Department of Homeland Security, 271
deployments, 256, 257, 261, 262, 267
depreciation, 18, 151
depth, 105, 212, 278
deregulation, 273
designers, 182, 310
deviation, 300
diffusion, 211, 286
discharges, 25, 30, 31, 54, 59, 72, 133, 138, 239, 249
distributed applications, 42
distribution, viii, 10, 11, 12, 16, 17, 18, 33, 34, 37, 49, 50, 60, 71, 72, 82, 95, 121, 122, 131, 133, 134, 140, 141, 142, 144, 146, 149, 150, 154, 164, 170, 171, 190, 194, 198, 202, 203, 234, 236, 237, 239, 240, 244, 246, 247, 249, 250, 252, 253, 255, 256, 257, 261, 맴264, 265, 284, 285, 294, 295, 300
District of Columbia, 237
diversity, 11, 15, 66, 112, 134, 143, 150, 178, 275, 276, 287
downsizing, 151
drawing, 142
dumping, 68, 94
durability, 214, 217, 250, 272, 287

E

economic activity, 79
economic consequences, 299
economic efficiency, viii, 261, 264
economic evaluation, 285
economic performance, 310
economics, 53, 194, 202, 204, 205, 228, 243, 284
economies of scale, 152, 286
education, 130

educational materials, 257
electric capacity, 151
electric current, 25
electrical resistance, 213
electrochemistry, 253
electrodes, 25, 206, 207, 208, 211, 217, 219, 222, 227, 277, 278, 305
electrolysis, 226, 227
electrolyte, 25, 206, 208, 211, 213, 214, 215, 216, 218, 219, 220, 221, 278
electromagnetic, 220, 221
electron, 211
electrons, 25
e-mail, 162
emergency, 11, 27, 71, 143, 170
emission, 126, 128
employees, 1, 193, 231, 291
EMS, 311
endothermic, 227
energy density, 28, 202, 208, 210, 211, 253, 277, 279, 304, 305, 306, 307
energy efficiency, 40, 82, 94, 129, 145, 148, 158, 251, 252, 277, 278, 279
energy input, 13, 246
energy prices, 5, 26, 40, 52, 53, 59, 78, 80, 86, 98, 99, 106, 108, 109, 111, 120, 122, 129, 130, 138, 152, 185, 187
energy supply, 78, 240, 245, 248
energy transfer, 28, 116
engineering, 36, 50, 96, 146, 147, 156, 157, 191, 217, 261, 267, 272, 284
environment, 145, 146, 147, 207, 214, 272, 311, 312
environmental impact, 150, 248, 285, 305
environmental issues, 303
environmental management, 306
environmental quality, 134, 146, 155
equity, 11, 12, 18
estimation process, 77
Europe, 279, 303, 307
evaporation, 31
evidence, 130, 252
evolution, 158
expertise, 197, 280, 284
exposure, 5
extraction, 144, 153

F

fabrication, 278
farms, 62, 199
financial, 3, 4, 11, 12, 13, 14, 15, 16, 17, 19, 20, 28, 50, 57, 65, 72, 79, 80, 81, 82, 85, 86, 88,

93, 95, 103, 105, 106, 112, 117, 121, 126, 128, 130, 133, 156, 157, 178, 181, 182, 183, 191, 194, 198, 202, 239, 248, 286, 287, 298, 301
fires, 207
flexibility, viii, 32, 60, 128, 137, 149, 153, 158, 234, 254, 261, 264, 297
flow field, 218
fluctuations, 43, 66, 67, 114, 120, 164, 244, 246, 247, 292, 296
force, 167
Ford, 277
forecasting, 258
formation, 283
friction, 26, 177, 221, 222
fuel cell, 195, 197, 202, 206, 224, 225, 226, 227, 243, 277
fuel efficiency, 125, 126
fuel prices, 106, 108
full capacity, 181
funding, 14, 124, 250, 257, 262, 268, 280, 286, 304
funds, 29, 280

G

Galaxy, 104
geography, 11
geology, 278
Germany, 200
global leaders, 279
goods and services, 12
grants, 271
graph, 108, 167, 296
greenhouse, 252
grid environment, 284
grid services, 285, 287
grids, 167, 236, 265, 295
growth, vii, viii, 49, 50, 65, 75, 82, 96, 97, 124, 149, 194, 196, 232, 234, 264, 285
growth rate, 50
guidance, 14, 27, 30, 156, 197, 235
guidelines, 184, 203, 244, 312

H

Hawaii, 284, 285
hazardous materials, 305, 306
healing, 13, 149
health, 51, 147, 190, 202, 218, 244, 286, 287
hedging, 149
homeowners, 293
homes, 295

host, 278
hub, 221, 222
hybrid, 5, 10, 79, 143, 152, 154, 200, 207, 223, 228, 278, 295, 302, 303, 307, 308, 309, 311
hydroelectric power, 26
hypothesis, 169, 184

I

ideal, 167, 210, 228, 247, 253, 303
imbalances, 275
Impact Assessment, 285
imports, 229, 257
improvements, vii, 3, 146, 148, 151, 153, 214, 218, 225, 226, 261, 267, 279, 288
impulses, 199, 241
impurities, 215, 218, 219
income, 12, 18, 129
income tax, 12, 18, 129
independence, 295
individuals, 271
industries, 147, 229, 258, 273, 274, 312
industry, viii, 17, 133, 194, 195, 196, 197, 201, 202, 203, 204, 217, 218, 224, 225, 228, 232, 233, 234, 235, 239, 243, 244, 245, 246, 247, 248, 249, 250, 251, 252, 253, 255, 256, 257, 265, 267, 271, 276, 277, 280, 286, 287, 288, 289, 292, 294, 303, 304, 310, 311, 312
infancy, 224, 225
inflation, 9, 12, 13, 23
information sharing, 271
infrastructure, vii, 3, 29, 99, 137, 145, 147, 150, 151, 152, 153, 158, 159, 170, 191, 194, 198, 201, 202, 228, 232, 236, 237, 239, 240, 242, 243, 246, 247, 250, 257, 265, 284, 285, 293, 294, 295, 296, 297, 300, 311
insertion, 215, 278
institutions, 197, 234
integration, 5, 16, 46, 47, 65, 66, 69, 70, 71, 114, 120, 144, 146, 148, 149, 152, 153, 154, 159, 190, 194, 203, 204, 218, 219, 220, 221, 232, 234, 238, 240, 244, 245, 249, 250, 251, 253, 254, 255, 256, 257, 262, 267, 278, 284, 297, 300, 303, 307, 308, 311, 312
integrity, 164
intelligence, 74, 77
interface, 220, 221, 295, 301
interoperability, 195, 205, 206, 251, 253, 254
investment, 5, 13, 18, 36, 50, 60, 96, 121, 123, 124, 156, 170, 194, 199, 201, 250, 252, 257, 265, 272, 279, 288

investments, vii, 18, 32, 36, 49, 50, 123, 134, 137, 147, 154, 156, 191, 194, 199, 228, 232, 257, 271, 277
investors, 241
ion transport, 212, 217
ions, 25, 278
iron, 206, 212, 217, 303, 304
issues, viii, 148, 198, 199, 205, 207, 210, 238, 240, 241, 245, 251, 252, 265, 274, 278, 280, 285, 292, 294, 309, 311

J

Japan, 200, 213, 278, 303, 306, 307
justification, 191, 248

K

kinetics, 224, 226

L

laptop, 210
laws, 16, 145
lead, 16, 25, 34, 49, 51, 58, 61, 71, 79, 86, 97, 124, 125, 126, 136, 141, 152, 153, 156, 157, 167, 168, 194, 197, 199, 206, 207, 208, 215, 217, 240, 251, 253, 262, 267, 268, 272, 274, 277, 278, 293, 302, 303, 304, 309, 312
lead-acid battery, 97, 302, 303
leadership, 261, 267
learning, 74
legislation, 82
life cycle, 272, 278, 279
life expectancy, 184
lifetime, 202, 207, 208, 212, 222, 243, 245, 246, 247, 251, 255, 256, 304, 309
light, 67, 121, 279, 280
lithium, 9, 25, 194, 199, 206, 210, 212, 240, 251, 253, 262, 267, 277, 303, 304
load balance, 165
lower prices, 152

M

magnesium, 206, 227
magnet, 26, 177, 222
magnetic field, 26, 167
magnitude, 12, 17, 28, 29, 56, 63, 73, 79, 141, 157, 183, 187, 278, 296
majority, 20, 51, 204

management, viii, 10, 42, 46, 50, 52, 53, 55, 56, 60, 65, 69, 70, 71, 79, 80, 82, 92, 94, 98, 99, 129, 136, 152, 170, 199, 203, 207, 208, 210, 211, 214, 220, 221, 227, 240, 244, 256, 271, 273, 276, 287, 288, 292, 294, 297, 299, 300, 301, 306, 307, 309
manufacturing, 79, 119, 195, 202, 205, 214, 216, 218, 244, 250, 257, 261, 265, 267, 273, 278, 279, 284, 286, 303, 306
market penetration, viii, 185, 217, 233, 256, 272, 278, 279, 286, 292, 294, 311, 312
market share, 184
market structure, 194, 232, 249, 252
marketing, 78
marketplace, vii, 3, 5, 20, 40, 55, 72, 75, 85, 88, 93, 94, 122, 123, 144, 145, 148, 149, 152, 153, 154, 156, 158, 280
Maryland, 230
mass, 28, 218
materials, 34, 194, 195, 197, 199, 202, 203, 204, 205, 206, 207, 211, 214, 215, 216, 217, 218, 220, 221, 222, 223, 225, 226, 227, 228, 229, 235, 244, 253, 255, 262, 267, 268, 272, 274, 277, 278, 279, 286, 304, 306
materials science, 197
matrix, 130, 278
matter, iv, 53, 100
measurements, 286
mechanical stress, 277
media, 31, 218
melt, 214
membranes, 206, 217, 218, 227, 277
MES, 9, 31, 278
metals, 227
methodology, 15, 114, 165, 236, 255, 284, 310
Mexico, 1, 197, 234, 291
military, 284, 305
mission, 265, 284
mixing, 208
mobile device, 311
models, viii, 5, 110, 157, 170, 212, 232, 250, 251, 255, 286, 287, 292, 294, 308, 310, 311
modernization, 196, 234
modules, 61, 111, 136, 137, 295
moisture, 214
momentum, 262, 268

N

nanomaterials, 206
natural disaster, 149
natural disasters, 149
natural gas, 26, 40, 67, 126, 179

negative consequences, 296
negative effects, 105, 145, 292, 295
negative outcomes, 81
neutral, vii, 2, 14, 267
New South Wales, 230
next generation, 303
nickel, 10, 25, 199, 240, 289, 303
nitrogen, 126, 199, 240, 279
nodes, 50
North America, 10, 43, 75, 160, 161, 177, 257, 264

O

obstacles, 204, 250, 279, 308
officials, 147
oil, 250
open markets, 151
operating costs, 23, 28, 29, 87, 97, 301
operations, 54, 66, 148, 152, 203, 207, 244, 245, 247, 248, 249, 250, 256, 280, 284, 296, 299, 301
opportunities, 3, 5, 61, 69, 72, 75, 128, 145, 149, 153, 157, 234, 249, 253, 257, 265, 285
optimization, 149, 150, 211, 273, 306, 309, 311
outreach, 233, 253, 257, 271
overlap, 145
overlay, 275, 276
oversight, 280, 283
ownership, 5, 18, 84, 88, 95, 158
oxidation, 25, 217, 278
oxide electrodes, 210
oxygen, 224, 226, 227

P

Pacific, 10, 52, 161, 162, 176, 185, 193, 194, 197, 230, 260, 299, 313
participants, 93, 129, 170, 197, 235
peer review, 271
penalties, 42, 46
percentile, 6, 24, 37, 38, 76, 83, 85, 96
performers, 271
permeability, 279
petroleum, 126
phosphate, 212, 303
photographs, 256
physics, 31
plants, 26, 86, 110, 115, 164, 177, 182, 237, 244, 246, 248, 275, 279
plastics, 218
polarization, 227

policy, 16, 144, 145, 274, 284, 285
policy makers, 274
policymakers, 2, 3, 5, 15, 16, 144, 145, 146, 148, 154, 155, 157, 158, 197, 234, 241, 257
pollution, 185, 244
polymer, 212, 226, 304
polymer electrolytes, 212
polymers, 212
portability, 214
portfolio, 170, 233, 265, 285, 287
portfolio management, 287
potential benefits, vii, 2
power generation, viii, 169, 237, 261, 264, 265, 284
power plants, 43, 78, 110, 126, 177, 236, 243, 246, 275
preparation, iv, 117
present value, 84
preservation, 286
President, 160
price signals, 34, 78, 133, 145, 148, 149, 151, 154
primary function, 164
principles, 261, 267, 286
probability, 110
probe, 212
producers, 257
professionals, 270, 286
profit, 18, 36, 71, 85, 128, 134, 146, 155, 178, 191
project, 2, 3, 14, 19, 21, 30, 123, 124, 134, 138, 146, 147, 150, 155, 157, 179, 182, 183, 184, 190, 250, 270, 283, 284, 285, 288
proliferation, 145, 152, 154
propagation, 279
property taxes, 12, 18
proposition, 5, 14, 78, 82, 92, 95, 129, 130, 133, 134, 138, 139, 140, 144, 145, 156, 157, 159, 178, 181, 184, 265, 267, 272
protection, 67, 246, 292, 293, 300, 301
prototype, 267, 303
prototypes, 274, 279, 307
public health, 190

Q

qualifications, 287
quality control, 195, 205

R

ramp, 31, 34, 45, 51, 60, 69, 208, 222, 244, 245, 286, 301
rate of return, 36
ratepayers, 16, 50, 133, 134, 146, 155, 159, 185
reaction time, 279
reactions, 25, 31, 212, 217, 227, 277
real estate, 302
real time, 165
reality, 234
recognition, 5, 145, 154, 195, 205, 272
recommendations, iv, 3, 292, 293, 309, 310
recovery, 43, 177, 245, 285, 286
regeneration, 226
regulations, 16, 18, 45, 73, 75, 78, 145, 147, 156, 190, 237, 300, 310, 311, 312
regulatory agencies, 146, 155
reimburse, 103
relief, 18, 40, 46, 49, 50, 55, 56, 60, 65, 69, 70, 94, 116, 117, 118, 133, 135, 136, 143, 148, 152, 154
renewable energy, vii, viii, 37, 49, 57, 58, 59, 60, 61, 62, 63, 65, 106, 109, 110, 111, 112, 126, 136, 143, 148, 151, 194, 197, 198, 199, 201, 228, 232, 234, 236, 237, 239, 240, 242, 243, 245, 248, 256, 257, 263, 265, 268, 285, 293
rent, 40, 60, 88, 89
reputation, 159
requirements, 11, 27, 28, 30, 33, 34, 40, 45, 49, 51, 52, 66, 67, 69, 90, 97, 98, 156, 181, 202, 203, 207, 210, 217, 225, 226, 240, 243, 244, 255, 272, 273, 275, 276, 279, 284, 287, 296, 297, 298, 299, 304, 305, 310, 311, 312
resale, 251
researchers, 3, 15, 147, 157, 197, 201, 203, 206, 232, 234, 241, 244, 253, 254, 255, 256
reserves, 11, 38, 45, 46, 91, 92, 135, 139, 149, 164, 165, 170, 244
resilience, 194, 198, 239, 243
resins, 219
resistance, 9, 12, 206, 215, 216, 218, 219, 227, 286, 306
response time, 30, 31, 45, 178, 202, 243, 244, 245, 246, 248, 254
retail, 52, 64, 104, 123, 151, 255
revenue, 4, 11, 12, 13, 16, 18, 23, 36, 81, 86, 93, 95, 123, 128, 129, 140, 155, 156, 170, 183, 252
risk, 5, 32, 50, 78, 92, 96, 121, 123, 124, 128, 137, 147, 150, 156, 158, 159, 190, 191, 207, 214, 256, 257, 273
robust design, 272

room temperature, 214
rules, 5, 18, 34, 45, 73, 75, 78, 147, 155, 156, 157, 158, 190, 218, 257, 286, 310, 311
Russia, 307

S

Sacramento Municipal Utility District, 260
safety, 146, 147, 155, 190, 200, 202, 206, 210, 211, 212, 221, 222, 224, 226, 244, 250, 251, 261, 267, 294, 300, 304, 306, 310, 311, 312
salt formation, 26
salts, 210
saturation, 279
scale system, 250, 262, 267
science, 215, 261, 267, 279, 284
scope, 4, 18, 20, 33, 73, 128, 150, 197, 235, 292, 293, 295
security, 144, 164, 165, 170, 246, 265, 284
seller, 164
sensors, 218
service provider, 3, 15
shape, 96, 214, 257, 265, 284
shelf life, 256, 305
shortage, 129
shortfall, 43, 45, 68, 69, 84, 94, 177
showing, 280, 302
signals, 34, 46, 47, 71, 164, 183, 184, 191, 245, 251, 257, 271
simulation, 251, 273, 310
simulations, 250
sine wave, 33
small businesses, 47, 271
smoothing, 245
societal cost, 124
society, 13, 16, 134, 144, 159
sodium, 10, 25, 194, 199, 206, 213, 214, 215, 216, 240, 250, 262, 267, 277, 303, 306
software, 60, 65, 310
solar collectors, 62
solution, 78, 81, 83, 93, 105, 125, 128, 133, 201, 208, 211, 228, 242, 257, 297
solvents, 210
species, 217
specifications, 33, 202, 204, 243, 251, 256, 286, 311
speculation, 19, 75, 181
spin, 26, 177
spot market, 106
stability, 46, 48, 67, 165, 169, 171, 181, 196, 214, 217, 218, 224, 225, 226, 232, 234, 257, 265, 287, 307, 311
stabilization, 38

staffing, 124
stakeholders, vii, 3, 4, 12, 16, 79, 85, 121, 130, 133, 134, 145, 146, 147, 155, 156, 157, 158, 190, 191, 197, 202, 232, 233, 234, 243, 250, 251, 256, 265, 270, 288, 311, 312
standard deviation, 164
state, 1, 48, 51, 68, 77, 88, 97, 146, 155, 158, 171, 181, 190, 193, 212, 215, 218, 232, 237, 265, 275, 279, 280, 286, 287, 292, 295, 308, 309
states, 34, 237, 259
steel, 144, 308
storage media, 29, 30, 125
stratification, 208
stress, 221, 222, 236, 238, 243, 246
structure, 19, 98, 99, 232, 250, 278, 299, 301
subjectivity, 15, 19
substitutes, 78, 194, 198
substitution, 203, 234, 244, 246, 247, 249, 253, 256, 257
sulfur, 10, 25, 126, 199, 213, 214, 240, 250, 277, 303
sulfuric acid, 207, 217
superconductivity, 26
supplier, 150
supply chain, 279
synchronize, 32

T

talent, 272
tanks, 26, 218, 219, 277
target, 77, 139, 262, 263, 265, 267, 268, 301
tariff, 38, 52, 53, 54, 55, 84, 98, 99, 100, 106, 112, 123
tax deduction, 129
taxes, 23, 95
teams, 274
technical assistance, 284
techniques, 215, 271
technological advances, 286
technology gap, 311
technology transfer, 271
telecommunications, 73, 199
telephones, 210
temperature, 26, 63, 121, 199, 211, 214, 216, 221, 222, 240, 252, 304, 306, 307
Tennessee Valley Authority, 161, 270
testing, 214, 215, 216, 222, 224, 225, 227, 233, 250, 251, 255, 256, 267, 272, 279, 283, 287, 308
thermal energy, 26, 35, 143, 151

time frame, 196, 197, 206, 208, 212, 217, 219, 223, 228, 232, 235
time periods, 52, 100
time series, 85
titanate, 212
trade, 1, 193, 197, 217, 231, 234, 291
trade-off, 217
training, 257, 312
transaction costs, 52, 74
transactions, 39, 85, 87, 152, 165
transparency, 286
transport, 169, 214, 217, 218, 246, 247
transportation, 129, 148, 153, 265
treatment, 20, 215
troubleshooting, 312

U

U.S. economy, vii, 2, 194, 232, 236
uniform, 255
unique features, 153
unit cost, 121
United, v, 1, 193, 195, 198, 205, 231, 235, 236, 241, 257, 258, 261, 279, 291, 292
United States, v, 1, 193, 195, 198, 205, 231, 235, 236, 241, 257, 258, 261, 279, 291, 292
universities, 274

V

vacuum, 222
validation, 208, 222, 255
valuation, 180
valve, 51, 278, 303
vanadium, 25, 199, 217, 218, 240, 310
variable costs, 108
variations, 43, 57, 62, 67, 114, 164, 165, 171, 177, 239, 240, 298, 300
vehicles, vii, viii, 5, 79, 148, 152, 154, 194, 197, 198, 200, 207, 210, 232, 234, 263, 265, 268, 303, 311
velocity, 68, 279
ventilation, 299
vision, 149, 261, 267
volatile organic compounds, 126
volatility, 66, 69, 114, 240

W

Washington, 163
water, 26, 224, 226, 279, 299, 303
wealth, 280

wear, 26, 28, 29, 44, 58, 86, 89, 124, 129, 153, 165, 177, 178
websites, 271
wetting, 215, 221, 222
wholesale, 39, 40, 42, 52, 55, 57, 64, 88, 93, 106, 108, 111, 122, 138, 143, 176, 185, 250, 251, 255
wind farm, 37, 61, 78, 112, 136, 246
wind gusts, 66
wind turbines, 66, 67, 68, 69, 115
wires, 17, 31, 60, 168
workers, 202, 244
worldwide, 200, 201, 241, 305, 306

X

X-axis, 108

Y

Y-axis, 139
yield, 19, 55, 85, 306

Z

zinc, 10, 25, 199, 217, 240